Performance-Based Seismic Design of Concrete Structures and Infrastructures

Vagelis Plevris
Oslo and Akershus University College of Applied Sciences, Norway

Georgia Kremmyda
University of Warwick, UK

Yasin Fahjan
Gebze Technical University, Turkey

A volume in the Advances in Civil and Industrial Engineering (ACIE) Book Series

IGI Global
DISSEMINATOR OF KNOWLEDGE

www.igi-global.com

Published in the United States of America by
IGI Global
Engineering Science Reference (an imprint of IGI Global)
701 E. Chocolate Avenue
Hershey PA, USA 17033
Tel: 717-533-8845
Fax: 717-533-8661
E-mail: cust@igi-global.com
Web site: http://www.igi-global.com

Library of Congress Cataloging-in-Publication Data

Names: Plevris, Vagelis, 1976- editor. | Kremmyda, Georgia, 1979- editor. | Fahjan, Yasin, 1967- editor.
Title: Performance-based seismic design of concrete structures and infrastructures / Vagelis Plevris, Georgia Kremmyda, and Yasin Fahjan, editors.
Description: Hershey, PA : Engineering Science Reference, 2017. | Includes bibliographical references.
Identifiers: LCCN 2016054200| ISBN 9781522520894 (hardcover) | ISBN 9781522520900 (ebook)
Subjects: LCSH: Concrete construction. | Earthquake resistant design. | Buildings--Performance.
Classification: LCC TA681.5 .P47 2017 | DDC 621.1/762--dc23 LC record available at https://lccn.loc.gov/2016054200

This book is published in the IGI Global book series Advances in Civil and Industrial Engineering (ACIE) (ISSN: 2326-6139; eISSN: 2326-6155)

British Cataloguing in Publication Data
A Cataloguing in Publication record for this book is available from the British Library.

Advances in Civil and Industrial Engineering (ACIE) Book Series

Ioan Constantin Dima
University Valahia of Târgovişte, Romania

ISSN:2326-6139
EISSN:2326-6155

Mission

Private and public sector infrastructures begin to age, or require change in the face of developing technologies, the fields of civil and industrial engineering have become increasingly important as a method to mitigate and manage these changes. As governments and the public at large begin to grapple with climate change and growing populations, civil engineering has become more interdisciplinary and the need for publications that discuss the rapid changes and advancements in the field have become more in-demand. Additionally, private corporations and companies are facing similar changes and challenges, with the pressure for new and innovative methods being placed on those involved in industrial engineering.

The **Advances in Civil and Industrial Engineering (ACIE) Book Series** aims to present research and methodology that will provide solutions and discussions to meet such needs. The latest methodologies, applications, tools, and analysis will be published through the books included in **ACIE** in order to keep the available research in civil and industrial engineering as current and timely as possible.

Coverage

- Optimization Techniques
- Hydraulic Engineering
- Coastal Engineering
- Productivity
- Production Planning and Control
- Operations Research
- Quality Engineering
- Earthquake engineering
- Materials Management
- Transportation Engineering

IGI Global is currently accepting manuscripts for publication within this series. To submit a proposal for a volume in this series, please contact our Acquisition Editors at Acquisitions@igi-global.com or visit: http://www.igi-global.com/publish/.

The Advances in Civil and Industrial Engineering (ACIE) Book Series (ISSN 2326-6139) is published by IGI Global, 701 E. Chocolate Avenue, Hershey, PA 17033-1240, USA, www.igi-global.com. This series is composed of titles available for purchase individually; each title is edited to be contextually exclusive from any other title within the series. For pricing and ordering information please visit http://www.igi-global.com/book-series/advances-civil-industrial-engineering/73673. Postmaster: Send all address changes to above address.

Titles in this Series

For a list of additional titles in this series, please visit: www.igi-global.com

Modeling and Simulation Techniques in Structural Engineering
Pijush Samui (National Institute of Technology Patna, India) Subrata Chakraborty (Indian Institute of Engineering Science and Technology (IIEST), Shibpur, India) and Dookie Kim (Kunsan National University, South Korea)
Engineering Science Reference • copyright 2017 • 524pp • H/C (ISBN: 9781522505884) • US $220.00 (our price)

Computational Modeling of Masonry Structures Using the Discrete Element Method
Vasilis Sarhosis (Newcastle University, UK) Katalin Bagi (Budapest University of Technology and Economics, Hungary) José V. Lemos (National Laboratory for Civil Engineering, Portugal) and Gabriele Milani (Technical University in Milan, Italy)
Engineering Science Reference • copyright 2016 • 505pp • H/C (ISBN: 9781522502319) • US $210.00 (our price)

Handbook of Research on Emerging Innovations in Rail Transportation Engineering
B. Umesh Rai (Chennai Metro Rail Limited, India)
Engineering Science Reference • copyright 2016 • 664pp • H/C (ISBN: 9781522500841) • US $235.00 (our price)

Advanced Research on Nanotechnology for Civil Engineering Applications
Anwar Khitab (Mirpur University of Science and Technology, Pakistan) and Waqas Anwar (Mirpur University of Science and Technology, Pakistan)
Engineering Science Reference • copyright 2016 • 339pp • H/C (ISBN: 9781522503446) • US $195.00 (our price)

Emerging Challenges and Opportunities of High Speed Rail Development on Business and Society
Raj Selladurai (Indiana University Northwest, USA) Peggy Daniels Lee (Kelley School of Business, Indiana University, USA) and George VandeWerken (Providence Bank, USA)
Engineering Science Reference • copyright 2016 • 289pp • H/C (ISBN: 9781522501022) • US $210.00 (our price)

Advanced Manufacturing Techniques Using Laser Material Processing
Esther Titilayo Akinlabi (Univeristy of Johannesburg, South Africa) Rasheedat Modupe Mahamood (University of Johannesburg, South Africa & University of Ilorin, Nigeria) and Stephen Akinwale Akinlabi (University of Johannesburg, South Africa)
Engineering Science Reference • copyright 2016 • 288pp • H/C (ISBN: 9781522503293) • US $175.00 (our price)

Handbook of Research on Applied E-Learning in Engineering and Architecture Education
David Fonseca (La Salle Campus Barcelona, Universitat Ramon Llull, Spain) and Ernest Redondo (Universitat Politècnica de Catalunya, BarcelonaTech, Spain)
Engineering Science Reference • copyright 2016 • 569pp • H/C (ISBN: 9781466688032) • US $310.00 (our price)

Table of Contents

Detailed Table of Contents

Recent developments in performance-based analyses and the high performance of computational facilities have led to an increased trend for utilizing nonlinear time-history analysis in seismic evaluation of the performance of structures. One of the crucial issues of such analysis is the selection of appropriate acceleration time histories set that satisfy design code requirements at a specific site. In literature, there are three sources of acceleration time histories: 1) recorded accelerograms in real earthquakes scaled to match design code spectrum/uniform hazard spectra/conditional mean spectrum, 2) artificial records generated from white noise spectra to satisfy design code spectrum, and 3) synthetic records obtained from seismological models. Due to the increase of available strong ground motion database, using and scaling real recorded accelerograms is becoming one of the most contemporary research issues in this field. In this study, basic methodologies and criteria for selecting strong ground motion time histories are discussed. Design code requirements for scaling are summarized for ASCE/SEI-7-10, EC8 and Turkish Seismic Codes. Examples for scaling earthquake records to uniform hazard spectra are provided.

Many existing reinforced concrete buildings were designed in Southern European countries before the introduction of modern seismic codes and thus they are potentially vulnerable to earthquakes. Consequently, simplified methodologies for the seismic assessment and retrofitting of existing structures are required. In this study, a displacement based procedure using non-linear static analyses is applied to a four-storey

RC frame in order to obtain an initial estimation of the overall inadequacy of the original structure as well as the extent of different retrofitting interventions. Accurate numerical models are developed to reproduce the seismic response of the RC frame in the original configuration. The effectiveness of three different retrofitting solutions countering structural deficiencies of the RC frame is examined through the displacement based approach. Non-linear dynamic analyses are performed to assess and compare the seismic response of the frame in the original and retrofitted configurations.

This chapter proposes a new single-run adaptive pushover method for the seismic assessment of shear wall structures. This method offers two main advantages: it does not require decomposing the structure in nonlinear domain and it avoids the pitfall of previous single-run adaptive pushover analyses in utilizing the modal combination in the determination of the applied loads instead of combining the response quantities induced by those loads in individual modes. After a brief review of the main adaptive pushover procedures, the proposed method is presented as well as its numerical implementation. The predictions of this method are compared to those of other recent adaptive pushover methods and as well as to the rigorous non-linear time history analysis. Analyses show the efficiency of the proposed method.

In many seismic countries in the world (e.g. Europe, Northern USA, Japan, Turkey, etc.), the assessment of existing structures is a priority, since the majority of the building heritage was designed according to out-of-date or even non-seismic codes. The uncertainties about the nonlinear behaviour of the structures are, therefore, important and the nonlinear response should be treated directly, with a correspondingly strong increase in complexity of the assessment procedure. The assessment of regular reinforced concrete frame buildings has been performed, according to the Italian Seismic Code, Eurocode 8 and the CNR DT-212 guideline. A lumped plasticity model has been used with the aim of quantifying the differences between a fixed and a continuously updated shear span and between the use of inelastic springs located at the member ends or continuously along the beam elements, and with the purpose of considering the influence of axial-bending-shear interaction on the global capacity of the buildings.

Strengthening of non-ductile public buildings is a never-ending issue. Selection of the suitable strengthening method and appropriate analysis type for the assessment of pre- and the post-intervention performances are still open to question. The displacement or drift limitations are crucial as well as demand capacity ratios for determination of such buildings performance under severe ground motion. In this chapter, an investigation of seismic performance focused on displacement criterion of strengthened non-ductile public RC buildings in Turkey is presented. Both the nonlinear static and response history analysis were conducted. Friction dampers which are fairly modern technique and conventional RC wall implementation method were introduced to as-is building. For the simplicity and the easy of the process, 2D frame selected for investigation. Comparison of the aforementioned techniques for non-ductile public RC buildings and performances particularly by means of displacement obtained using different methods for those investigated schemes are carried out and presented in the chapter.

High-rise structures exposed to a strong vertical component of an earthquake excitation are endangered by auto-parametric resonance effect. While in a sub-critical state, the vertical and horizontal response components are independent. Exceeding a certain limit causes the vertical response to lose stability and induces dominant horizontal response. This effect is presented using two mathematical models: (1) the non-linear lumped mass model; and (2) the one dimensional model with continuously distributed parameters. Analytical and numerical treatment of both leads to three different types of the response: (1) semi-trivial sub-critical state with zero horizontal response component; (2) post-critical state (auto-parametric resonance) with a periodic or attractor type chaotic character; and (3) breaking through a certain limit, the horizontal response exponentially rises and leads to a collapse. Special attention is paid to transition from a semi-trivial to post-critical state in case of time limited excitation period as it concerns the seismic processes.

This chapter focuses on the comparison of the conventional linear force-based method with the advanced nonlinear deformation-based method that are commonly preferred to investigate the seismic performances of the existing RC school buildings. School buildings which have different structural characteristics and RC infill wall index are generated from an existing school's layout plan. During the nonlinear dynamic analysis, seven recorded earthquake motions which are scaled in accordance with the Turkish Earthquake Code are employed. Seismic performances of the school buildings against the two different earthquake hazard level are evaluated considering not only various RC infill wall indexes but also different material strengths and number of stories in terms of limit states specified in the code. In order to determine the most appropriate method related to material strength, floor level and RC infill wall index for the seismic strengthening of the existing RC school buildings, the obtained linear forced and nonlinear deformation based analyses results are compared to each other.

Seismic isolation of structures is becoming a more common method of providing protection from earthquake damage. Starting from 1994, 53 buildings and structures have been designed by the author of this chapter with application of base or roof isolation systems. Of these designed buildings, the total number of already constructed and retrofitted buildings has reached 45. The number of seismically isolated buildings per capita in Armenia is one of the highest in the world. Together with that seismic isolation laminated rubber-steel bearings (SILRSBs) different by their shape and dimensions, as well as by damping (low, medium and high) were designed and about 5000 SILRSBs were manufactured in the country, tested locally and applied in construction. since 2003 seismic isolation technologies were designed and then extensively applied in construction of multi-story buildings. these are: base isolated residential complexes, business centers, hotels, schools, and hospital buildings. Original and innovative structural concepts, including the new approach on installation of seismic isolation rubber bearings by clusters, were developed and designed for construction of these base isolated buildings and they are described in the given chapter. The advantages of this approach are listed and illustrated by the examples. It is mentioned that suggested new structural solutions in seismic isolation are bringing to significant savings in construction cost. The earthquake response comparative analyses were carried out for some of the considered buildings in two versions (i.e., when the buildings are base isolated and when they are fixed base). Several histories were used in the analyses and for both cases the buildings were analyzed also according to the requirements of the Armenian Seismic Code. Comparison of the obtained results indicates the high effectiveness of the proposed structural concepts of isolation systems and the need for further improvement of the Seismic Code provisions regarding the values of the reduction factors. A separate section in the chapter dedicated to the design of high damping laminated rubber-steel bearings and to results of their tests.

In this chapter, it is expressed how performance based design criteria are modeled with fuzzy logic and why it needs to be modeled with fuzzy logic. Firstly, a brief information is given about current theories and techniques such as Fuzzy Logic, Chaos, Fractal Geometry, Artificial Neural Networks; and it is mentioned about advantages of applying these techniques to the problems. Then graphical inference deduction technique is expressed by giving a summary info about Fuzzy Set Theory, membership functions, clustering, rule-based systems and defuzzification. Then the divergence of performance based design criteria is set forth by making curvature evaluations for various regulations on a column section as a structure element. Finally, it is shown in detail how to use the fuzzy set theory for solution of this disharmony by using the clustering technique on a univariate sample first and then on a multivariate sample.

This chapter defines resilience in different contexts comprehensively, and organizes the mathematical theory of network resilience by providing a generalization in order to create a quantitative framework for resilience characterization of an infrastructure network. At this point, a new performance index measuring delivery importance was employed for an applied purpose and an industrial example using realistic data was solved to evaluate the resilience of the entire network. It can be utilized for any type of hazard which might lead to the disruption of the system. The principles and theory in this study can also be applied to other infrastructures that are interconnected and operate as a network, such as transporting systems, electrical power, water supply and distribution systems.

Considering the fact that similar structural and construction deficiencies which are exposed during the recent destructive earthquake events are existing in many southern European, Middle Eastern and west Asian countries settling on highly seismic zones, designating the seismic adequacy of the existing building stock for providing structural safety is a significant necessitation in the mitigation of losses during the future seismic events. In most of these regions, a clear majority of the building stock has not been adequately designed or constructed in terms of the seismic regulations of the design codes, while some have even not benefitted from engineering services. Post-earthquake site observations demonstrate the insufficient capacities in these buildings depending on different structural and construction deficiencies. Within the scope of this research, it is aimed to investigate and compare the analytically calculated structural performances of a building ensemble consists of 3~5 story structures with known damage level by utilizing different procedures.

Preface

Earthquakes, and particularly structural response to earthquakes, are complex phenomena depending on uncertain physical conditions, and gauged by social needs and expectations for which there is no universal standard. Earthquake occurrence, earthquake characteristics, and structural response characteristics are not deterministic, and need to be viewed as complex probabilistic phenomena. Designs are implemented by mere humans, and the design product may be altered by natural or human influences throughout the life of the structure. In the recent years, significant research has been undertaken to improve former methods applied for analyzing the seismic reliability of structures. The traditional design methods were merely based on strength or force considerations. When the importance of introducing performance measures and acceptance criteria became appreciated, it was attempted to modify and enhance the existing force-based approaches to include considerations of structural performance.

During the last twenty years, several researchers worked on proposing performance-based approaches for earthquake engineering evaluation and design, with the aim of providing improved reliability in the engineering process by directly relating computed response and expected structural performance. Traditionally, seismic design was performed by using linear elastic approaches, while recent code guidelines suggest the use of nonlinear analysis procedures for the design and assessment against seismic actions. Such analysis methods provide a highly efficient framework since they allow more direct design criteria and less simplifying assumptions. Today the use of static nonlinear analysis (pushover) and nonlinear time-history analysis, has become a common practice for seismic analysis and design, especially in the case of special structures.

"Performance-based engineering" has become a standard norm for research, development, and practice of earthquake engineering, particularly after the nineties. The challenges related to this topic vary from the characterization of strong motions and their effects on the structural response, quantification of multiple levels of performance associated with the functionality, damage, and safety limit states, examinations into the interaction of various nonstructural components and building contents with building performance, among many others. In most cases, real data obtained from observations and experiments are essential, to verify individual research findings and assure the expected performance of new, innovative practices. The goal of performance based seismic design is to assist in the engineering of cost-effective facilities, whose safety and resistance to damage from earthquakes meet the needs and expectations of key stakeholders and society at large more effectively and reliably than can be achieved with simple codes that use prescriptive design rules.

Performance-based earthquake engineering has emerged as the most important development in the field of earthquake engineering during the last three decades. Since then, it has started penetrating codes and standards on seismic assessment and retrofitting while making headway towards seismic design standards for new structures. This constitutes a major shift from the traditional structural design concept and represents the future of earthquake engineering. These new procedures help assure that the design will reliably meet a desired level of performance during a given earthquake. The fundamental components of Performance-Based Design (PBD) are nonlinear static or dynamic analyses where attempts are made to capture the real behavior of the structure by explicitly modeling and evaluating post-yield ductility and energy dissipation when subjected to actual earthquake ground motions.

Performance-based design uses nonlinear methods of analysis, which generally involve tedious and intensive computational effort. It is a highly iterative process needed to meet designer-specified and code requirements. The purpose of using PBD is to: (1) acquire realistic data about performance, progress of damage, and final failure of the concerned structure in deformation ranges that are far beyond those considered in contemporary seismic design; (2) examine the interaction between the local damage induced into individual elements and the global damage sustained by the whole structure; (3) examine the interaction between the structural system and exterior finishes; and (4) calibrate the capacity of numerical analyses to trace the behavior to collapse.

This book reports on the outline of the concepts and definitions of performance-based engineering of concrete structures. Interest is placed on specific requirements that correspond to a wide range of potential seismic actions. Other issues of interest, i.e., interaction between the local damage and global behavior, and behavior to final collapse are also explored, and relevant findings are presented. Performance-based approaches may be also used for evaluating relevant concrete material properties under seismic loading conditions using relevant test methods or model predictions. In combination with, and eventually as a replacement for traditional prescriptive design methods, performance concepts offer powerful means to improve the design for concrete durability and mechanical properties. Traditional design methods often fail to consider modern and innovative concrete technology, including high performance concrete, self-compacting concrete, light- or heavy-weight concrete, sustainable mixes with minimized cement contents, concretes for aggressive environments or extreme loading conditions, etc. In these fields, performance concepts offer powerful tools for the application and development of modern concrete technology.

The book aims to contribute to the afore-mentioned aspects involved in the performance-based seismic design of concrete structures and infrastructures. In this volume, we have tried to expand the research frontier in the subject area by presenting leading research work organized into 11 chapters authored by 25 researchers from various countries around the world. The chapters present analytical and numerical methodologies, formulations, applications and case studies as well as state-of-the-art developments on the performance-based design of concrete structures and infrastructures. A brief description of each of the eleven chapters follows:

In the first chapter, the authors identify the existing challenges in selecting and scaling time history records for linear and nonlinear analyses as a part of performance based design. The basic methodologies for selecting strong ground motion time histories are categorized as: (1) recorded accelerograms from real earthquakes scaled to match the design code spectrum/uniform hazard spectra/conditional mean spectrum; (2) artificial records generated from white noise spectra to satisfy the design code spectrum, and (3) synthetic records obtained from seismological models. Comparisons of the different methodologies are explained with a detailed procedure and examples.

In Chapter 2, the authors examine the seismic assessment and retrofitting of an under-designed reinforced concrete (RC) frame. A displacement based procedure using non-linear static analyses is applied to a four-story RC frame to obtain an initial estimation of the overall inadequacy of the structure as well as the extent of different retrofitting interventions. Accurate numerical models are developed to reproduce the seismic response of the RC frame in the original configurations. The effectiveness of three different retrofitting solutions is examined through a displacement based approach using non-linear dynamic analyses.

In the third chapter, the authors propose a new single-run adaptive pushover method for the seismic assessment of shear wall structures. The main adaptive pushover procedures in the literature are reviewed and the proposed method is presented together with its numerical implementation. The main advantages of the method are explained and its predictions are compared to those of other recent adaptive pushover methods as well as to the rigorous non-linear time history analysis.

In the fourth chapter, the authors examine the influence of the shear-bending interaction on the global capacity of RC frames. The assessment of regular RC frame buildings is performed in accordance to the provisions of the Italian Seismic Code, Eurocode 8 and the CNR DT-212 guideline. A lumped plasticity model has been used with the aim of quantifying the differences between a fixed and a continuously updated shear span and between the use of inelastic springs at the member ends or continuously along the beam elements, with the purpose of considering the influence of axial-bending-shear interaction on the global structural capacity.

Chapter 5 discusses the selection of the suitable strengthening method and appropriate analysis type for the assessment of pre- and the post-intervention structural performance. The importance of displacement and drift limitations is emphasized for the determination of building performance under severe ground motion. Two strengthening techniques for non-ductile public RC buildings are utilized. Friction dampers and conventional RC-wall implementation methods are compared in terms of seismic performance particularly by means of displacement obtained using different methods.

The sixth chapter discusses the problem of the dynamic stability and post-critical processes of slender auto-parametric systems. High-rise structures exposed to a strong vertical component of an earthquake excitation are endangered by auto-parametric resonance effect. While in a sub-critical state, the vertical and horizontal response components are independent, exceeding a certain limit causes the vertical response to lose stability and induces dominant horizontal response. This effect is presented using two mathematical models: the non-linear lumped mass model; and the one-dimensional model with continuously distributed parameters. Special attention is given to the transition from a semi-trivial to a post-critical state in case of time limited excitation period as it concerns the seismic processes.

In the seventh chapter, the conventional linear force-based method is compared with the advanced nonlinear deformation-based method that is commonly preferred to investigate the seismic performances of the existing RC school buildings. The authors introduce the RC infill wall index parameter based on structural characteristics and layout plan of school buildings. The material strength and "RC infill wall index" for the seismic strengthening of existing RC school buildings are studied in terms of their influence on the seismic performance of the buildings.

In Chapter 8, the author presents some innovative structural concepts for the non-linear analysis, design and construction of multi-story base isolated buildings, including a new approach on installation of seismic isolation laminated rubber-steel bearings (SILRSBs) by clusters. Advantages of this approach are listed and illustrated by examples, while earthquake time histories analyses of base isolated vs fixed base buildings are carried out for comparison purposes.

In the ninth chapter, the author introduces a fuzzy logic concept for the performance based design criteria. Current theories and techniques such as Fuzzy Logic, Chaos, Fractal Geometry, and Artificial Neural Networks are reviewed. The implementation of Fuzzy Set Theory to performance-based design utilizing the graphical inference deduction technique is summarized taking into account the membership functions, clustering, rule-based systems and defuzzification. Finally, the divergence of performance based design criteria is expressed by making curvature evaluations for various regulations on a column section as a structural element. The clustering technique is applied on a univariate sample first and then on a multivariate one.

Chapter 10 defines resilience in different contexts in a comprehensive way and organizes the mathematical theory of network resilience by providing a generalization to create a quantitative framework for resilience characterization of an infrastructure network. A new performance index measuring delivery importance is employed and an industrial example using realistic data is solved to evaluate the resilience of the entire network. The presented principles and theory can be also applied to other infrastructures that are interconnected and operate as a network, such as transportation systems, electrical power, water supply and distribution systems.

In the eleventh chapter, the authors address the issue of seismic performance of buildings with construction deficiencies exposed to destructive earthquakes. Post-earthquake site observations in developing countries demonstrate insufficient capacities in these buildings depending on different structural and construction deficiencies. Structural Performance procedures in different earthquake codes are summarized with an aim to investigate and compare the analytically calculated structural performances of a building ensemble which consists of 3-5 story structures with known damage level by utilizing different procedures.

The book is intended for all those who wish to find out the results of the latest research in this scientific area of performance-based design of earthquake resilient concrete structures, in order to enhance their knowledge, provide a foundation for further study, indicate areas of new developments and/or areas where further developments are necessary. It intends to be a coherent and accessible manuscript summarizing the available knowledge and it can be efficiently used by researchers and professionals working in the field of earthquake engineering with emphasis on performance-based analysis and design. It could also be used for educational purposes at a postgraduate level on courses focusing on contemporary methods for reinforced concrete design.

Vagelis Plevris
Oslo and Akershus University College of Applied Sciences, Norway

Georgia Kremmyda
The University of Warwick, UK

Yasin Fahjan
Gebze Technical University, Turkey
October 10, 2016

Acknowledgment

The editors of the book would like to express their deep gratitude to all the contributors for their most valuable support during the preparation of this volume, for their time and effort devoted to the completion of their contributions, and for their great expert help in the review process. In addition, we are most appreciative to the members of the Editorial Advisory Board (EAB) of the book for their constructive comments and suggestions offered during the preparation process and for their contribution in reviewing some of the book chapters. We are also grateful to all the colleagues who, although did not contribute chapters to the book, were kind enough to offer their expert help during the review process. Finally, we would also like to thank the personnel of IGI Global, especially Ms. Jan Travers, Ms. Marianne Caesar and Ms. Katherine Shearer for their most valuable continuous support for the publication of this book.

Vagelis Plevris
Oslo and Akershus University College of Applied Sciences, Norway

Georgia Kremmyda
The University of Warwick, UK

Yasin Fahjan
Gebze Technical University, Turkey
October 10, 2016

Chapter 1
Selection and Scaling Time History Records for Performance-Based Design

Yasin M. Fahjan
Gebze Technical University, Turkey

F. İlknur Kara
Gebze Technical University, Turkey

Aydın Mert
Boğaziçi University, Turkey

ABSTRACT

Recent developments in performance-based analyses and the high performance of computational facilities have led to an increased trend for utilizing nonlinear time-history analysis in seismic evaluation of the performance of structures. One of the crucial issues of such analysis is the selection of appropriate acceleration time histories set that satisfy design code requirements at a specific site. In literature, there are three sources of acceleration time histories: 1) recorded accelerograms in real earthquakes scaled to match design code spectrum/uniform hazard spectra/conditional mean spectrum, 2) artificial records generated from white noise spectra to satisfy design code spectrum, and 3) synthetic records obtained from seismological models. Due to the increase of available strong ground motion database, using and scaling real recorded accelerograms is becoming one of the most contemporary research issues in this field. In this study, basic methodologies and criteria for selecting strong ground motion time histories are discussed. Design code requirements for scaling are summarized for ASCE/SEI-7-10, EC8 and Turkish Seismic Codes. Examples for scaling earthquake records to uniform hazard spectra are provided.

DOI: 10.4018/978-1-5225-2089-4.ch001

INTRODUCTION

Seismic analysis and design of engineering structures resist to earthquakes effects requires determination of seismic loads. Seismic design codes generally define ground shaking or seismic loading in the form of a design response spectrum and/or probabilistic uniform hazard response spectrum (UHS) obtained from probabilistic seismic hazard analysis (PSHA). Utilizing acceleration design spectrum or uniform hazard spectrum, seismic design is traditionally performed for most common engineering structures by the means of equivalent lateral static loading or modal spectrum analyses.

Nowadays, so many different advanced engineering applications such as tall buildings, critical facilities or irregular structures require more complex dynamic linear or non-linear analysis. For this reason, dynamic response history analyses (DRHA) are becoming more common and more popular methodology in the area of earthquake engineering with the help of recent advances in computer technologies and computational technologies and calculation techniques. This rigorous and sophisticated engineering computations require input motion in terms of acceleration time series consistent with design codes of such structures.

Basically, in the literature, there are three different sources available for design engineers to obtain acceleration time series. The first is to use real accelerograms recorded during past earthquakes. Due to the increase of available strong ground motion databases, real strong motion records are easily accessible in large numbers and their manipulation is relatively straightforward. Design engineer can be able to prepare a suite of time histories but the consistency with the design codes requirements should be taken into account.

The second is to use artificial accelerograms that are generated to match a target response spectrum by obtaining a power spectral density function from the smoothed response spectrum, and then to drive sinusoidal signals having random phase angels and amplitudes. The sinusoidal motions are then summed and an iterative procedure can be invoked to improve the match with the target response spectrum, by calculating the ratio between target and actual response ordinates at selected frequencies (Boomer and Acevedo, 2004).

The third is to use simulated accelerograms that is the acceleration time history obtained through earthquake simulations generated from seismological source models and accounting wave propagation path and site effects. These model range from point source stochastic simulations through their extension to finite source, to fully dynamic models of stress release

SOURCE OF ACCELEROGRAMS

There are three sources of acceleration time histories:

- Accelerograms recorded in real earthquakes,
- Artificial acceleration time series compatible with design response spectrum and
- Synthetic records obtained from seismological models (Figure 1).

Figure 1. Source of strong motion records

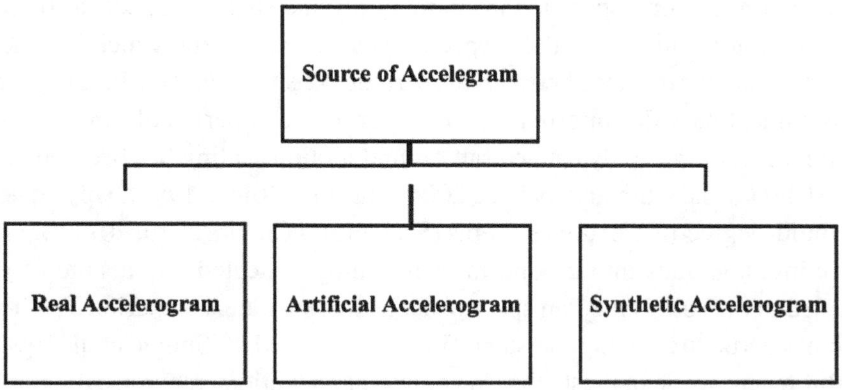

Real Accelerograms

Real Strong ground motion accelerograms contain a wealth of information about the nature of the ground shaking and carry all the ground-motion characteristics (amplitude, frequency, and energy content, duration and phase characteristics), and reflect all the factors that influence accelerograms (characteristics of the source, path, and site).

Due to the increase of available strong ground motion records, using and scaling real recorded accelerograms become one of most referenced contemporary research issues in this field. Despite the continued growth of the global strong motion databank, there are many combinations of earthquake parameters such as magnitude, rupture mechanism, source to site distance and site classification that are not well represented, which can make obtaining suitable records difficult in some circumstances (Boomer et al., 2003).

With the aim of accurate estimation of structural response, scaling of properly selected real ground-motion records can be used to warrant the accelerograms having similar strong-motion characteristics. To ensure correct estimation of median structural response with minimum dispersion for nonlinear Response History Analysis (RHA) scaling procedures must be used with sufficient number of accelerograms (Ay & Akkar, 2014). The most convenient method for selecting and scaling ground motions will depend on the structural response parameter(s) of interest, whether record-by-record variability in structural response is to be predicted and whether maximum responses or collapse responses are to be predicted.

The best method for selecting and scaling ground motions will depend on the type of assessment being performed. Generally, there are three types of performance assessment:

- Intensity,
- Scenario, and
- Time-based.

Intensity-based assessments are the most common of the three types and compute the response of a building and its components for a specified intensity of ground shaking. A scenario-based assessment computes the response of a building to a user specified earthquake event, which is typically defined by earthquake magnitude and the distance between the earthquake source and the building site. A risk-based or time-based assessment provides information on response over a period of time.

In the literature, most of the study carried out ground motion scaling has been on single-degree-of-freedom systems (Chopra and Chinatanapakdee 2004; Martinez-Rueda 1998). Up to date, there are only a few studies on multi-degree-of-freedom systems (Kalkan and Chopra 2010, 2011; Shome et al., 1998). Early quantitative investigations into ground motion scaling indicated that a suite of ground motions may be safely scaled to the suite's median spectral acceleration value, at a period T, without biasing the median response of a structure having the same first-mode period T (Shome et al. 1998, Iervolino and Cornell 2005). But recent work suggests that in some other situations record scaling may induce some bias in structural response (Baker and Cornell 2006b, Luco and Bazzurro 2007).

Artificial Accelerograms

Artificial accelerograms are generated to match a target response spectrum by obtaining a power spectral density function from the smoothed response spectrum, and then to derive sinusoidal signals having random phase angles and amplitudes. The sinusoidal motions are then summed and an iterative procedure can be invoked to improve the match with the target response spectrum, by calculating the ratio between the target and actual response ordinates at selected frequencies (Boomer and Acevedo, 2004). In order to get other characteristics of artificial spectrum compatible record, such as duration, it is necessary to obtain supplementary information about the expected earthquake motion apart from the response spectrum. Even though, it is possible to obtain acceleration time-series that are almost completely compatible with the elastic design spectrum, the generated accelerograms often have an excessive number of cycles of strong motion, and consequently have unrealistically high energy content.

The difficulty of the artificial time history generation methods lies in trying to match a single ground motion to a design response spectrum that is not intended to represent the motion from an individual earthquake (Naeim and Kelly, 1999). The design response spectrum is generally a result of a statistical analysis that considers the influence of several seismic sources simultaneously, whence the response at different periods may be driven by earthquakes in different sources and the spectrum is the envelope of spectra corresponding to scenarios in each of the sources (Reiter, 1990, Bommer et al., 2000).

Synthetic Accelerograms (Simulation of Acceleration Time History)

Different ground motion simulation methods have been developed by researchers. Especially after the adoption of performance-based design approach with the Earthquake resistant design of engineering structures, accurate estimation or simulation of strong motion time histories have been essential and one of the most popular study area in engineering seismology and earthquake engineering. In order to obtain necessary simulated ground motions with the intention of dynamic analysis of engineering structures and earthquake response, synthetic accelerograms can be generated from seismological source models considering the fault rupture processes and accounting for path and site effects.

In general, there are actual difficulties in defining appropriate input parameters such as the source, path, and site characteristics. To generate synthetic accelerograms there is a need for a definition of a specific earthquake scenario in terms of magnitude, rupture mechanism in addition to geological conditions and location of the site. Generally, most of these parameters are not often available, particularly when using seismic design codes (Boomer and Acevedo, 2004).

In recent years, numerous studies have been undertaken by seismologists particularly regarding the modelling of earthquake rupture processes in order for accurate prediction of ground motions generated by large earthquakes. There are numerous methods in the literature dealing with earthquake ground motion simulations. These methods can be mainly classified as

- Deterministic,
- Stochastic,
- Green Function and
- Hybrid methods (Figure 2).

Deterministic Earthquake Simulations Models

Deterministic earthquake simulations, utilizing kinematic source models, require the knowledge of fault slip distribution. In principle, deterministic methods combine Green's Function (GF) with the source function to generate surface ground motion during an earthquake (Kramer, 1996). In order to determine the ground motion at any point by this method, the earthquake source geometry, slip functions for the entire source, and the GF should be defined than the source needs to be divided into a finite number of discrete elements (Krinitzsky and Chang, 1977).

- Finite difference,
- Finite element, and
- Discrete wave number methods are the most commonly used approaches used in the literature in generating deterministic strong ground motion simulations.

Aki (1968), and Bouchon and Aki (1977) are two of most significant studies in this field that developed and utilized this model.

Figure 2. Ground motion simulation methodologies

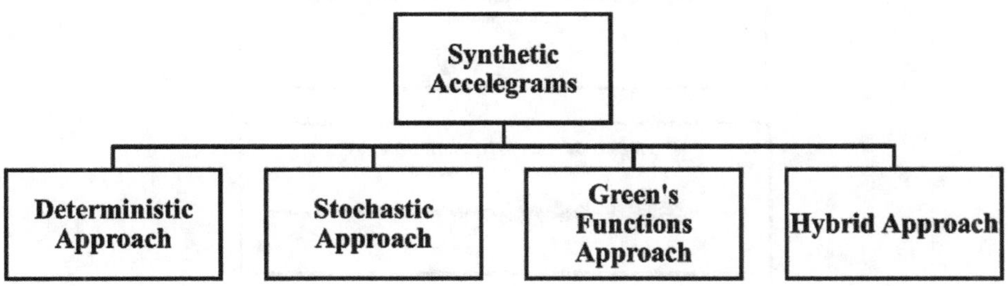

Stochastic Earthquake Simulation Models

Stochastic method is one of the most widely utilized strong ground motion simulation models. It is based on the assumption that the high frequency components of earthquake motion can be represented by an omega-squared (ω2) average spectrum obtained band-limited Gaussian noise. The Fourier amplitude spectra of Brune (1970) are satisfied by the main parameters of high frequency ground motion for earthquakes within a wide magnitude range (Trifunac and Brady, 1975).

Basic steps of the earthquake simulations using the stochastic method are shown in Figure 3. In general, two different source models, point source and finite source, are used in stochastic simulation techniques (Boore, 2000). The point source model (SMSIM - Stochastic Point Source Simulation) proposed by Boore (1983, 2003), in which the source is localized into a single point, is based on the assumption that the obtained waveform simulations involves both deterministic and random processes. Deterministic part that is basically function of both magnitude and distance and stochastic part is obtained by white noise. The point source model yields highly accurate results when the distance between the source and the receiver is significantly larger than the dimensions of the source itself.

Figure 3. Basic step in Stochastic earthquake simulations

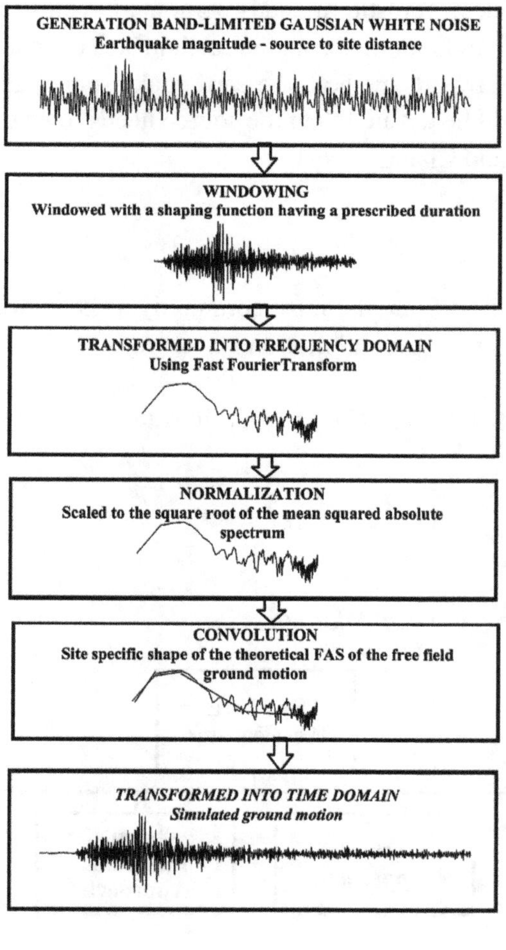

Beresnev and Atkinson (1997, 1998) developed a method for finite-fault strong ground motion simulation (FINSIM – Finite Fault Simulation) by modeling the fault rupture plane as a subfault matrix and by considering propagation originating from each subfault as a subfault point source with ω2-model spectrum (in conjunction with the principles of). The finite source model solved an important problem in stochastic earthquake simulation techniques, since it involves finite-fault rupture effects (e.g., source geometry, uncertainty in rupture parameters, non-homogeneity, etc.), which cannot be modeled by stochastic point source models. Motazedian and Atkinson (2005) introduced the concept of dynamic corner frequency to reduce the dependencies to sub source size. This, in turn, eliminated multiple triggering of sub-events. The sub source dependency was terminated by Boore (2009) with modification to sub-event normalization. This methodology was named as Stochastic Extended Simulation (EXSIM). The major advantage of EXSIM over FINSIM is its insensitivity to sub-fault size, conservation of energy radiated during the rupture process and only a portion of fault will be active at any given point of time (Atkinson and Boore, 2006).

Green's Function-Based Earthquake Simulations Models

In general terms, the Green's Function (GF) can be defined as the response of the ground to an impulsive point source. This definition is also an expression of the theoretical "Synthetic Green's Function" (SGF). In order to synthetically calculate the SGF it is necessary to know the crustal velocity structure of the region between the earthquake source and the recording station. Synthetic Green's functions at low frequencies (< 1 Hz) can be accurately calculated because the geology is often known well enough to be accurately modelled with a coarse spatial resolution. However, it is not possible to determine the heterogeneous nature of the ground with sufficient resolution, especially for the high-frequency or short wavelength components of ground motion.

Simulation of high frequency ground motion is still a difficult problem in seismology due to its random nature. Probably the most important restriction for that kind of simulation algorithm is related to the calculation of source and propagation path characteristics. Using the small magnitude earthquake as an Empirical Green's Functions (EGF) together with appropriate source function to simulate high frequency ground motion is the most promising idea to cope with this problem. Practically, the problems that stem from this heterogeneity can be best overcome by utilizing the EGF, which includes the effects of crustal velocity structure. EGFs are used to obtain not only the effect of the free surface and attenuation but also refractions, reflections and scattering due to heterogeneities along the propagation path. In addition, EGFs inherently include linear site response at the site where they are recorded.

The EGF method, which was first proposed by Hartzell (1978) and Wu (1978), is based on the principle that earthquake rupture and resulting ground motion can be modeled based on the elasto-dynamic representation theory and using the ground motion information recorded from small earthquakes. This method was later modified, improved, and utilized by some other researchers in their studies such as Hadley and Helmberger (1980), Irikura (1983), Hutchings and Wu (1990), and Hutchings (1991). The basis of this method is to utilize the observed small earthquakes originating from the rupture area of the simulated large earthquake to account on the actual propagation path and local site effects, and thereby determine, with great accuracy, the asperity fields in the source and the heterogeneity. A schematic presentation of earthquake simulations developed using GF is provided in Figure 4.

Figure 4. Schematic illustration of Green's Function simulations

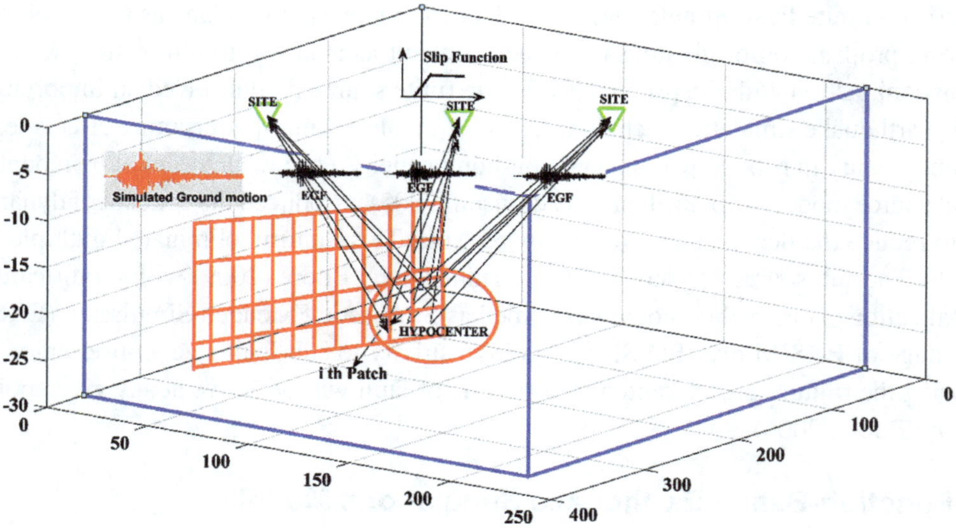

Hybrid Broadband Earthquake Simulations Models

During generation of broadband strong ground motion simulations by the hybrid approach, first the ground motion components from different frequency bands of the spectrum are determined individually, and then these data are combined together. Recent seismologic developments make the mathematical solution of the SGFs possible for the large wavelength (or low frequency) components, for which the geological heterogeneities can be modeled. High frequency components of ground motion, on the other hand, cannot be mathematically solved due to uncertainties related to the earthquake source and the geological heterogeneities. As a result, these components should be calculated either by stochastic simulation techniques of high-frequency components of the strong ground motion or by utilizing an EGF-based method that uses small earthquake records as a GF. Both low and high frequency band components of earthquake ground motion are determined through different methods, and these are later combined via specific filtering functions to obtain ground motion simulations within a wide frequency range. Designing of the filtering functions that will be used during combination processes is a particularly important subject as energies obtained from different parts of the spectrum need to be combined properly.

SELECTION OF TIME HISTORY RECORDS

Real earthquake records can be selected to match specific features of the ground motion, generally based on elastic response spectrum. The selection criteria of proper time history records to fit the design code spectrum are also taking into account the geological and seismological conditions at a specific site. The seismological and geological parameters can be classified in terms of

- Magnitude,
- Faulting type,

- Distance to fault,
- Rupture directivity,
- Site condition and
- Spectral content.

Both record selection and scaling are equally important processes for success of any nonlinear Response History Analysis applied. Appositely selecting records considering the hazard conditions for a given site helps to reduce the dispersion of Engineering Demand Parameters (EDPs) and increase accuracy by achieving better estimates of the "true median". Before scaling ground motions, one needs to define the hazard conditions associated with a given site either through deterministic or probabilistic site-specific hazard analysis or alternatively from the USGS seismic hazard maps (Kalkan and Chopra, 2010).

Guidance given in seismic design codes on how to select appropriate real records is usually focused on compatibility with the response spectrum rather than seismological parameters. Therefore, most of the time, acceleration time history records are selected on the basis of strong-motion parameters such as:

- Peak ground acceleration,
- Peak ground velocity, and
- Duration, to match a design response spectrum.

The ground motion time histories used to represent an intensity measure corresponding to a particular hazard level (or return period) should reflect the magnitude, distance, site condition, and other parameters that control the ground motion characteristics. Selection of records having appropriate magnitudes is important because magnitude strongly influences frequency content and duration of ground motion. In the literature, there are different studies point out that magnitude differences how to influence the response spectral shape (Kalkan and Chopra, 2010). In general, maximum amplitudes of the average spectral shape are relatively stable but events with larger magnitude yield wider response. It is desirable to use earthquake magnitudes within ± 0.25 magnitude units of the target magnitude (Stewart et al., 2001).

Selection of records having appropriate fault-site distances is important especially for near-fault sites, because the characteristics of near-fault ground motions differ from those of other ground motions. Previous studies reported that predominant period shifts to higher values with increase in distance from the fault for a given earthquake (Abrahamson and Silva, 1997).

Site conditions have a major effect on the characteristics and frequency content of the strong ground motion records. Even though the ground motions are amplified in soft soils, the high frequency motions are attenuated. Predominant period of spectral shape from a rock site is generally lower than that of a soil site. Unlike predominant period shifting to higher values with reduction in VS30, peak intensity of spectral shapes remains similar. Generally, the ground motions amplification effects can be observed in spectral acceleration of the records at intermediate to long period.

Fault type is not important criterion in the initial time series selection but near-source effects such as directivity is an important parameter to select initial time series especially if site is close to the fault and in the forward directivity zone. One another important issue is that if multiple recordings is used initial time series should be selected considering to nonstationary characteristic of ground motion. The easiest way to represent nonstationary characteristic of ground motion, initial time series should be selected from multiple earthquakes or at least multiple azimuths for a single earthquakes (Al Atik, 2010).

METHODS OF GROUND MOTION SCALING AND MATCHING

Generally, international seismic codes specify design spectra to identify earthquake loads to the engineering structures and recommend modifying selected ground motions to represent the seismic hazard associated with a design response spectrum and matching spectral acceleration within the period range of interest to use in the dynamic time history analysis.

Rarely, some region in the world, for a given site, it is possible to obtain recorded ground motion that represent important earthquake scenarios considering magnitude, distance to fault, fault characteristics and site characteristics. Most of the time, it is impossible to have sufficient recorded ground motion which match the controlling earthquake scenarios. In this condition, it may be desirable to modify some recorded acceleration time histories to obtain modified time histories that their response spectrum ordinates match the design response spectral ordinates over a period range of interest.

In the literature, basically two methods can be applied to modify actual time history record to be consistent with the design response spectrum:

- Scaling and
- Spectral matching (Al Atik, 2010).

To scale initial time history, multiply it with a constant to obtain scaled time series which is equal or exceeds the target spectrum over a specified period range. On the other hand, spectral matching requires modifying the frequency content of the recorded acceleration time series to match the design spectrum at all spectral periods (Figure 5).

Figure 5. Schematic illustration for ground motion scaling and matching

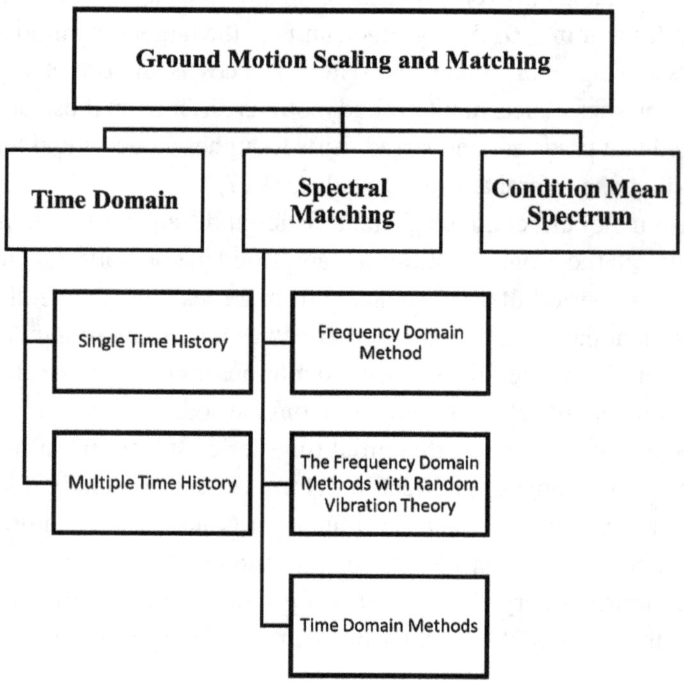

Ground Motion Scaling

In this approach, recorded motion is simply scaled up or down uniformly to best match the target spectrum within a period range of interest, without changing the frequency content.

General Procedure for a Single Time History

The procedure is based on minimizing the differences between the scaled motion's response spectrum and target spectrum in a least-square sense. The methodology proposed herein considers as the squared scaled-to-target "Difference", evaluated by the integral

$$\left|\text{Difference}\right| = \int_{T_S}^{T_F} \left[\alpha \; S_a^{actual}(T) - S_a^{\text{target}}(T) \right]^2 dT \tag{1}$$

where

S_a^{target} target acceleration response spectrum,
S_a^{actual} acceleration spectrum of the given (actual) time history,
α scaling factor,
T period of oscillator,
T_S lower period of scaling, and
T_F upper period of scaling

In order to minimize the difference, the first derivative of the "Difference" function with respect to the scaling factor has to be zero:

$$min \left|Difference\right| \quad \Rightarrow \quad \frac{d \left|Difference\right|}{d \; \alpha} = 0 \tag{2}$$

By combining Equations 1 and 2, we get Equation 3 in a discrete form in terms of initial (T_S) and final (T_F) periods and step increment (ΔT) of the response spectra range:

$$\alpha = \frac{\sum_{T=T_A}^{T_B} (S_a^{actual} \; S_a^{target})}{\sum_{T=T_A}^{T_B} (S_a^{actual})^2} \tag{3}$$

Scaling Procedure for Three-Dimensional Analysis

Scaling multi-component record suggested to have the same scale factor by in ASCE 7-10 (American Society of Civil Engineers, 2010). So ground motions consist of pairs of appropriate horizontal ground motion acceleration time-history components should be selected and used. The acceleration time history records have to be consistent with magnitudes, fault distance, and source mechanism that control the maximum considered earthquake. For each pair of horizontal acceleration time history components the square root of the sum of the squares (SRSS) of the five percent-damped spectrum of the scaled horizontal components is to be constructed. Each pair of ground motions shall be scaled over a period range from 0.2 T to 1.5T and the average of the SRSS spectra from all horizontal component pairs does not fall below the design spectrum.

General Procedure for Multiple Time History

When dealing with more than one input time history, one can either use the same procedure to fit each record separately, or try to best-fit the average of the produced spectra to the target spectrum. Multiple time history problems can be solved in three distinct ways:

1. The average of the (N) time histories is fitted to the target using a single scaling factor for all time histories. In this case, the matching of the produced average spectrum to the target one is quite good, since all-time histories are amplified by the same factor.
2. The single time history procedure is used for each one of the (N) time histories individually. Although each motion is "best-scaled", the resulting average does not perfectly match the target spectrum.
3. The (N) input motions are fitted using different scaling factors for each time history. In this method, a set of scaling factors are found such that the average spectrum of the scaled motions best fits the target. It should be noted that when using this method, even though an optimal average spectrum can be achieved, the outcome scaling factors for the different input time history may be very large or very small even negative values can be achieved.

Spectral Matching

In the literature several different algorithm was developed to generate artificial spectrum compatible acceleration time history (Preumont, 1984; Boomer and Acevedo, 2004; Adekrist and Eatherton, 2016). In general, spectral matching method can be categorized into three groups:

- The Frequency domain methods,
- The Frequency domain methods with random vibration theory and
- Time domain methods (Atik and Abrahamson, 2010).

Early spectral matching approaches used the Frequency domain methods. This method is based on the concept of using actual records to generate time histories that fit a given target response spectrum. In this method, an actual motion is filtered in the frequency domain by its spectral ratio with the design target spectrum. The Fourier phases of the motions remain unchanged during the entire procedure. The technique is repeated iteratively until the desired matching is achieved for a certain range of periods. The resulting time histories should be investigated in terms of suitability as input for linear and nonlinear time history analyses of engineering structures. For example, in real earthquake records, average of the ductility factor is expected to be equal to structural behavior factor at longer periods (equal displacement rule) especially for velocity and displacement sensitive spectral regions.

Although Frequency domain spectral matching was a common approach, there are two disadvantages; first, the methodology does not have good convergence properties in general. Namely, multiplicative scale functions applied to the Fourier amplitude spectra to match in the frequency domain disrupt the velocity and displacement time series obtained by integrating the acceleration record. That kind of corruptions can be clearly observed both in the velocity time series as a nonzero value offset at the end of motions and in the displacement time series as a linearly increasing or decreasing displacement. Second, the method alters the nonstationary character of the real recorded ground motion and increases the total energy in the ground motion finally it no longer looks like a time history record from an earthquake (Atik and Abrahamson,2010).

The Frequency domain methods with random vibration theory use the same algorithm with the Frequency domain methods. The only difference in this approach is that initial large adjustment to the Fourier amplitude spectrum was made using by the random vibration theory. In this approach, power spectral density function calculated from smoothed response spectrum and sinusoidal signals with random phase angles and amplitudes are used together with the iterative procedure to improve the matching level with the target response spectrum and recorded acceleration response. If you consider the acceleration and velocity time history record this approach produces good results. But frequently the character of displacement time series changes even if several base line correction methods can be applied (Atik and Abrahamson, 2010).

The third approach, the time domain spectral matching is based on adding wavelets to the initial time series in order to adjust it in the time domain. The wavelet is a mathematical function described effectively limited duration waveform with a zero average and its amplitude starts at zero, increases and then decreases again to zero.

Since the character of a real ground motion is better preserved (Lilhanand and Tseng, 1988) this approach is generally accepted as an improved approach in the literature (Adekrist and Eatherton, 2016). Time domain spectral matching was first proposed by Kaul (1978) and then extended by Lilhanand and Tseng (1987, 1988) adding with a tool to simultaneously match spectra at different damping values. This primitive time domain spectral matching algorithm which used an wavelet adjustment function provide numerical stability but does not preserve the nonstationary character of the initial acceleration time series and introduces drift to the resulting velocity and displacement time series.

Abrahamson (1992) developed the RspMatch program using by Lilhanand and Tseng algorithm and suggest a new adjustment wavelet (cosine wavelet) in order to preserve the nonstationary character of the time series and to ensure stability of the numerical solution. However, even he used a new adjustment wavelet it does not enough to solve drift problem not only the resulting velocity but also displacement time series. For that reason this version of RspMatch program requires baseline correction to correct the drift in the corresponding velocity and displacement time series.

Hancock et al. (2006) modify RspMatch to eliminate this drift and used to update RspMatch (2005). For this purpose, Hancock et al. (2006) propose a new adjustment wavelets and improve the methodology to obtain numerical stability by adding correction wavelet or implementing a reduction factor to the target response spectrum. Different wavelet adjustment function proposed by Al Altik and Abrahamson (2010) to obtain zero velocity and displacement time series without including any baseline correction.

The basic steps of this new methodology are summarized below;

1. For each period and damping ratio, compute the response of an elastic SDOF system.
2. To determine the inconsistency Compare the resulting response spectrum for each period and damping ratio with the target spectrum.
3. Calculate the spectral sensitivity matrix.
4. Calculate the set of wavelet magnitudes.
5. Add wavelets to the acceleration time histories with the appropriate phase and amplitude to modify the spectral ordinates.
6. To obtain spectral match with acceptable tolerance iterate the procedure repeating by the above steps.

CONDITIONAL MEAN SPECTRUM

Dynamic analysis of structures requires accurately estimating the structural response. General procedure to predict dynamic behavior of engineering structure based on selecting suites of ground motions according to a target design spectrum. Design engineer mostly use UHS as a target spectrum. UHS is determined by enveloping the spectral accelerations (Sa) at each periods that are exceeded with a given probability depending on PSHA procedure (Cornell, 1968). However, those spectral acceleration at each period are frequently very high values because, they were calculated corresponding to low probability exceedance such as 2% in 50 years. One another important consideration regarding to UHS, moderate earthquakes at short distances control the spectral response at short periods whereas large and far away earthquakes control the response at long periods (Bommer et al., 2000) so that it is not characterize the ground motion from a single earthquake

Alternative method named conditional mean spectrum (CMS) is proposed by Baker and Cornell (2006), Baker (2011) to reduce the UHS into realistic target spectra considering to the magnitude (M), distance (R) and epsilon (ε) which is a measure of the difference between spectral acceleration of a record and the mean of ground motion prediction equation at the given period. The CMS is a spectrum that matches the UHS at a conditioning period but that also represents the response from a single earthquake scenario more realistically than the UHS.

The UHS remains at a constant ε standard deviation from the median response spectrum at all periods. Considering a high positive ε value, the UHS represents an expected ground motion spectrum that all of its points take simultaneously rare values. This issue is in a significant conflict with the nature of real ground motion records (e.g., Haselton et al. 2011). The rate of observing a high positive ε at all periods is much lower than the rate of observing a high ε at any single period. Thus it can be concluded that the UHS represents a nearly impossible earthquake scenario, especially in higher levels of hazard (Baker and Cornell 2006a). The conservatism of the UHS has been addressed also by other researchers (e.g., Reiter 1990, Naeim and Lew 1995).

Previous studies in the literature clearly indicate that consideration of ground motion parameter ε values selecting earthquake record is important because ε is related to spectral shape and thus predictor of structural response. It has been shown that when selecting earthquake motions for dynamic structural analysis based on their ε values is more effective than selecting records based on magnitude and distance (Baker and Cornell 2006). While it is not appropriate the nature of real ground motion records the UHS has a constant ε standard deviation from the median response spectrum at all periods thus UHS represents a nearly impossible earthquake scenario, especially in higher levels of hazard.

CMS maintains the probabilistic rigor of PSHA, so that consistency is achieved between the PSHA and the ground-motion selection. This enables one to make quantitative statements about the probability of observing the structural response levels obtained from dynamic analyses that utilize this spectrum; in contrast, the UHS does not allow for such statements. In Probabilistic engineering assessments, this rigor is a significant benefit that likely justifies the slight changes in the analysis approach relative to traditional ground motion selection approaches.

The calculations involved in obtaining the CMS can be summarized as step-by-step calculation procedure is presented in Figure 6 and can be described as follows:

1. As a result of probabilistic hazard analysis, determine the target spectral acceleration at the dominant period of the structure model (T*). Perform de-aggregation process for the spectral acceleration at T* to determine the associated Magnitude (M), closest distance (R) and epsilon (ε)
2. Compute the median and logarithmic standard deviation of the response spectrum at T* given M and R resulted from de-aggregation process and using the same attenuation relationship used in probabilistic hazard analysis.
3. Knowing ε at the dominant period (T*), compute ε for the period range required for the structural analysis (Figure 7). Correlation of epsilons for the NGA ground motion models is given by Baker and Jayaram (2008)
4. Using median and logarithmic standard deviation of spectral acceleration at T* and correlation of epsilons, compute conditional mean spectrum (CMS) for T* using the equations suggested in Baker (2011). An example for CMS at T*=0.85 is plotted in Figure 8.

Figure 6. Simple procedure for computing CMS

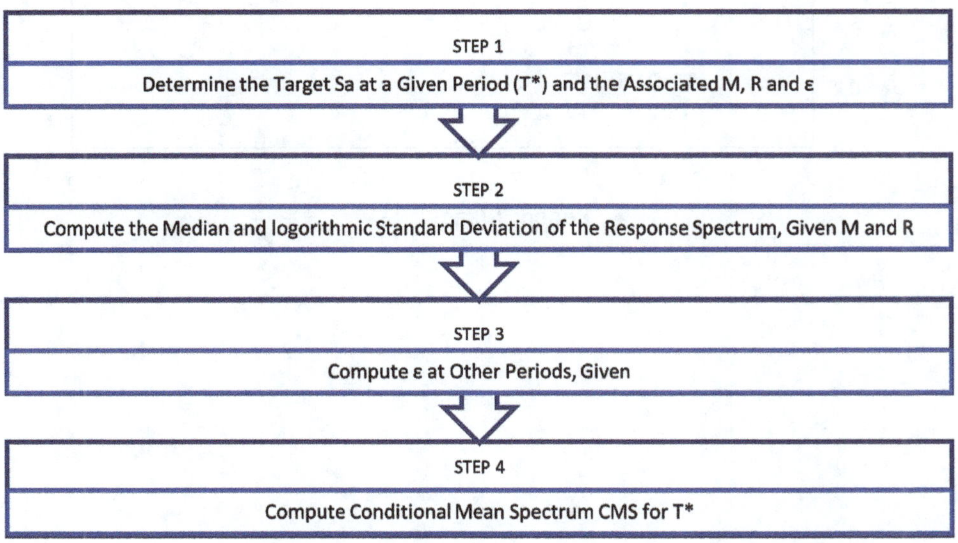

Figure 7. De-aggregation process for the spectral acceleration at T to determine the associated Magnitude (M), closest distance (R) and epsilon (ε)*

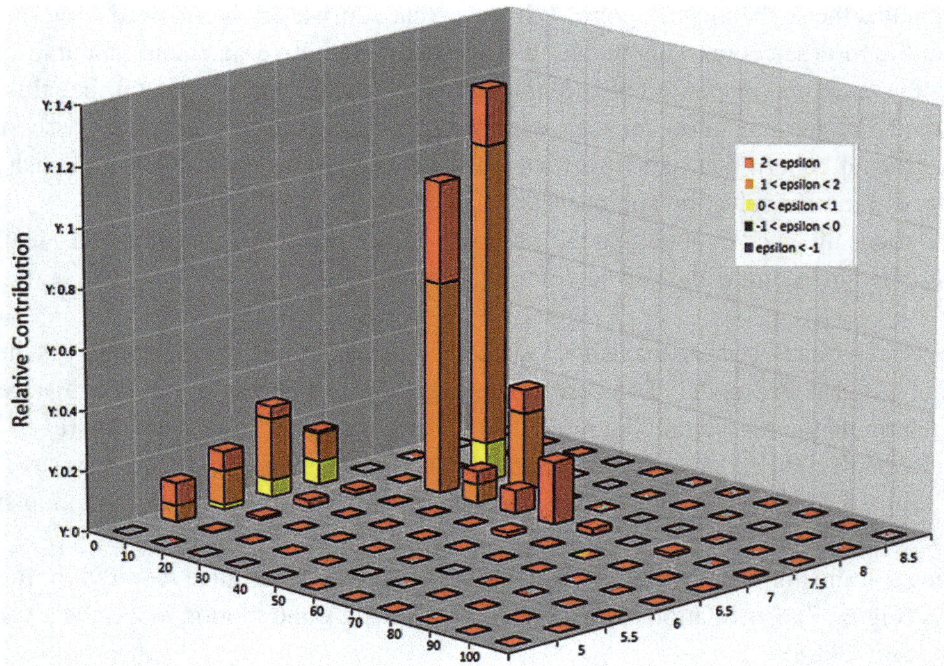

Figure 8. Correlation of epsilons for CMS

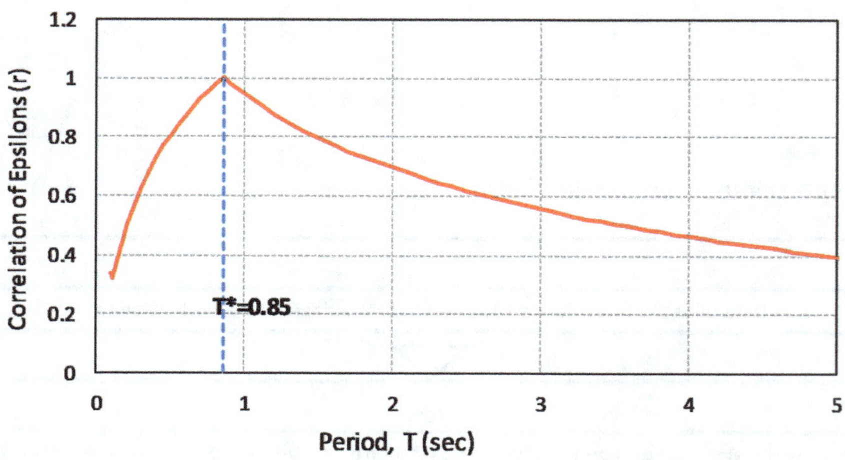

Figure 9. Condition mean spectrum curves compared with uniform hazard spectrum

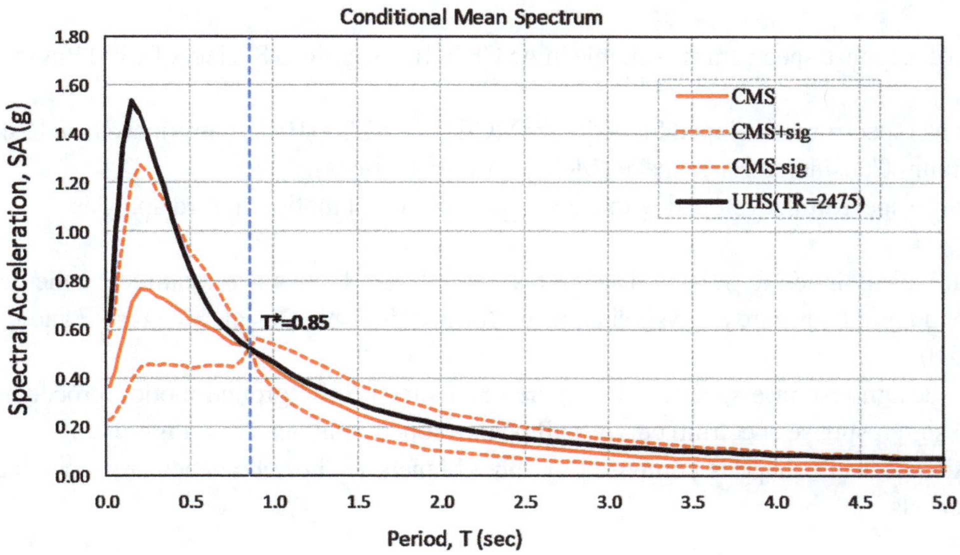

$$\mu_{lnSa(T_i)|lnSa(T^*)} = \mu_{lnSa}\left(M,R,T\right) + \rho\left(T_i,T^*\right)\varepsilon\left(T^*\right)\sigma_{lnSa}\left(T_i\right) \tag{4}$$

$$\sigma_{lnSa(T_i)|lnSa(T^*)} = \sigma_{lnSa}\left(T_i\right)\sqrt{1-\rho\left(T_i,T^*\right)^2} \tag{5}$$

where, in equation 4 and 5, $\mu_{lnSa}\left(M,R,T\right)$ and $\sigma_{lnSa}\left(T_i\right)$ are the predicted mean and standard deviation, respectively, of *lnSa* at given period, and *lnSa(T)* is the log of the spectral acceleration of interest. Note that $\varepsilon\left(T\right)$ is formulated in terms of *lnSa* values because *Sa* values are well represented by lognormal distributions. $\rho\left(T_i,T^*\right)$ *is the* correlation coefficient between the ε values at the two periods that have been calculated in previous studies. The parameter ε is defined as the number of standart deviations by which a given *lnSa* value differs from the mean predicted *lnSa* value for a given magnitude and distance. The relationships between amplitude, distance and ε value is shown in Figure 7. More detail explanation of the formulations can be found in Baker (2011).

RESPONSE SPECTRUM DEFINITION AND SELECTION AND SCALING GROUND MOTIONS ACCORDING TO SEISMIC DESIGN CODES

Seismic design codes generally define ground shaking in the form of a response spectrum of acceleration and permit to use spectrally matched scaled real/ artificial/ synthetic accelerograms recorded during earthquakes.

Response Spectrum Definition in ASCE/SEI-7-10

Two Types of response spectrum is available in ASCE 7-10 (American Society of Civil Engineers, 2010)

1. Response spectrum developed for bedrock (MCE_R). (ASCE 7-10 has introduced new Risk-Targeted Maximum Considered Earthquake (MCE_R) ground motions).
2. Response spectrum developed with a site-specific ground motion hazard analysis

For buildings assigned to Seismic Design Category E and F, or when required by the building official, a ground motion hazard analysis shall be performed. Seismic Design Category E and F is defined in ASCE 7-10.

Where a design response spectrum is required and site-specific ground motion procedures are not used, the design response spectrum curve shall be developed as indicated in Figure 10.

The basic spectrum shape is composed by three branches. These branches are separated by three "corner" periods:

- T_0,
- T_S,
- T_L, which are based on S_{DS} and S_{D1}.

S_{DS} is the design earthquake spectral response acceleration parameter at short period, S_{D1} at 1 sec period. S_{DS} and S_{D1} shall be determined from Equations (4) and (5), respectively.

$$S_{DS} = \frac{2}{3} S_{MS} \tag{4}$$

$$S_{D1} = \frac{2}{3} S_{M1} \tag{5}$$

S_{MS} and S_{M1} are the MCE_R spectral response acceleration parameter for short periods and at 1 s, respectively. S_{MS} and S_{M1} adjusted for site class effects and shall be determined by Equations (6) and (7).

$$S_{MS} = F_a S_S \tag{6}$$

$$S_{M1} = F_v S_1 \tag{7}$$

where F_a is the short-period site coefficient (at 0.2 sec-period) and, F_v is the long-period site coefficient (at 1.0 s-period). The parameters S_S and S_1 shall be determined from the 0.2 and 1.0 sec spectral response accelerations shown on figures which are prepared for MCE_R and Site Class B. Based on the site soil properties, the site shall be classified as Site Class; A, B, C, D, E, or F in ASCE7-10.

Figure 10. Developing a response spectrum based on ASCE 7-10

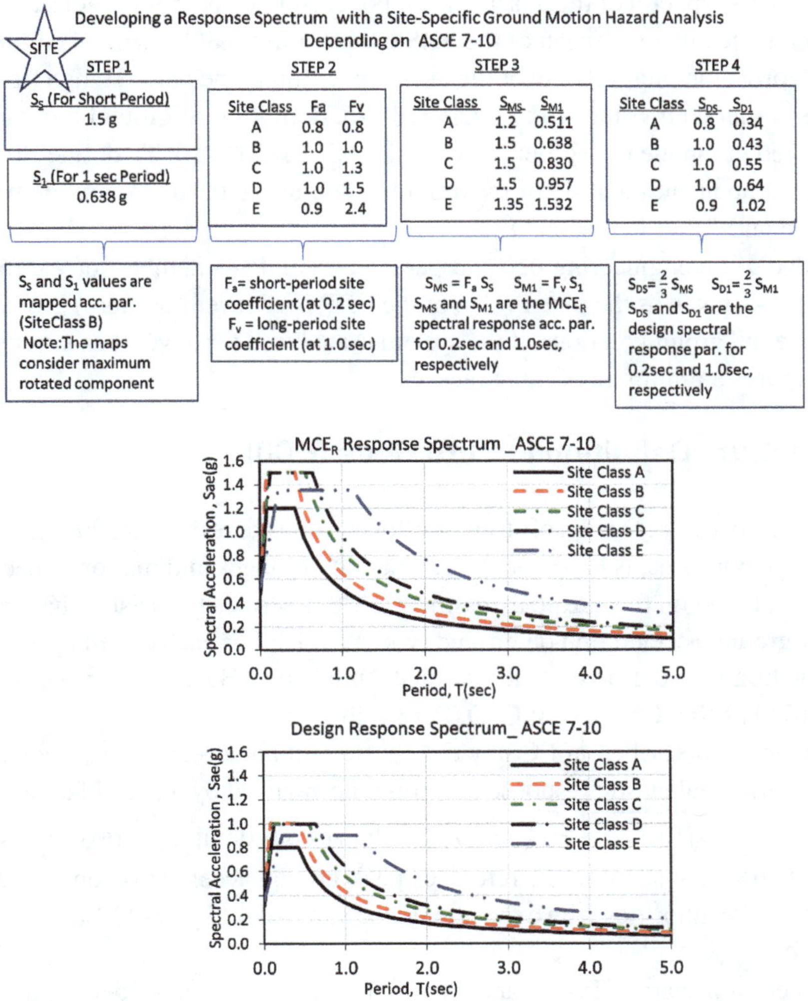

Selection and Scaling Ground Motions According to ASCE/SEI-7-10

According to ASCE/SEI-7-10 earthquake records should be selected from events of magnitudes, fault distance and source mechanisms that are consistent with the maximum considered earthquake (MCE_R). The risk-targeted motions MCE_R represent the ground motion levels associated with 1% probability of collapse in 50 years.

For two-dimensional analysis of symmetric-plan buildings, ASCE/SEI-7-10 requires intensity-based scaling of ground motion records using appropriate scale factors so that the mean value of the 5%-damped response spectra for the set of scaled records is not less than the design response spectrum over the period range from $0.2T_n$ to $1.5T_n$. (Where T_n is the elastic first-"mode" vibration period of the structure).

For three-dimensional analyses, ground motions should consist of pairs of appropriate horizontal ground motion acceleration components (the horizontal ground motions are to be determined for the orientation that provides the maximum structural response). For each pair of horizontal components, a square root of the sum of the squares (SRSS) spectrum should be constructed by taking the SRSS of the 5%-damped response spectra of the unscaled components. Each pair of motions are then scaled with the same scale factor such that the mean of the SRSS spectra from all horizontal component pairs does not fall below the corresponding ordinate of the target spectrum in the period range from $0.2T_n$ to $1.5T_n$.

It is worth to say that comparing with ASCE/SEI 7-5, definition of ground motion was changed to the maximum rotated component of ground motion in ASCE/SEI 7-10. To define the spectrum as the maximum rotated component, the ratio of maximum to geometric mean spectral demand can be taken as 1.1 at 0.2 second and 1.3 at 1.0 second [2].

The design value of an engineering demand parameter (EDP)—member forces, member deformations or story drifts—is taken as the mean value of the EDP over seven (or more) ground motions, or its maximum value over all ground motions, if the system is analyzed for fewer than seven ground motions. This procedure requires a minimum of three records.

Response Spectrum Definition in Euro Code (EC8)

According to EC8 (Eurocode 8: Design of structures for earthquake resistance, 2005), the seismic action, considered for design purposes, is based on the estimation of the ground motion expected at each location in the future, i.e. based on the hazard assessment. EC8 describes the seismic hazard by the value of the reference peak ground acceleration on ground type A, (a_gR). For each country, the seismic hazard is described by a zonation map defined by the National Authorities. Based on the site soil properties, the site shall be classified as Site Class A, B, C, D, E, or F in EC8.

The ground motion is described in EC8 by the elastic ground acceleration response spectrum. The basic shape of the horizontal elastic response spectrum, normalized by ag. Soil factor "S" and damping correction factor "η" $\eta = \sqrt{10 / (5 + \xi)} \geq 0,55$, ξ is the viscous damping ratio of the structure) are the main parameters of spectrum shapes. The reference peak ground acceleration on type A ground, a_gR is defined by the National Authorities with a recommended value of 475 years (10% probability of exceedance in 50 years) for T_{NCR}.

Two types of spectrum shapes, Type 1 and Type 2, (Figure 11) are considered for seismicity conditions. In this regard, provisions of EC8 state:

If the earthquakes that contribute most to the seismic hazard defined for the site for the purpose of probabilistic hazard assessment has a surface wave magnitude, M_s, not greater than 5.5, it is recommended that the Type 2 spectrum is adopted.

Type 1: High and moderate seismicity regions ($M_s > 5,5$).
Type 2: Low seismicity regions ($M_s \leq 5,5$); near field earthquakes.

The basic spectral shape is composed by four branches. These branches are separated by three "corner" periods: T_B, T_C and T_D which are Nationally Determined Parameters (NDPs), allowing the adjustment of the spectral shape to the seismo-genetic specificities of each country.

Figure 11. Developing a response spectrum based on EuroCode EC-8

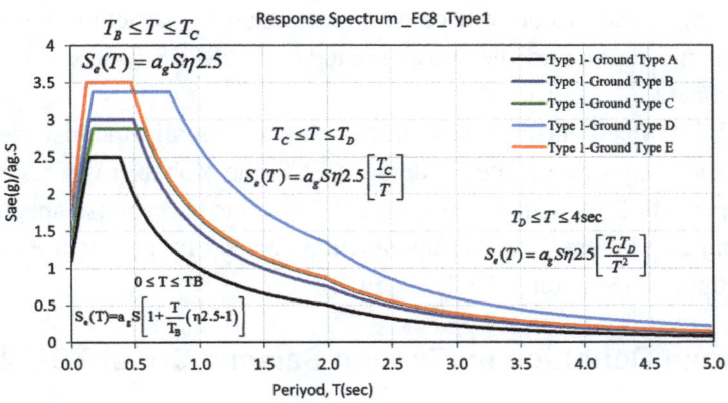

(a) *Type 1; High and moderate seismicity regions (Ms > 5,5)*

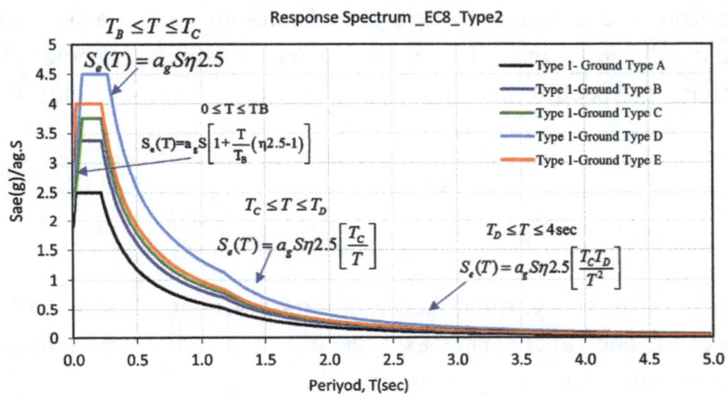

(b) *Type 2; Low seismicity regions (Ms ≤ 5,5); near field earthquakes*

Selection and Scaling Ground Motions According to Euro Code (EC8).

According to EN 1998-1 (EuroCode 8 (EC8) -Part1), selection of real ground motion records should consider the requested criteria such as:

- Range of magnitude,
- Fault distance,
- Site classes etc.

To perform the seismic analysis, recorded ground accelerations, artificial or synthetic accelerograms can be used. For the analysis, minimum three accelerograms should be used. When 3 different accelerograms are used, EDP is taken from the unfavorable value of three dynamic analyses. When at least seven different (real or simulated) records are used, EDP can be derived from the average of the response quantities resulting from all analyses (Section 4.3.3.4.3, EC8). The mean of the spectral acceleration values at zero-period, as calculated from the individual records selected, must exceed the value of $a_g S$ for the site considered (S – soil amplification factor). EC8 states that seismic motion should consist of three simultaneously acting accelerograms representing the two horizontal and the vertical component of strong ground motion. The same accelerograms may not be used simultaneously along both horizontal directions. The maximum value of each action effect on the structure due to the two horizontal components of the seismic action may then be estimated by the SRSS values of the action effect due to each horizontal component. (4.3.3.5.1).

The mean of the 5% damped elastic spectrum calculated from all time histories should not be less than 90% of the corresponding value of the 5% damped EC8 (the probability of exceedance of the design earthquake within a period of 50 years is 10%. (2.1, EC8)) elastic response spectrum, in the range of periods from $0.2T_1$ and $2T_1$, where T_1 is the fundamental period of the structure in the direction where the accelerograms is applied (Section 3.2.3.1.2.4, EC8).

Response Spectrum Definition in (Turkish Seismic Code)-TSC 2007

When required, elastic spectral acceleration S_{ae} (T) which is defined as the ordinate of 5% damped elastic design acceleration spectrum may be determined through special investigations by considering local seismic and site conditions (Figure 12). S_{ae}(T) is equal to spectral acceleration coefficient times the acceleration of gravity, g. Spectral acceleration coefficient A(T) can be determined with Equation 8.

$$\begin{aligned} A(T) &= A_0 IS(T) \\ S_{ae}(T) &= A(T)g \end{aligned} \tag{8}$$

In Equation 8, A_0 is the Effective Ground Acceleration Coefficient, I is the building importance factor, and S(T) is the spectrum coefficient. The basic spectral shape is composed by three branches. These branches are separated by three "corner" periods: T_A, T_B.

Selection and Scaling Ground Motions According to (Turkish Seismic Code): TSC 2007

Real earthquake records are selected to match specific features of the ground motion, generally based on either elastic response spectrum, or an earthquake scenario with the minimum parameters being the magnitude, distance and site classification. Turkish seismic code (Code, T. E.,2007) allow the use of artificially generated, previously recorded or simulated accelerograms as input ground motions for linear and nonlinear seismic analyses. Mean spectral acceleration of generated ground motions for zero-periods shall not be less than (A_0 g) and the mean spectral accelerations of artificially generated acceleration records for 5% damping ratio shall not be less than 90% of the elastic spectral accelerations, S_{ae} (T), in the period range between $0.2T_1$ and $2T_1$ with respect to dominant natural period, T_1. of the building in the earthquake direction considered. The design earthquake considered in the Turkish specification corresponds to high intensity earthquake for residential buildings where the probability of exceedance of the design earthquake within a period of 50 years is 10%.

At least three ground motions shall be used where the maximum of the results, and if at least seven ground motions are used and the mean values of the results shall be considered for design.

Figure 12. Developing a response spectrum depending on TSC-2007

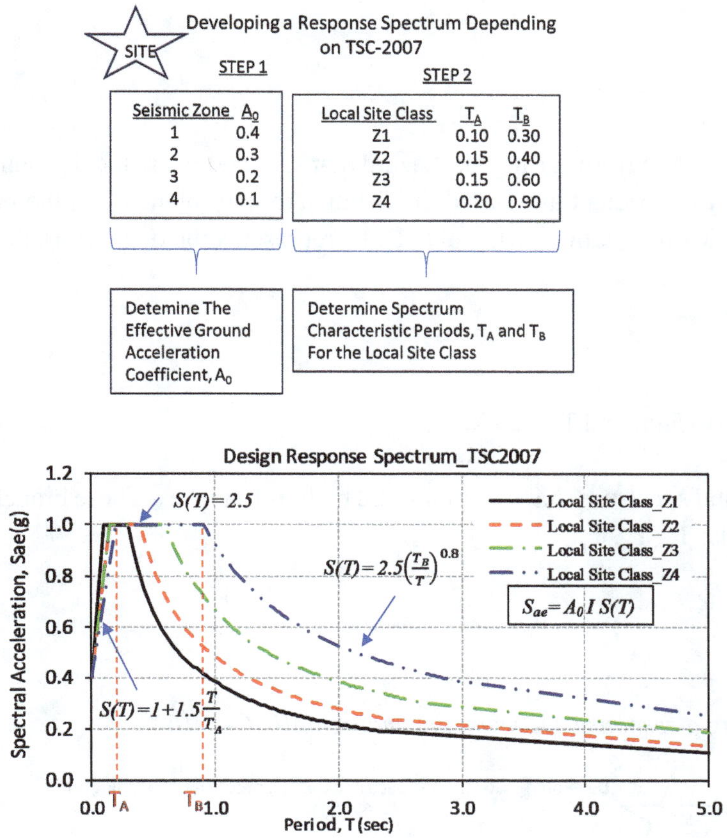

Response Spectrum Definition in (Turkish Seismic Code): TSC 2016

According to TSC-2016 (Code, T. E., 2016, daft version of Seismic Code is available in date), the seismic action is based on the hazard assessment. The seismic hazard is described by Earthquake Hazard Maps. Hazard maps (www.deprem.gov.tr/) are prepared for four ground motion levels and reference soil ($V_{s,30}$=760 m/sec). Based on the site soil properties, the site shall be classified as Site Class ZA, ZB, ZC, ZD, ZE, or ZF in TSC-2016.

For a given ground motion state and reference soil properties, response spectrums, with %5 damping ratio, defined with;

- Map spectral acceleration coefficients,
- Fault proximity coefficient and
- Local ground effect coefficients.

When required, spectrums can be defined based on site specific seismic hazard analysis.

Design earthquake spectral response acceleration parameter at short period, S_{DS}, and at 1 sec period, S_{D1}, shall be determined from Equations (8) and (9), respectively. (Figure 13)

$$S_{DS} = S_S \; F_S \tag{9}$$

$$S_{D1} = S_1 \; \gamma_F \; F_1 \tag{10}$$

where S_S and S_1 can be taken from earthquake hazard maps. F_S and F_1 are local ground effect coefficients are the fault proximity coefficient and based on distance to fault plane. γ_F is the near-fault coefficient and can be calculated with Equation (10) and (11). L_F represents the distance to the fault plane (km).

$$\gamma F = 1.2 \; LF < 15 \text{ km} \tag{11}$$

$$\gamma F = 1.2 - 0.02(LF - 15) \; 15 \text{ km} < LF \leq 25 \text{ km} \tag{12}$$

The basic horizontal spectrum shape is composed by four branches. These branches are separated by three "corner" periods: T_A, T_B, T_L.

Figure 13. Developing a response spectrum depending on TSC-2016

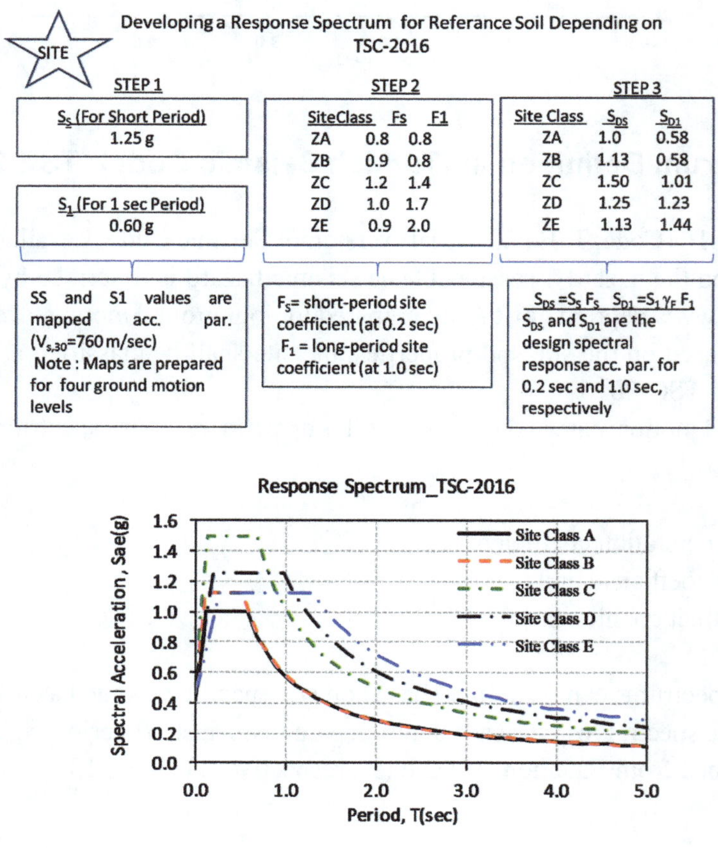

Selection and Scaling Ground Motions According to TSC 2016

According to TSC-2016 earthquake records should be selected from events of magnitudes, fault distance and source mechanisms that are consistent with the design earthquake level. According to the TSC 2016 the eleven ground motion sets are used The number of records from same earthquake will not exceed three. If available for the site real records should be used, if not synthetic record can be used.

For two-dimensional analysis, the set of scaled records is not less than the design response spectrum over the period range from $0.2T_p$ to $1.5T_p$ (where T_p is the elastic first-"mode" vibration period of the structure).

For three-dimensional analyses, each pair of horizontal components should be combined with square root of the sum of the squares (SRSS) methodology. Average of the combined spectrums of all the selected records are then scaled. The ratio of scaled spectrum and design spectrum not fall below 1.3 in the period range from $0.2T_p$ to $1.5T_p$. Each pair of ground motions will be scaled with the same scale factor.

Examples on Selection and Scaling Ground Motions

Source mechanisms, path attenuation and site effects are considered in the selection of ground motions. The selection of ground motions is performed based on magnitude, distance to source and site soil conditions. Earthquakes with large magnitudes have, in general, long effect duration. Distance to source has a high impact on site ground motion through path attenuation. Based on the criteria discussed above, the selected ground motion records are selected to cover a wide range of ground motions with different natural characteristics, and details of the selected ground motions. With de-aggregation analyses the moment magnitude and closest distance to the fault of the predominant scenario for MCE_R, seismic levels are computed. Based on ASCE 7-10, 2010, a general procedure for selecting and scaling pairs of horizontal ground motion time histories components for three-dimensional analyses of structures can be summarized as follows:

- Real earthquake ground motion records (EQGMs) are used for selecting each ground motion compatible with the design spectra/uniform hazard spectrum for specified seismic levels (MCE_R) corresponding to return period $T_R = 2475$ years.
- The ground motion duration in accelerograms between the points where the acceleration exceeds 5% g for the first and last times equals, at least, to 15 sec or five times the fundamental period of vibration of the structure (T). (ASCE 7.05)
- For each pair of horizontal ground motion components, a square root of the sum of the squares (SRSS) spectrum shall be constructed by taking the SRSS of the 5 percent-damped response spectra for the scaled components (where an identical scale factor is applied to both components of a pair).
- The ground motions are scaled such that the average value of the SRSS spectra does not fall below 1.3 times the 5% damped spectrum of the design earthquake between 0.2T and 1.2T. For isolated structures, the average of the SRSS spectra from all horizontal component pairs does not fall below the corresponding ordinate of the response spectrum used in the design (S_{Max}) between the periods $0.5T_D$ and $1.25T_M$, respectively, where T_D is the effective period of the isolated structure at design displacement and T_M is the effective period of the isolated structure at maximum displacement.

- The best fitted 7 ground motions to design response spectrum are selected. PEER WEST2 NGA (2014) database ground motions can be used for the selection.

As an example the best fitted ground motions to a design spectrum is shown in terms of SSRS of acceleration and displacement response spectrum in Figure 14. Site specific uniform hazard spectrum resulted from probabilistic hazard analysis can be also used as a target spectrum with the same scaling procedure. Geometric means of acceleration and displacement response spectra of the horizontal components of scaled records are shown Figure 15.

In the case of scarcity of number of records to fit the design/uniform hazard spectrum for a specific period range especially long periods for tall structures, spectral matching of the ground motion records can be applied. The best scaled records are selected as initial step before the matching process. Spectral matching of the ground motion records can be achieved utilizing RSPMatch09 (Al Atik & Abrahamson

Figure 14. SRSS of acceleration and displacement spectra of horizontal components for best fitted records of design spectrum using time domain scaling method

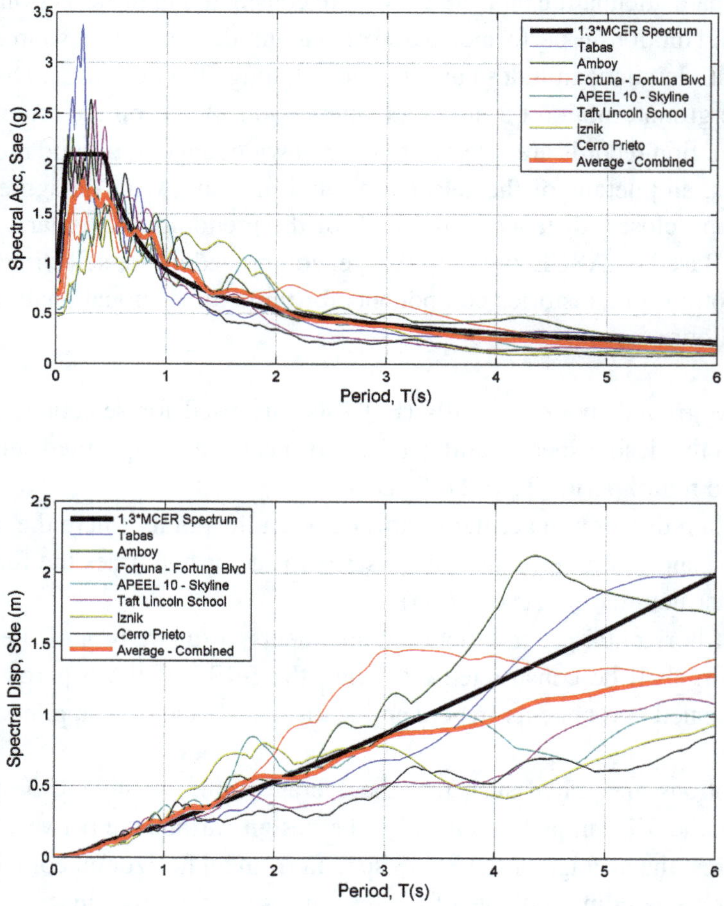

Figure 15. Geometric means of acceleration and displacement horizontal components spectra for best fitted records of uniform hazard spectrum using time domain scaling method

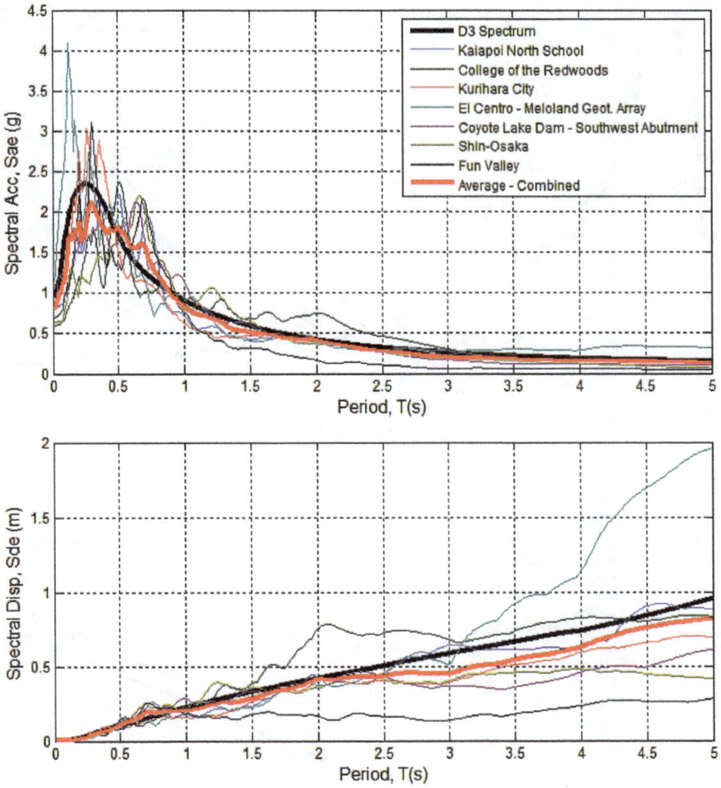

2010) using improved tapered cosine model. The improved algorithm utilized in RSPMatch09 extends the seed time histories by zero padding in the beginning and adding long period wavelets at the end of the time histories to achieve the zero drift in the time histories. Synchronizing the different padding zeros for the same record's components should be performed (2 horizontal, 1 vertical). Example for spectral matching results is shown in Figure 16 in terms of acceleration and displacement response spectrum.

- Acceleration, velocity, displacement time histories;
- Normalized Arias intensity, spectral acceleration, displacement and design and combined spectra;
- Average spectral acceleration and displacement combined spectra for the scaled ground motions for MCE_R seismic hazard levels are given in Figure 17 and Figure 18.

Conditional mean spectrum (CMS) methodology can be applied for selecting and scaling the horizontal components of the records as discussed though the text. An example for geometric mean of the scaled records' acceleration and displacement response spectra is shown Figure 19.

Figure 16. SRSS of acceleration and displacement spectra of horizontal components for best fitted records of design spectrum using spectral matching method

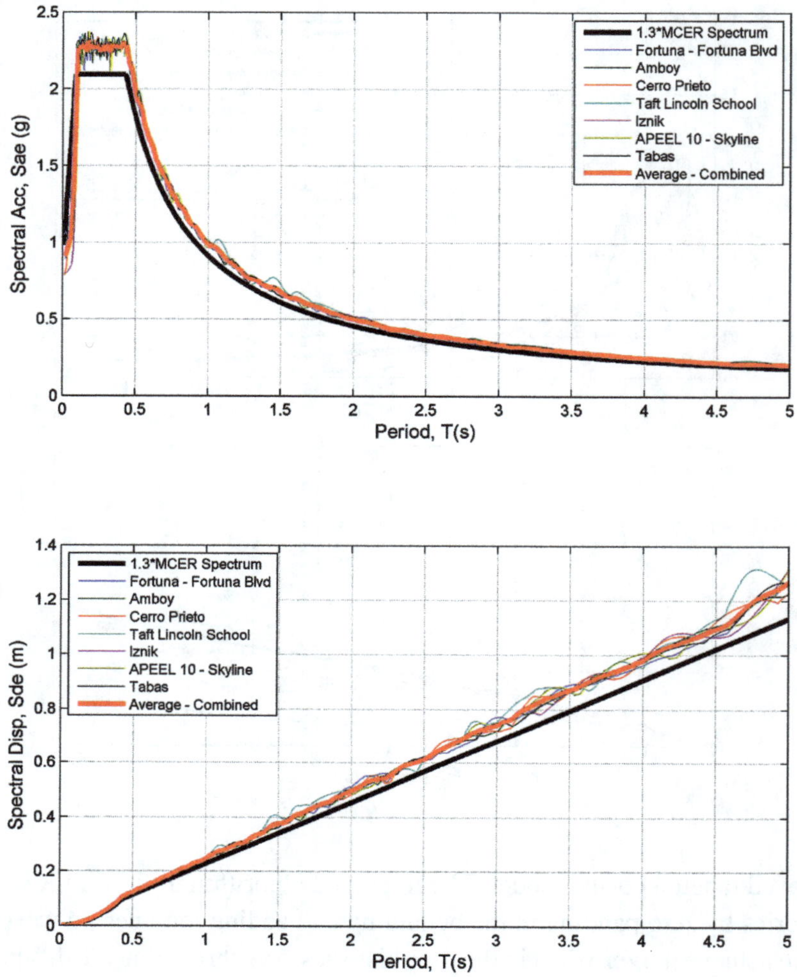

Figure 17. Acceleration, velocity and displacement time histories of original and scaled records using spectral matching method

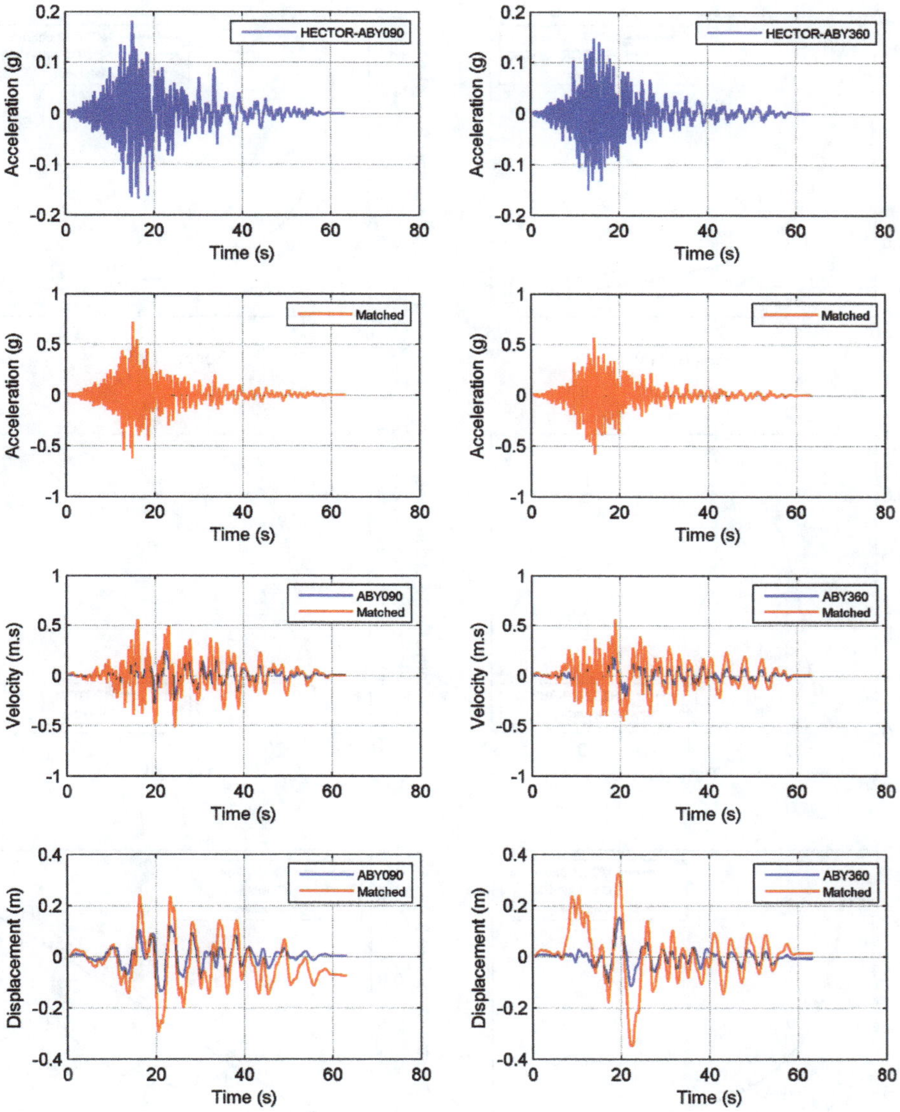

Figure 18. Spectral acceleration and Arias intensity for original and scaled components using spectral matching method

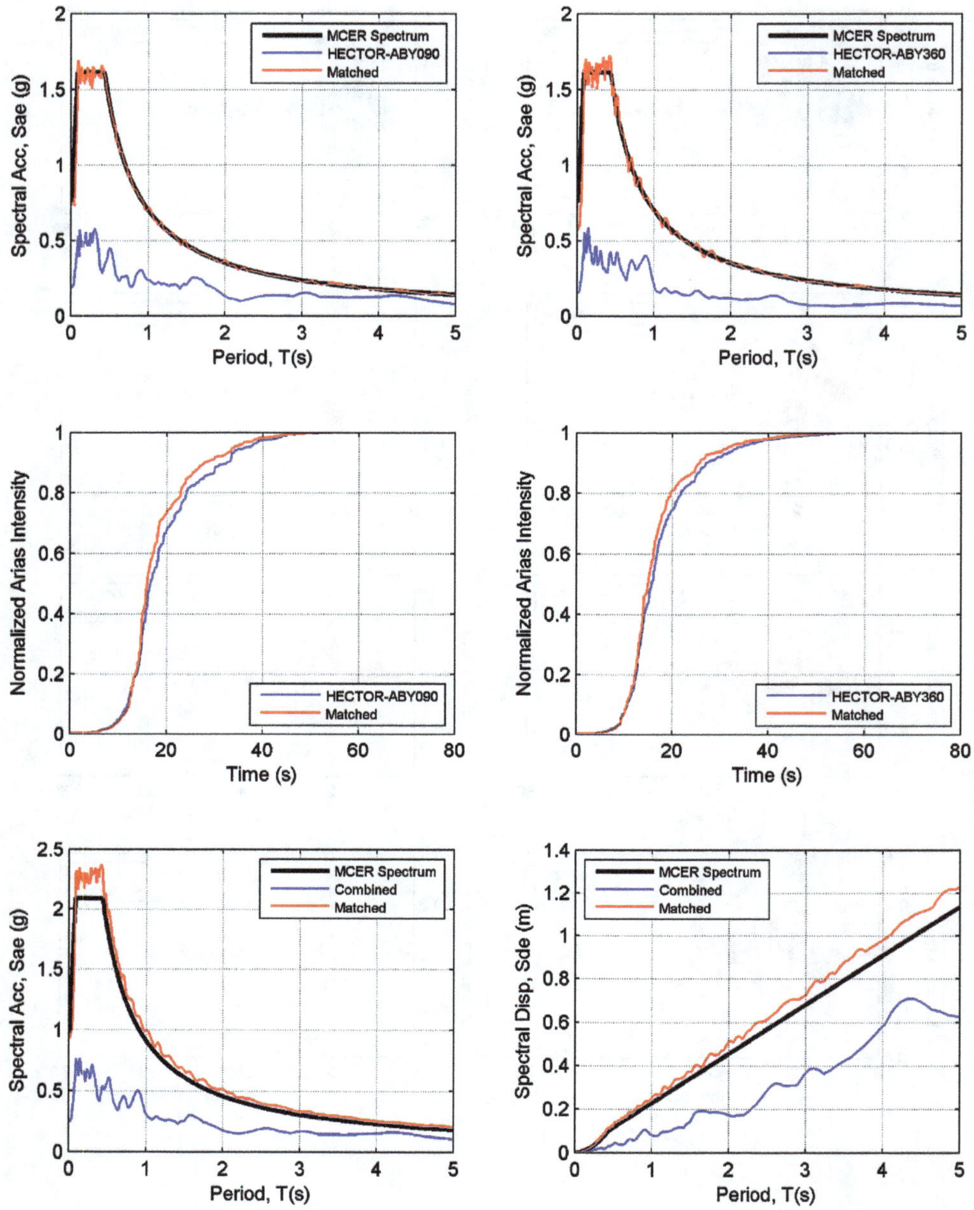

Figure 19. Spectral acceleration scaling using Conditional Mean Spectrum Methodology

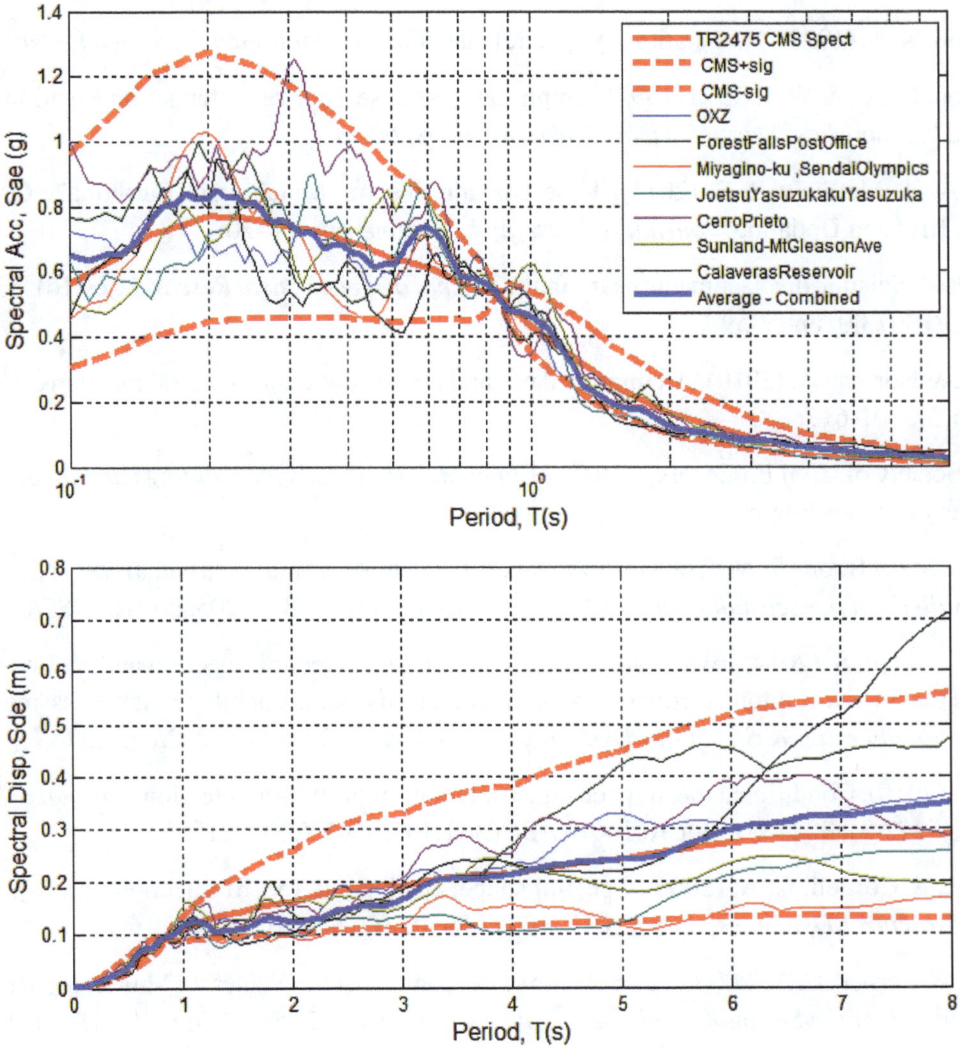

REFERENCES

Abrahamson, N. A. (1992). Non-stationary spectral matching. *Seismological Research Letters, 63*(1), 30.

Abrahamson, N. A., & W. J. Silva (1997). Empirical Response Spectral Attenuation Relations for Shallow Crustal Earthquakes. *Seismological Research Letters, 68*(1).

Adekristi, A., & Matthew, R. E. (2015). Time-Domain Spectral Matching of Earthquake Ground Motions using Broyden Updating. *Journal of Earthquake Engineering*, 1–20.

Aki, K. (1968). Seismic displacements near a fault. *Journal of Geophysical Research, 73*(16), 5359–5376. doi:10.1029/JB073i016p05359

Al Atik, L., & Norman A. (2010). An improved method for nonstationary spectral matching. *Earthquake Spectra, 26*(3), 601-617.

American Society of Civil Engineers. (2010). *Minimum design loads for buildings and other structures*. Amer Society of Civil Engineers.

Atkinson, G. M., & Boore, D. M. (2006). Earthquake ground-motion prediction equations for Eastern North America. *Bulletin of the Seismological Society of America, 96*(6), 2181–2205. doi:10.1785/0120050245

Ay, B. Ö., & Akkar, S. (2014). Evaluation of a recently proposed record selection and scaling procedure for low-rise to mid-rise reinforced concrete buildings and its use for probabilistic risk assessment studies. *Earthquake Engineering & Structural Dynamics, 43*(6), 889–908. doi:10.1002/eqe.2378

Baker, J. W. (2010). Conditional mean spectrum: Tool for ground-motion selection. *Journal of Structural Engineering, 137*(3), 322–331. doi:10.1061/(ASCE)ST.1943-541X.0000215

Baker, J. W., & Cornell, C. A. (2006). Spectral shape, epsilon and record selection. *Earthquake Engineering & Structural Dynamics, 35*(9), 1077–1095. doi:10.1002/eqe.571

Baker, J. W., & Cornell, C. A. (2006b). Correlation of Response Spectral Values for Multicomponent Ground Motions. *Bulletin of the Seismological Society of America, 96*(1), 215–227. doi:10.1785/0120050060

Baker, J. W., & Jayaram, N. (2008, February). Correlation of Spectral Acceleration Values from NGA Ground Motion Models. *Earthquake Spectra, 24*(1), 299–317. doi:10.1193/1.2857544

Beresnev, I. A., & Atkinson, G. M. (1997). Modeling finite-fault radiation from the ωn spectrum. *Bulletin of the Seismological Society of America, 87*(1), 67–84.

Beresnev, I. A., & Atkinson, G. M. (1998). FINSIMa FORTRAN program for simulating stochastic acceleration time histories from finite faults. *Seismological Research Letters, 69*(1), 27–32. doi:10.1785/gssrl.69.1.27

Bommer, J. J., & Acevedo, A. B. (2004). The use of real earthquake accelerograms as input to dynamic analysis. *Journal of Earthquake Engineering, 1*(1), 43–91. doi:10.1080/13632460409350521

Bommer, J. J., Acevedo, A. B., & Douglas, J. (2003). The Selection and Scaling of Real Earthquake Accelerograms for Use in Seismic Design and Assessment.*Proceedings of ACI International Conference on Seismic Bridge Design and Retrofit.*

Bommer, J. J., Scott, S. G., & Sarma, S. K. (2000). Hazard-consistent earthquake scenarios. *Soil Dynamics and Earthquake Engineering*, *19*(4), 219–231. doi:10.1016/S0267-7261(00)00012-9

Boore, D. M. (1983). Stochastic simulation of high-frequency ground motions based on seismological models of the radiated spectra. *Bulletin of the Seismological Society of America*, *73*, 1865–1894.

Boore, D. M. (2000). *SMSIM – FORTRAN Programs for Simulating Ground Motions from Earthquakes: Version 2.0 – A Revision of OFR 96-80-A*. USGS Open File Report OF 00-509.

Boore, D. M. (2003). Simulation of ground motion using the stochastic method. *Pure and Applied Geophysics*, *160*(3), 635–676. doi:10.1007/PL00012553

Boore, D. M. (2009). Comparing stochastic point-source and finite-source ground-motion simulations: SMSIM and EXSIM. *Bulletin of the Seismological Society of America*, *99*(6), 3202–3216. doi:10.1785/0120090056

Bouchon, M., & Aki, K. (1977). Discrete wave-number representation of seismic wavefields. *Bulletin of the Seismological Society of America*, *67*, 259–277.

Brune, J. N. (1970). Tectonic stress and the spectra of seismic shear waves from earthquakes. *Journal of Geophysical Research*, *75*(26), 4997–5009. doi:10.1029/JB075i026p04997

Chang, F. K., & Krinitzsky, E. L. (1977). *Duration, spectral content, and predominant period of strong motion earthquake records from western United States*. US Army Engineer Waterways Experiment Station.

Chopra, A. K., & Chinatanapakdee, C. (2004). Inelastic deformation ratios for design and evaluation of structures: Single-degree-of-freedom bilinear systems. *Journal of Structural Engineering*, *130*(9), 1309–1319. doi:10.1061/(ASCE)0733-9445(2004)130:9(1309)

Code, T. E. (2007). *Specifications for buildings to be built in seismic areas*. Ankara, Turkey: Ministry of Public Works and Settlement. (in Turkish)

CodeT. E. (2016). http://www.deprem.gov.tr/belgeler2016/tbdy.pdf

Cornell, C. A. (1968). Engineering seismic risk analysis. *Bulletin of the Seismological Society of America*, *58*(5), 1583–1606.

Eurocode 8: Design of structures for earthquake resistance. (2005).

Fahjan, Y. M. (2008). Selection and scaling of real earthquake accelerograms to fit the Turkish design spectra. *Teknik Dergi*, *19*(3), 4423–4444.

Hadley, D. M., & Helmberger, D. V. (1980). Simulation of strong ground motions. *Bulletin of the Seismological Society of America*, *70*(2), 617–630.

Hancock, J., & Bommer, J. J. (2006). A state-of-knowledge review of the influence of strong-motion duration on structural damage. *Earthquake Spectra*, *22*(3), 827–845. doi:10.1193/1.2220576

Hartzell, S. H. (1978). Earthquake aftershocks as Greens functions. *Geophysical Research Letters*, *5*(1), 1–4. doi:10.1029/GL005i001p00001

Haselton, C. B., Baker, J. W., Liel, A. B., & Deierlein, G. G. (2011). Accounting for ground-motion spectral shape characteristics in structural collapse assessment through an adjustment for epsilon. *Journal of Structural Engineering*, *137*(3), 332–344. doi:10.1061/(ASCE)ST.1943-541X.0000103

Hutchıngs, L. (1991). Prediction" of strong ground motion for the 1989 Loma Prieta earthquake using empirical Green's functions. *Bulletin of the Seismological Society of America*, *81*, 88–121.

Hutchıngs, L., & Wu, F. (1990). Empirical Greens functions from small earthquakes: A waveform study of locally recorded aftershocks of the San Fernando earthquake. *Journal of Geophysical Research*, *95*(B2), 1187–1214. doi:10.1029/JB095iB02p01187

Iervolino, I., & Cornell, C. A. (2005). Record selection for nonlinear seismic analysis of structures. *Earthquake Spectra*, *21*(3), 685–713. doi:10.1193/1.1990199

Irikura, K. (1983). Semi-empirical estimation of strong ground motions during large earthquakes. *Bull. Disaster Prevention. Res. Inst Kyoto Univ*, *33*, 63–104.

Kalkan, E., & Chopra, A. K. (2010). *Practical Guidelines to Select and Scale Earthquake Records for Nonlinear Response History Analysis of Structures*. U.S. Geological Survey Open-File Report.

Kalkan, E., & Chopra, A. K. (2011). Modal-Pushover-based Ground Motion Scaling Procedure. *Journal of Structural Engineering*, *137*(3), 298–310. doi:10.1061/(ASCE)ST.1943-541X.0000308

Kaul, M. K. (1978). Spectrum-consistent time-history generation. *Journal of the Engineering Mechanics Division*, *104*(4), 781–788.

Kramer, S. L. (1996). *Geotechnical Earthquake Engineering*. Prentice Hall.

Krinitzsky, E. L., & Chang, F. K. (1977). *Specifying Peak Motions for Design Earthquakes, State-of the-Art for Assessing Earthquake Hazards in the United States, Report 7, Miscellaneous Paper S-73-1*. Vicksburg, MS: US Army Corps of Engineers.

Lilhanand, K., Tseng, W. S. (1987). Generation of synthetic time histories compatible with multiple-damping design response spectra. *Structural Mechanics in Reactor Technology.*

Lilhanand, K., & Tseng, W. S. (1987). Development and application of realistic earthquake time histories compatible with multiple-damping design spectra. *Development*, *3*, 7–8.

Luco, N., & Bazzurro, P. (2007). Does amplitude scaling of ground motion records result in biased nonlinear structural drift responses? *Earthquake Engineering & Structural Dynamics*, *36*(13), 1813–1835. doi:10.1002/eqe.695

Martinez-Rueda, J. E. (1998). Scaling procedure for natural accelerograms based on a system of spectrum intensity scales. *Earthquake Spectra*, *14*(1), 135–152. doi:10.1193/1.1585992

Motazedian, D., Atkinson, G. M. (2005). Stochastic finite-fault modeling based on a dynamic corner frequency. *Bulletin of the Seismological Society of America, 95*(3), 995-1010.

Naeim, F., & Kelly, J. M. (1999). *Design of Seismic Isolated Structures: From Theory to Practice*. John Wiley & Sons. doi:10.1002/9780470172742

Naeim, F., & Lew, M. (1995). On the use of design spectrum compatible time histories. *Earthquake Spectra*, *11*(1), 111–127. doi:10.1193/1.1585805

Preumont, A. (1984). The generation of spectrum compatible accelerograms for the design of nuclear power plants. *Earthquake Engineering & Structural Dynamics*, *12*(4), 481–497. doi:10.1002/eqe.4290120405

Reiter, L. (1990). *Earthquake Hazard Analysis: Issues and Insights*. Columbia University Press.

Reyes, J. C., & Kalkan, E. (2012). How many records should be used in an ASCE/SEI-7 ground motion scaling procedure? *Earthquake Spectra*, *28*(3), 1223–1242. doi:10.1193/1.4000066

Shome, N., Cornell, C. A., Bazzurro, P., & Carballo, J. E. (1998). Earthquakes, records, and nonlinear responses. *Earthquake Spectra*, *14*(3), 469–500. doi:10.1193/1.1586011

Stewart, J. P., Chiou, S. J., Bray, J. D., Graves, R. W., Somerville, P. G., & Abrahamson, N. A. (2001). *Ground Motion Evaluation Procedures for Performance-Based Design, PEER Report 2001/09*. Berkeley, CA: Pacific Earthquake Engineering Research Center, University of California.

Trifunac, M. D., & Brady, A. G. (1975). A Study on the Duration of Strong Earthquake Ground Motion. *Bulletin of the Seismological Society of America*, *65*, 581–626.

Venture, N. C. J. (2011). NIST GCR 11-917-15–selecting and scaling earthquake ground motions for performing response-history analyses. US National Institute for Standards and Technology.

Wu, F. (1978). Prediction of strong ground motion using small earthquakes, *Proceedings of the 2nd International Conference on Microzonation*, *2*,701-704.

Chapter 2
Seismic Assessment and Retrofitting of an Under-Designed RC Frame Through a Displacement-Based Approach

Marco Valente
Politecnico di Milano, Italy

Gabriele Milani
Politecnico di Milano, Italy

ABSTRACT

Many existing reinforced concrete buildings were designed in Southern European countries before the introduction of modern seismic codes and thus they are potentially vulnerable to earthquakes. Consequently, simplified methodologies for the seismic assessment and retrofitting of existing structures are required. In this study, a displacement based procedure using non-linear static analyses is applied to a four-storey RC frame in order to obtain an initial estimation of the overall inadequacy of the original structure as well as the extent of different retrofitting interventions. Accurate numerical models are developed to reproduce the seismic response of the RC frame in the original configuration. The effectiveness of three different retrofitting solutions countering structural deficiencies of the RC frame is examined through the displacement based approach. Non-linear dynamic analyses are performed to assess and compare the seismic response of the frame in the original and retrofitted configurations.

INTRODUCTION

Several reinforced concrete (RC) buildings were designed according to older Codes in high-seismicity European countries. These structures are very likely to experience severe damage or collapse during moderate seismic events. Therefore, simplified procedures to evaluate the seismic performance of existing and retrofitted buildings are needed, along with reliable and effective seismic upgrading techniques, (CEB-FIB, 2003; ATC, 1996; FEMA, 2000; Fajfar, 1999, 2000; Fajfar et al. 2005; Masi, 2003; Fardis et

DOI: 10.4018/978-1-5225-2089-4.ch002

al., 2013; Rozman & Fajfar, 2009; Antonopoulos & Triantafillou, 2002; Valente, 2012, 2013a, 2013b). This study presents the results of numerical investigations on the seismic assessment and retrofitting of a RC structure, designed mainly for gravity loads without specific earthquake-resistant provisions, that was pseudo-dynamically tested at the JRC ELSA laboratory in Italy. Detailed numerical models were developed to reproduce the seismic response of the RC structure in the original configuration and non-linear dynamic analyses were performed. A displacement based procedure using non-linear static pushover analyses was used to evaluate the seismic performance of the original structure and the most relevant results are discussed. The effectiveness of the use of different approaches for the seismic performance assessment of other typologies of existing structures is presented by the authors in Milani and Valente (2015a, 2015b) and Valente and Milani (2016a, 2016b).

The displacement based procedure is also used for a preliminary design and evaluation of viable retrofitting interventions. In particular, three alternative retrofitting solutions for the RC structure are proposed and verified through a displacement based approach.

The first intervention scheme was defined on the basis of the results of the numerical analyses on the bare frame, adopting two different design strategies, like strength-only and ductility-only solutions. These selective interventions were applied to different members of the structure in order to improve its global and local seismic behavior. A strength-only intervention using RC jacketing was implemented in the strong column at the third and fourth stories of the structure to reduce the large difference in terms of flexural capacity. Moreover, a ductility-only intervention was accomplished at the first three stories of the structure, where a large inelastic deformation demand was expected. The intervention required the arrangement of external fiber reinforced polymer (FRP) composites, which provided an increase in concrete confinement of the columns.

The second retrofitting intervention was based on the addition of steel bracing, which can be considered as a very effective method for global strengthening of buildings, (Youssefa et al., 2007; Mazzolani, 2008; Di Sarno & Elnashai, 2009; Giannuzzi et al., 2014). In particular, the use of eccentric steel bracings in the rehabilitation of existing RC structures is efficient in limiting inter-storey drifts and can provide a stable energy dissipation capacity.

The third intervention was carried out by casting a concrete shear wall to the full width of the short bay of the frame. This solution leads to significant increases in overall strength and stiffness for the retrofitted frame, when compared to that of the initial frame configuration. This intervention is efficient in controlling global lateral drift and thus reducing damage in structural members.

The efficacy of the three retrofitting solutions adopted for the RC structure was evaluated by non-linear dynamic analyses.

Retrofitting Strategies

A displacement based procedure is used to evaluate the seismic performance of existing and retrofitted RC structures. The procedure is based on a simplified approach using non-linear static pushover analyses and allows comparing alternative retrofitting strategies countering different structural deficiencies. The base shear-top displacement relationships obtained from pushover analysis were transformed into capacity curves in the acceleration-displacement (AD) format. Several target displacements and capacity curves were estimated assuming that different strategies could be used to retrofit the structure. In fact, different types of structural interventions on existing structures can be classified according to their effect on the behavior of the structure and can be represented by the capacity curves presented in Figure 1.

Figure 1. Capacity curves, demand spectra and different retrofitting strategies in the acceleration-displacement (AD) format

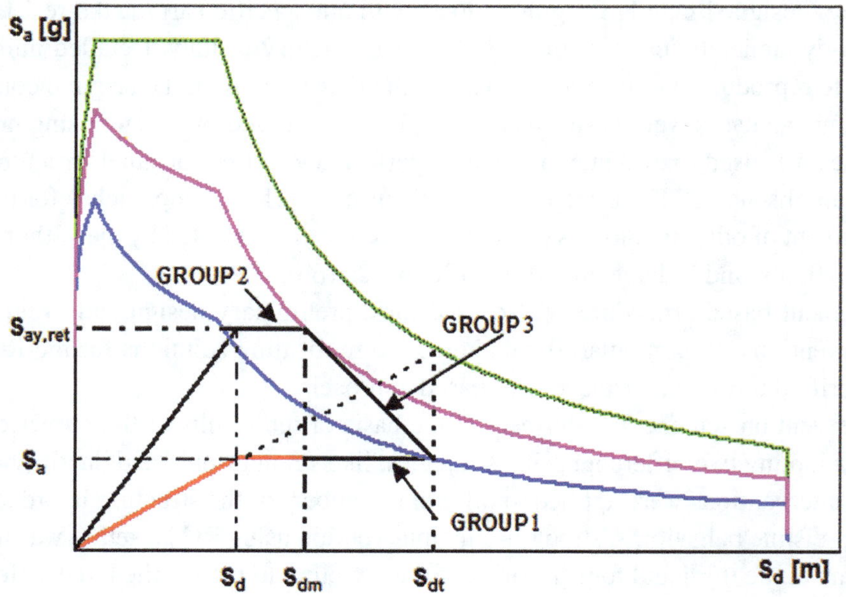

- The first group (group 1) includes strategies whose basic aim is to improve the overall ductility of the structure;
- The second group (group 2) represents interventions resulting in system strengthening and stiffening;
- The intermediate group (group 3) comprises solutions increasing both strength and ductility of the structure.

The range of available retrofitting solutions is bounded by assuming the two alternative structural intervention techniques corresponding to groups 1 and 2.

In Figure 1 the bilinear curve (red line) shows the capacity spectrum for the unretrofitted equivalent system. S_{ay} and S_{dy} are the spectral acceleration and the spectral displacement, respectively, at the yield point of the unretrofitted equivalent system. S_{dm} is the spectral displacement corresponding to the point at the end of the capacity spectrum before retrofitting and S_{dt} is the target spectral displacement after retrofitting according to the strategies of group 1. The extension of the horizontal (red) line represents the spectral displacement required to achieve the target spectral displacement. The green curve is the elastic demand spectrum and the blue curve is the inelastic spectrum corresponding to the ductility demand for the strategies of group 1.

The bilinear curve (black line) shows the capacity spectrum for the retrofitted equivalent system according to the strategies of group 2. The velvet curve is the inelastic spectrum corresponding to the ductility demand for the strategies of group 2. $S_{ay,ret}$ is the spectral acceleration at the yield point of the retrofitted equivalent system and S_{dm} is the target spectral displacement corresponding to the performance point after retrofitting.

The line joining the two points with coordinates ($S_{ay,ret}$, S_{dm}) and coordinates (S_{ay}, S_{dt}) represents the location of the target spectral displacements for the different alternative effective retrofit solutions belonging to group 3.

RC Structure Under Investigation

The seismic assessment and retrofitting procedure based on a displacement based approach was applied to a four-storey RC frame, designed only for gravity loads, that was subjected to real-scale pseudo-dynamic tests at the JRC ELSA laboratory at Ispra (the frame was braced with lateral steel pinned bars in order to avoid out-of-plane deformation during the experimental tests). The frame was intended to be representative of the typical design and construction practice in many Southern European countries in the 1960's. It was also representative of a number of rather recent buildings, designed without the application of the capacity design principles and without up-to-date detailing. The elevation of the RC frame, the geometric characteristics and the reinforcement details of the columns are shown in Figure 2. The longitudinal reinforcement consisted of smooth round bars, as usually found up to the early seventies in many European countries. The reinforcement of the two internal columns (columns 2 and 3) was reduced in the top two storeys and only column 2 worked along its strong axis, as a result of the adopted non-seismic design philosophy. Column 2 had a rectangular cross-section with dimensions of 0.60 m x 0.25 m in the first and second storeys, and 0.50 m x 0.25 m in the third and fourth storeys. The other columns (columns 1 and 4) had cross-sections of 0.20 m x 0.40 m and 0.20 m x 0.30 m, the same in all floors. The longitudinal reinforcement of all columns had a lapped splice (70 cm) at the base of the first storey and another at the base of the third storey. All the beams were 0.25 m wide and 0.50 m deep and the slab thickness was 0.15 m.

The concrete grade was C16/20 (according to Eurocode 2 CEN, 2004) and the class of the smooth reinforcement was FeB22k (according to Italian standards). Of course, tests on samples of the materials (steel reinforcement and concrete) were carried out to provide the input data required by numerical analysis. The actual values of concrete strength in compression and steel strength at yielding were approximately 16 MPa and 344 MPa, respectively. A more detailed description of the structure can be found in an ELSA Report, (Pinto et al, 2002).

Figure 2. Geometric characteristics and reinforcement details of the columns of the test structure

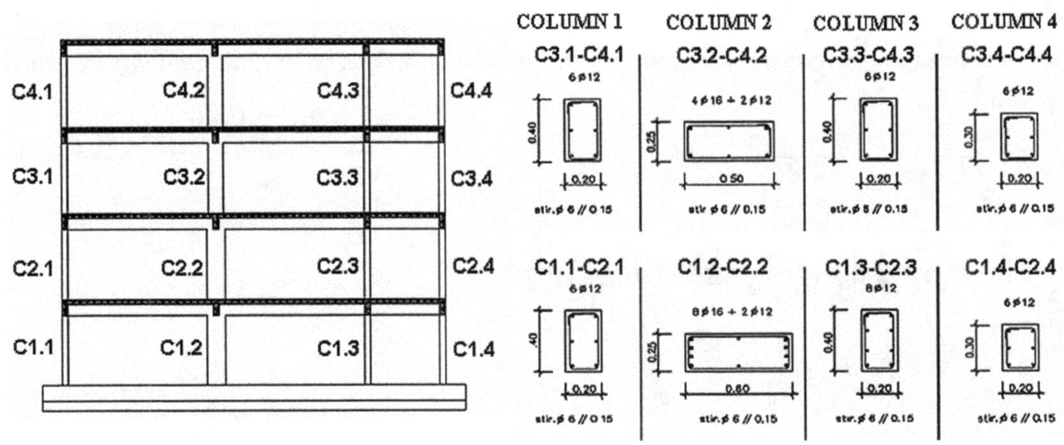

The experimental activity carried out at the JRC ELSA Laboratory consisted of several pseudo-dynamic tests on the bare and retrofitted RC frames for different earthquake-intensity levels. The seismic motions used as input in the tests represented a moderate/high European seismic hazard scenario. A set of hazard- consistent acceleration time-histories (15s duration) was artificially generated and three time-histories with different return periods (475, 975 and 2000 years) were chosen for the tests (Pinto, 2002): Acc-475 (peak acceleration equal to 0.23g), Acc-975 (peak acceleration equal to 0.3g) and Acc-2000 (peak acceleration equal to 0.38g). The acceleration time-history corresponding to a 475-year return period is shown in Figure 3.

The full-scale RC frame was modelled by using both the code SeismoStruct (Seismosoft, 2007), based on a fibre-modelling approach, and the code Ruaumoko (Carr, 2006), based on a lumped plasticity approach.

In Seismostruct, concrete was modelled by using a uniaxial constant-confinement model based on the constitutive relationship proposed by Mander et al. (1988), and later modified by Martinez-Rueda and Elnashai (1997). The confinement effects, provided by the transverse reinforcement, were taken care of as proposed by Mander, whereby a constant confining pressure was assumed in the entire stress-strain range. The longitudinal reinforcement was modelled via the Menegotto–Pinto model, (Menegotto & Pinto, 1973).

In Ruaumoko, beams and columns were modelled by one-dimensional elastic elements with inelastic behaviour concentrated at the edges in plastic hinge regions (Giberson model). The Modified Takeda hysteresis model (Otani, 1981), widely used for RC sections, was used to represent the moment-curvature behaviour in the hinge region of the member.

The actual materials properties measured during the experimental tests were introduced into the numerical models. The comparison of the numerical predictions with the experimental test results allowed calibrating the main parameters of the numerical models developed in this study.

The pseudo-dynamic tests performed on the bare frame at the JRC Ispra were numerically simulated through non-linear dynamic analyses. The same artificial accelerograms with increasing intensities adopted in the pseudo-dynamic tests were used in the numerical analyses, which were performed sequentially, according to the various steps of the experimental campaign, in order to better reproduce the laboratory tests. The comparison with the test results made it possible to calibrate and improve analytical modelling and to check the accuracy of the numerical models.

Figure 3. Acceleration time-history and acceleration response spectrum (5% damping) of the input ground motion (Acc-475)

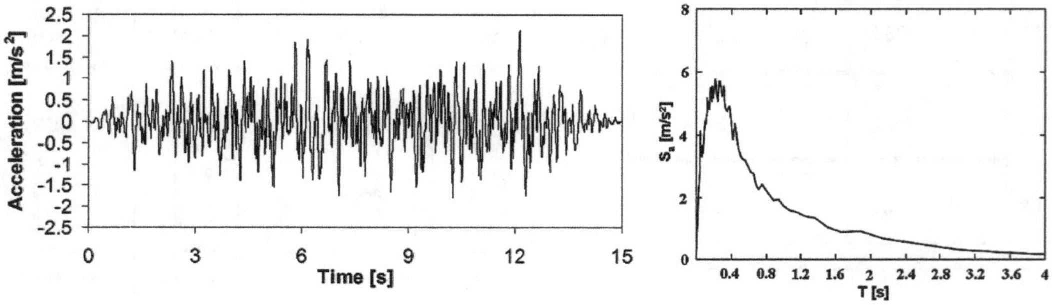

Non-Linear Dynamic Analysis With Acc-475 Record

The numerical results obtained with Acc-475 artificial motion are reported in terms of top storey displacement time-history, maximum inter-storey drift and storey shear profiles. Comparisons are made between the model based on a fibre-modelling approach and the experimental results. In Figure 4 the top storey displacement time-history shows that the numerical model fits well the tests in terms of phase and peak values (5.8 cm vs 6 cm). As for the drift profiles, the model is able to identify the maximum drift at the third level, Figure 5. Furthermore, Figure 6 shows that the maximum base shear is close to 210 kN and that the agreement between the experimental and numerical storey shear profiles is very satisfactory.

Non-Linear Dynamic Analysis With Acc-975 Record

The numerical predictions and comparisons with the experimental results under Acc-975 record are reported in Figures 7 to 9. The experimental test was stopped after 7.5 seconds, when the third storey almost collapsed, in order to allow for repair and subsequent strengthening of the frame. Therefore, the

Figure 4. Top storey displacement time-history under Acc-475 record

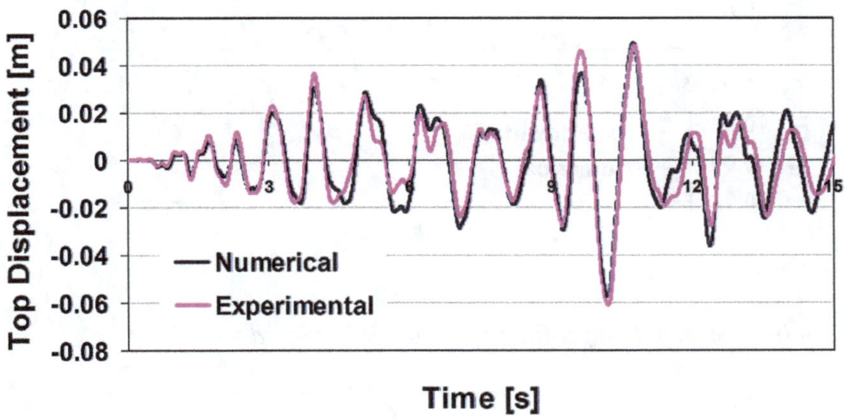

Figure 5. Maximum inter-storey drift profile under Acc-475 record

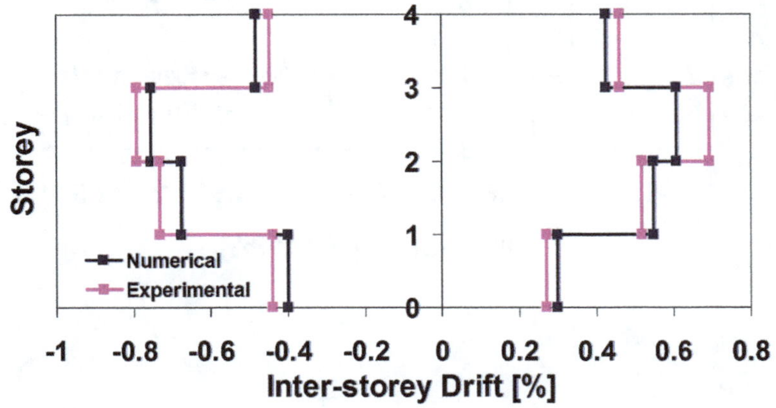

Figure 6. Maximum storey shear profile under Acc-475 record

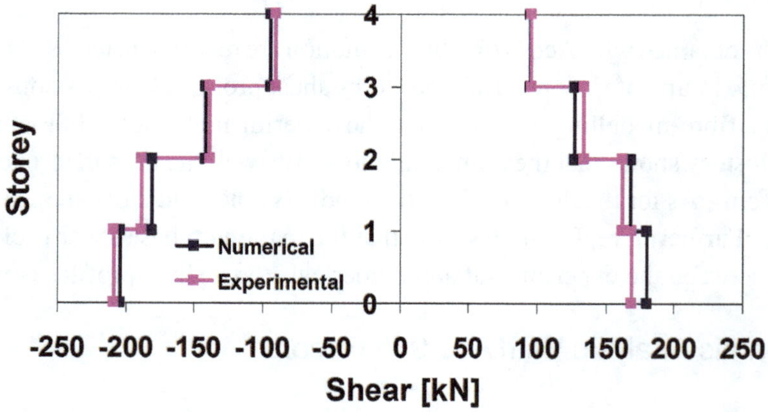

Figure 7. Top displacement time-history under Acc-975 record

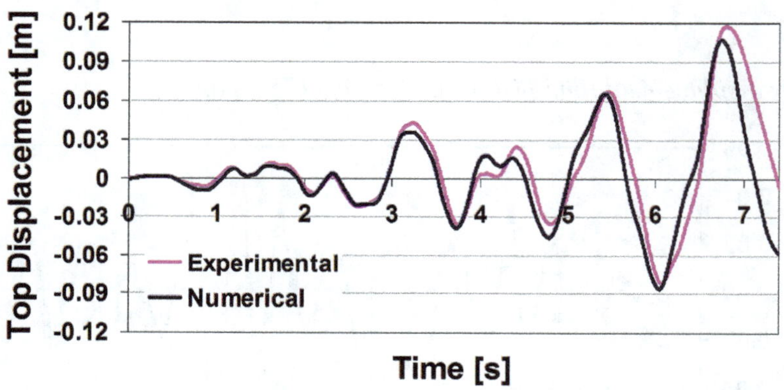

Figure 8. Maximum inter-storey drift profile under Acc-975 record

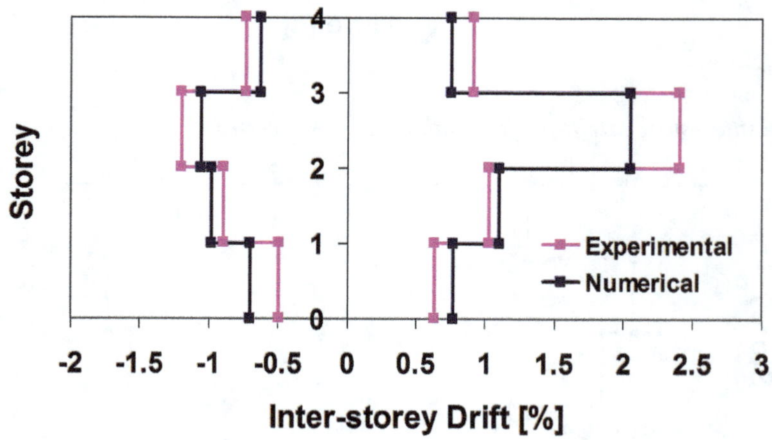

Figure 9. Maximum storey shear profile under Acc-975 record

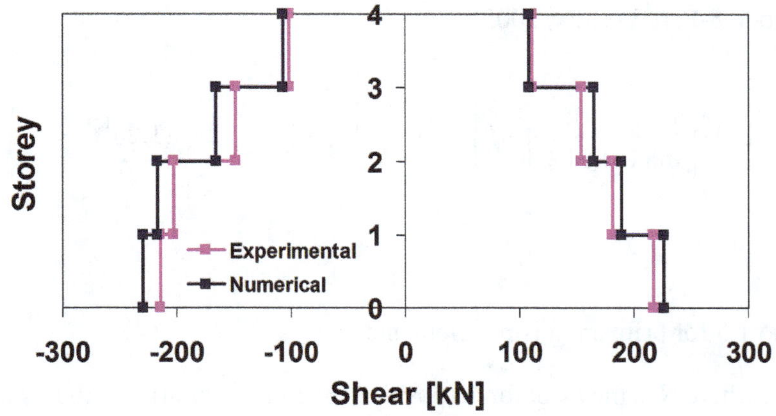

results of the numerical analyses are reported only for the first 7.5 seconds of the original Acc-975 record. The numerical model of the bare frame was able to quite well reproduce the experimental results in terms of top-storey displacement time-history and peak values, (10.7 cm vs 11.6 cm), Figure 7. The soft-storey drift at the third floor was fully captured by the numerical model, even if some differences were observed in the prediction of the peak drift value at the soft-storey (Figure 8). Figure 9 shows a close agreement between the experimental and numerical storey shear profiles.

The numerical results confirmed the high vulnerability of the frame already experimentally observed. In fact, it was shown that in spite of the limited inter-storey drifts for the Acc-475 earthquake, the demands for a slightly higher intensity earthquake (1.3 times the Acc-475 earthquake in terms of peak acceleration) led to much larger inter-storey drifts. This result was due to the excessive strength difference between the second and third storeys caused by the abrupt changes in dimension and reinforcement detailing of the strong column, leading to the formation of an undesirable soft-storey failure mechanism.

Simplified Assessment Procedure for the Bare Structure

Non-linear static pushover analyses *were performed* to obtain the capacity of the structure represented by a force-displacement curve. The base shear force and the roof displacement were converted to the spectral accelerations and to the spectral displacements of an equivalent SDOF system, and these spectral values defined the capacity spectrum. The bilinear (elastic-perfectly plastic) idealization of the pushover curve was defined on the basis of the "equal-energy" concept (the areas underneath the actual and bilinear curves are approximately the same, within the range of interest). In Figure 10 the seismic demand for the equivalent SDOF system was determined for the Limit State of Significant Damage (LSSD). The elastic acceleration and the corresponding elastic displacement demand were computed by intersecting the radial line corresponding to the elastic period of the idealized bilinear system with the elastic demand spectrum. The inelastic demand in terms of accelerations and displacements was provided by the intersection point of the capacity curve with the demand spectrum corresponding to the ductility demand. In this study, the seismic demand was computed with reference to the general Eurocode 8 response spectrum (Type 1, soil type A), (Eurocode 8-Part 1, CEN, 2005). The theoretical predictions were performed for a PGA level equal to 0.3g. The value of the total ultimate chord rotation capacity, ϑ_u, of concrete members under cyclic loading may be calculated from the following expression, according to Eurocode 8-Part 3 (CEN, 2005):

$$\theta_u = \frac{1}{\gamma_{el}} \cdot 0.016 \cdot \left(0.3^{\nu}\right) \cdot \left[\frac{\max(0.01;\omega')}{\max(0.01;\omega)} \cdot f_c\right]^{0.225} \cdot \left(\frac{L_V}{h}\right)^{0.35} \cdot 25^{\left(\alpha \cdot \rho_{sx} \cdot \frac{f_{yw}}{f_c}\right)} \cdot (1.25^{100 \cdot \rho_d}) \quad (1)$$

where:

- γ_{el} is equal to 1.5 for primary seismic elements,
- $\nu = \dfrac{N}{b \cdot h \cdot f_c}$, where N is the axial force (positive for compression), b is the width of compression zone and h is the depth of the cross-section,
- $\omega' = \dfrac{A'_s f_y}{bhf_c}$, $\omega = \dfrac{A_s f_y}{bhf_c}$ are the mechanical reinforcement ratios of the longitudinal reinforcement in compression and in tension, respectively,
- L_V is the shear span, i.e. the moment/shear ratio at the end section,
- $\alpha = \left(1 - \dfrac{s_h}{2 \cdot b_0}\right) \cdot \left(1 - \dfrac{s_h}{2 \cdot h_0}\right) \cdot \left(1 - \dfrac{\sum b_i^2}{6 \cdot b_0 \cdot h_0}\right)$ is the confinement effectiveness factor (b_i is the spacing of longitudinal bars laterally restrained by a stirrup corner or a cross-tie along the perimeter of the cross-section, b_0 and h_0 are the width and the depth of confined core, s_h is the stirrup spacing),
- $\rho_{sx} = \dfrac{A_{sx}}{b_w s_h}$ is the ratio of the transverse reinforcement parallel to the direction x of loading,
- ρ_d is the steel ratio of diagonal reinforcement (if any), in each diagonal direction,
- f_y (or f_{yw}) and f_c are the mean values of the steel yield strength and the concrete compression strength, respectively, both in MPa, as obtained from *in-situ* tests and from any additional sources of information, appropriately divided by the confidence factors, accounting for the level of knowledge attained. A knowledge level equal to 3 (according to Eurocode 8-Part 3, CEN, 2005) was assumed corresponding to a confidence factor of 1 on the assumption that the original construction drawings were available, together with full information on the material properties. As a consequence, the mean values were adopted for the materials.

The values of the total chord rotation capacity given by expression (1) should be multiplied by 0.825 in those members that have not been designed according to any seismic provisions and multiplied by 0.575 in those members that are reinforced with smooth (plain) longitudinal bars not lapping in the vicinity of the end regions, where steel yielding is expected.

The Limit State of Significant Damage (LSSD) corresponds to the attainment of $0.75 \cdot \theta_u$. Figure 10 shows that the bare frame lacked the appropriate capacity to resist the 0.3g PGA seismic intensity level at the LSSD. The displacement demands in Figure 10 refer to the equivalent SDOF system. The displacement demands of the MDOF system were obtained by multiplying the SDOF system demand by the

Figure 10. Demand spectra and capacity curve in AD format at the LSSD for the bare frame

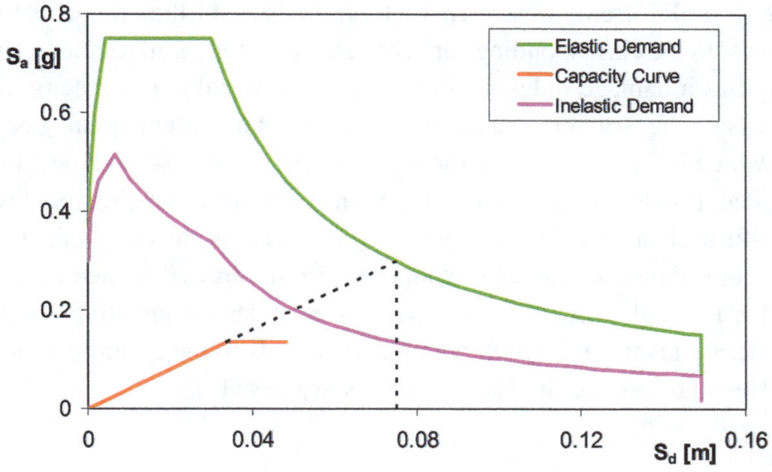

transformation factor $\Gamma = \sum m_i\phi_i / \sum m_i\phi_i^2$, where m_i is the mass in the ith storey and ϕ_i is the component of the normalized displacement shape. A gap in terms of maximum top displacement was observed at the LSSD; the difference between the seismic demand and the displacement capacity was 0.036m (0.098m vs 0.062m).

The results of the simplified procedure showed that the attainment of the LSSD occurred at column 2 of the third floor, where the most significant damage was observed in the laboratory tests and the highest value of the Demand-to-Capacity Ratio was registered during non-linear dynamic analyses, see Figure 11. The developed numerical models provided reliable predictions of the behavior of the test specimen identifying the main structural deficiencies.

Non-linear dynamic analyses by using seven artificial accelerograms with PGA=0.3g were carried out. The suite of artificial accelerograms was generated using the computer code SIMQKE (Vanmarcke

Figure 11. Non-linear dynamic analyses: Demand-to-Capacity Ratio for the bare frame

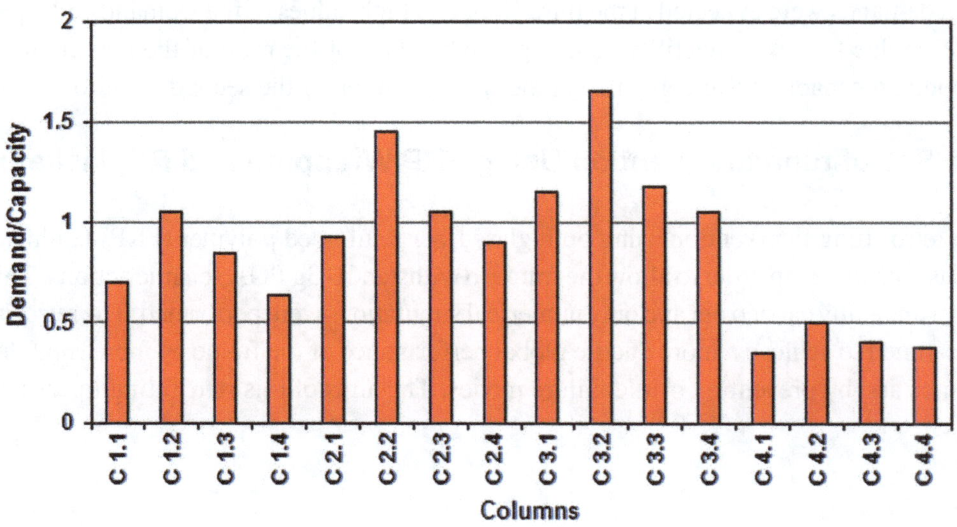

et al., 1976) so as to match a Type 1 response spectrum for soil class A, according to Eurocode 8. The seismic assessment of the RC frame was carried out on the basis of the ratio of the chord rotation demand (at member ends) to the corresponding capacity. The member chord rotation Demand-to-Capacity Ratio was considered as a damage index against the loss of lateral load capacity of the member and was investigated to assess the structural capacity. The chord rotation demand may be taken equal to the element drift ratio, which is the deflection at the end of the shear span with respect to the tangent to the axis at the yielding end, divided by the shear span. In the columns belonging to a frame under seismic action, the lateral drifts at shear span ends are generally much larger than the nodal rotations at columns ends. The nodal rotations of the columns can be neglected in structures designed without capacity design procedures, where the flexural stiffness of the beams is much larger than that of the columns. For the Limit State of Significant Damage (LSSD), which corresponds to the attainment of $0.75 \cdot \theta_u$ according to Eurocode 8, the Demand-to-Capacity Ratio, DCR, is expressed as:

$$DCR = \frac{\theta_{demand}}{\frac{3}{4} \cdot \theta_u} \tag{2}$$

For each dynamic analysis, the ratio of the maximum demand to the corresponding capacity was evaluated for each structural member at each time step, because the capacities of the ductile and brittle mechanisms are both a function of the demand and, in particular, of the axial load, according to both considered Codes.

Figure 11 provides the values of the Demand-to-Capacity Ratio for all the columns of the frame. The values of the Demand-to-Capacity Ratio reproduced quite well the behaviour observed during the experimental tests. The maximum value of the Demand-to-Capacity Ratio was registered in correspondence with column 2, as confirmed by the experimental evidence. In fact, the 975-yrp test was stopped at 7.5 seconds because of the impending collapse of the third storey and of the severe damage detected at the strong column of the third storey. High deformation demands were registered at the base and at the top of the strong column of the third storey. Column 2 had significantly higher stiffness and strength than the other columns, since it was the only one with the strong axis in the loading direction. High deformation demands were expected at the third floor and high values of the Demand-to-Capacity Ratio were registered due to lack of ductility detailing. High values of the ratio of the maximum demand to the corresponding capacity were computed at the strong column of the second floor too.

Selective Retrofitting Intervention Using FRP Wrapping and RC Jacketing

A selective retrofitting intervention using both glass fiber reinforced polymer (FRP) laminates and RC jacketing was considered in order to allow the frame to withstand 0.3g PGA seismic actions. This solution was not experimentally carried out and only numerical simulations were performed. The retrofitting intervention was aimed to achieve a more ductile global performance of the frame by increasing the ductility of the columns and by preventing brittle failure modes. The aims of this retrofitting intervention were:

1. To prevent the soft-storey mechanism at the third floor by mitigating the strength difference involved by the change of section of the strong column between the second and the third floor;
2. To improve the global deformation capacity of the frame by increasing the ductile resources of the columns, preventing brittle failure modes.

According to a selective retrofitting scheme, a ductility-only intervention was applied to the first three storeys and a strength-ductility intervention was applied to the strong column at the top two storeys (Figure 12). A mixed intervention was carried out using FRP, whose principal characteristics are reported in Table 1, and RC jacketing.

The FRP consisted of a 3-layer wrapping applied to the columns of the first three floors; the RC jacketing consisted of 4+4 steel bars on two opposite sides of the strong column at the top two storeys.

A numerical model of the frame retrofitted with FRP wrapping and RC jacketing was developed by using the SeismoStruct FE code (SeismoSoft 2007). The non-linear variable confinement model that includes the constitutive relationship and cyclic rules proposed by Mander et al. (1988) in compression, and those of Yankelevsky and Reinhardt (1989) in tension, was adopted to model RC sections retrofitted

Figure 12. Selective retrofitting intervention with FRP wrapping and RC jacketing

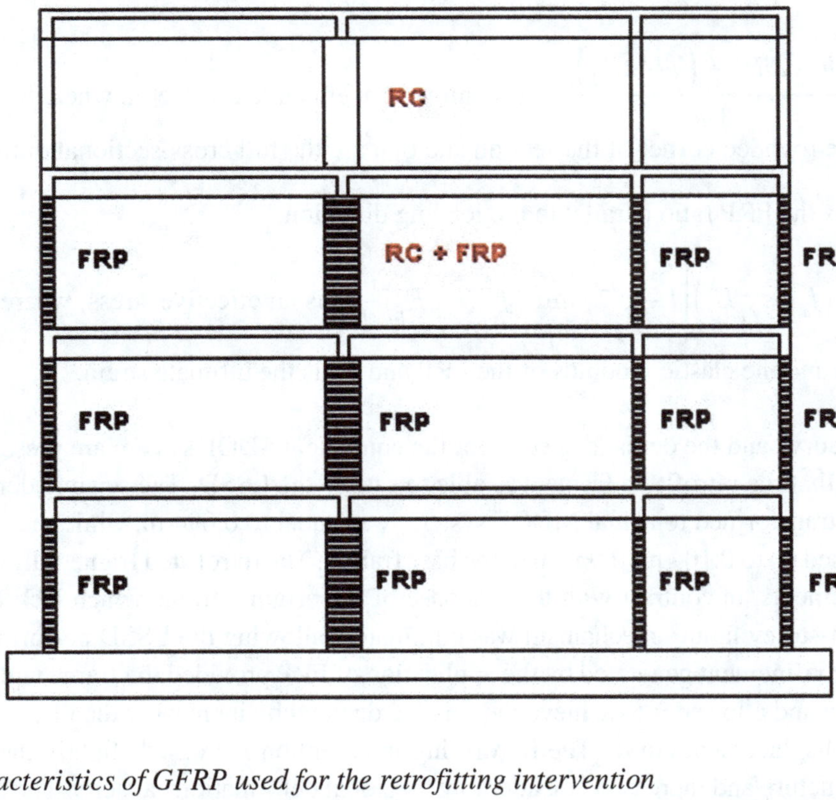

Table 1. Characteristics of GFRP used for the retrofitting intervention

Young Modulus (GPa)	Ultimate Tensile Stress (MPa)	Ultimate Tensile Deformation	Layer thickness (mm)
65	1700	0.026	0.23

by FRP. The confinement effect introduced by the FRP wrapping was modelled by means of the rules proposed by Spoelstra and Monti (1999). The model is based on modified Mander's model (Mander et al., 1988), where maximum strength and corresponding strain of the confined concrete are defined as a function of confinement pressure. A constant lateral pressure, depending on steel yielding stress, is considered for steel confinement, whereas confinement pressure is linearly varying with concrete lateral dilation in the case of external FRP wrapping. An iterative procedure is adopted to obtain the axial stress corresponding to a given value of axial strain, taking into account the effect of confinement, according to Spoelstra and Monti (1999). The RC jacketed rectangular section available in Seismostruct libraries was used for the modelling of rectangular columns retrofitted by means of RC jacketing. Different confinement levels for the internal (pre-existing) and the external (new) concrete materials were defined. To evaluate the properties of the retrofitted elements, the following assumptions were adopted according to Eurocode 8: the jacketed column behaves monolithically with full composite action between old and new concrete, the concrete properties of the jacket apply over the full section of the element, the axial load is considered acting on the full composite section.

The enhancement of the deformation capacity of the member, ϑ_u, was determined by adding a term due to FRP to the term describing the confinement provided by the transverse reinforcement. The total chord rotation capacity was calculated from expression (1) with the exponent of the term due to confinement increased by ($\alpha^* \rho_f f_{f,e} / f_c$), where:

- $\alpha^* = 1 - \dfrac{\left(b-2R\right)^2 + \left(h-2R\right)^2}{3bh}$ is the confinement effectiveness factor, where R = 20 mm is the radius of the rounded corner of the section and b, h are the full cross-sectional dimensions,
- $\rho_f = \dfrac{2t_f}{b}$ is the FRP ratio parallel to the loading direction,
- $f_{f,e} = \min\left(f_{u,f}; \varepsilon_{u,f} E_f\right)\left(1 - 0.7 \cdot \min\left(f_{u,f}; \varepsilon_{u,f} E_f\right)\dfrac{\rho_f}{f_c}\right)$ is an effective stress, where $f_{u,f}$ and E_f are the strength and the elastic modulus of the FRP and $\varepsilon_{u,f}$ is the ultimate strain.

The capacity curve and the demand spectra for the equivalent SDOF system are presented in Figure 13, which shows that the retrofitted frame was able to satisfy the LSSD. The seismic demand in terms of displacement, transformed to actual MDOF system, was equal to 0.094 m, while the capacity of the frame was increased up to 0.104 m (0.062 m in the bare frame). The retrofitted frame fully complied with the LSSD requirements, in contrast with the response of the original frame which lacked the required ductility. The soft-storey failure mechanism was eliminated, allowing the LSSD performance target to be met. Column confinement generated by the application of FRP provided the frame with significantly enhanced ductility and allowed it to achieve the seismic demand by increasing the plastic branch of the base shear - top displacement curve. The retrofitting intervention increased slightly the stiffness and strength of the structure and increased considerably its global deformation capacity. The adopted selective scheme proved to be effective for seismic upgrading of the RC frame.

Non-linear dynamic analyses by using seven artificial accelerograms with PGA=0.3g were carried out to verify the validity of the simplified procedure and the effectiveness of the retrofitting intervention. The storey drift profile of the retrofitted model exhibited an increase of the drift value at the first two floors,

Figure 13. Demand spectra and capacity curve in AD format at the LSSD for the frame retrofitted with FRP wrapping and RC jacketing

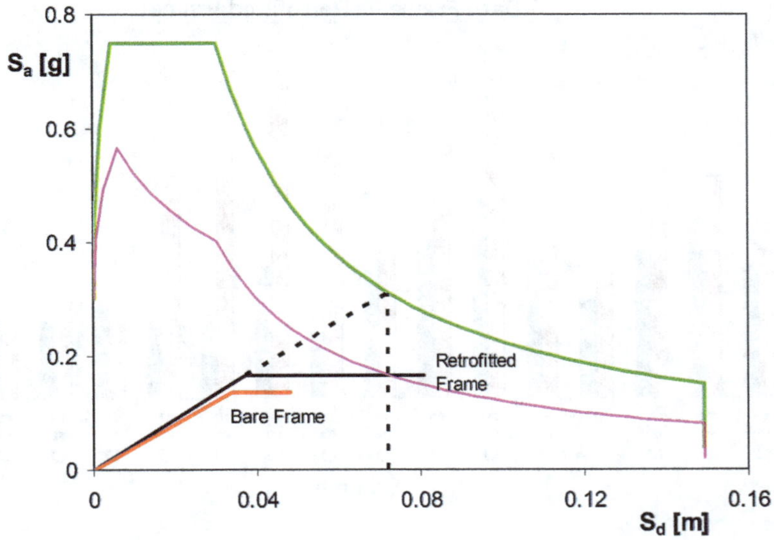

while a significant reduction of the storey drift occurred at the third floor (Figure 14). The increase of the flexural strength of column 2 and the confining effect of the FRP prevented the development of a soft-storey behavior at the third floor. Figure 15 provides the values of the Demand-to-Capacity Ratio in terms of chord rotation obtained from the dynamic analyses for the columns of the bare and retrofitted models. A significant reduction of the Demand-to-Capacity Ratio can be observed for all the columns of the first three floors, above all at the third floor. Numerical analyses indicated high values of deformation demand in the strong column, but in this case the column was detailed for ductility due to high level of confinement provided by FRP. A considerable improvement in deformation capacity was obtained and a significant decrease of the Demand-to-Capacity Ratio was observed for the retrofitted model.

Figure 14. Non-linear dynamic analyses: inter-storey drift profiles for the bare and retrofitted frames

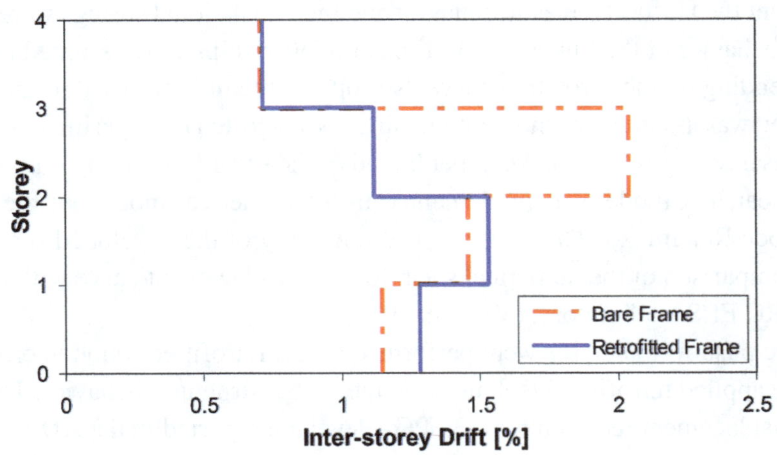

Figure 15. Non-linear dynamic analyses: Demand-to-Capacity Ratio for the bare and retrofitted frames

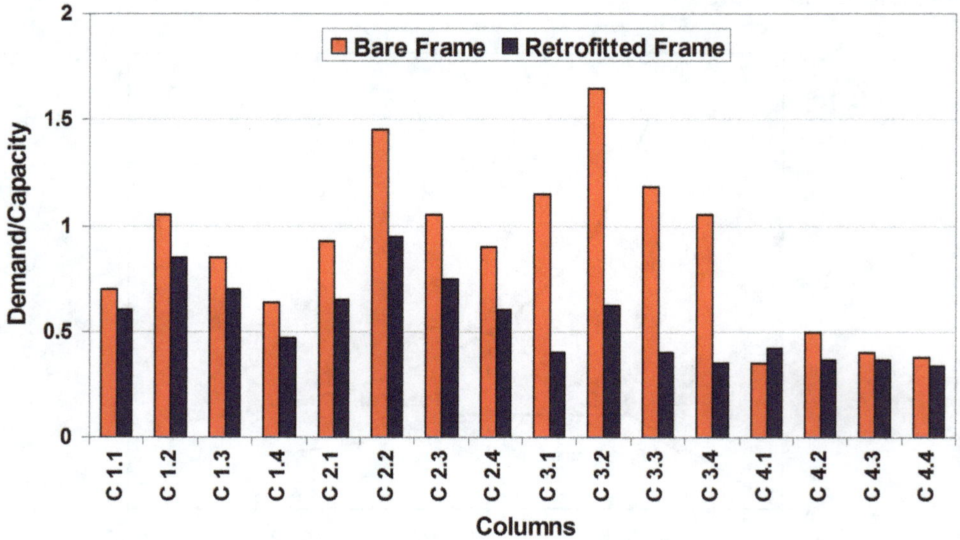

Retrofitting Intervention Using Eccentric Steel Bracing

A retrofitting solution involving the introduction of eccentric steel bracing (chevron bracing with a vertical shear link) was also investigated. The steel bracing assembly was inserted in the middle bay of all the floors. The vertical shear link was located at the mid-span underneath the floor beams and connected to a chevron bracing. For the link specimen a European HE120A section was adopted. Vertical steel straps were connected to the adjacent columns and two horizontal steel beams were anchored to the floor beams. Figure 16 shows a schematic view of the proposed eccentric bracing system inserted in the middle bay of the RC frame.

A link model was developed for the numerical analyses of the retrofitted frame, based on the approach proposed by Ricles and Popov (1987) for horizontal shear link elements. Steel links are subjected to high levels of shear forces and bending moments in the active link regions and elastic and inelastic deformations of both the shear and flexural behaviors have to be taken into account. The link was modelled as a linear beam element with non-linear rotational and translational springs at the ends. The rotational spring was used to represent the flexural inelastic behavior and the translational springs were used to represent the inelastic shear behavior of the link web. Multilinear relationships were assumed for the shear force-deformation and bending moment-rotation curve. Isotropic hardening was used in shear yielding, while kinematic hardening was assumed for moment yielding, as suggested by experimental evidence, (Ricles & Popov, 1987; Kasai & Popov, 1986). An upper bound of the shear force after complete hardening was considered. The shear link model was implemented in the numerical models of the frame developed by the computer code Ruaumoko. The accuracy and reliability of the developed shear link model was verified through comparison of the numerical shear force – displacement curve with experimental tests performed at the JRC ELSA laboratory, (Valente, 2011).

Non-linear static pushover analyses were performed on the retrofitted frame in order to estimate the effectiveness of the applied retrofitting technique on the global structural behavior. The results in terms of maximum top displacement required for 0.3g PGA level are reported in the AD format in Figure 17.

Figure 16. Elevation view of the RC frame retrofitted with eccentric steel bracing

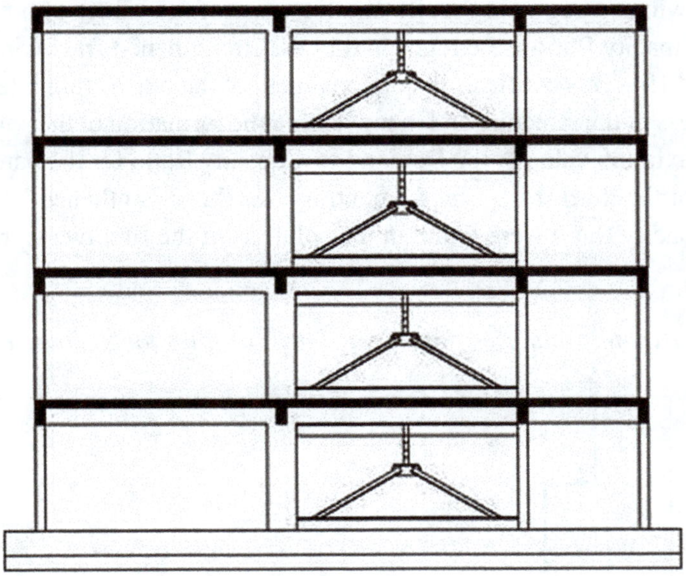

Figure 17. Demand spectra and capacity curve in AD format at the LSSD for the frame retrofitted with eccentric steel bracing

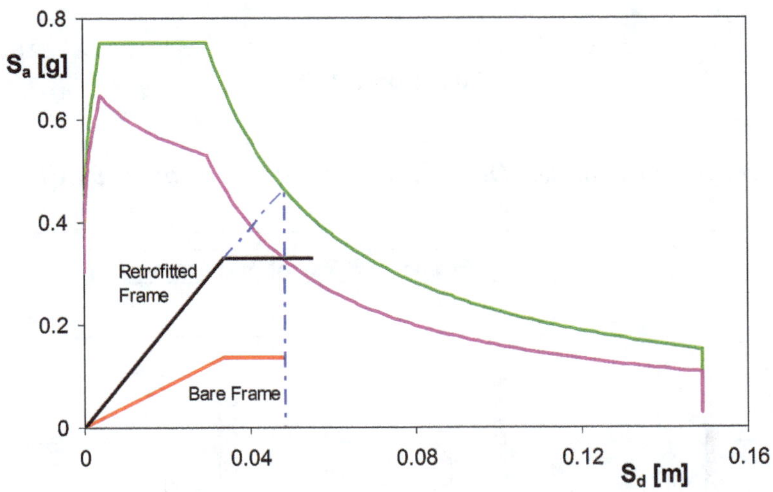

The retrofitting intervention considerably increased the stiffness and strength of the structure. The retrofitted frame was able to satisfy the LSSD and the capacity exceeded the demand. The seismic demand in terms of displacement, transformed to actual MDOF system, was equal to 0.06 m (0.098 m for the bare frame), while the capacity of the frame was increased up to 0.071 m (0.062 m in the bare frame). The procedure confirmed the effectiveness of the retrofitting intervention in both reducing the displacement demand and increasing the global deformation capacity of the bare frame. The retrofitting intervention eliminated the irregularities of the frame and the global response of regular structures may be captured more accurately by pushover analyses.

Non-linear dynamic analyses were carried out using seven artificial accelerograms with PGA=0.3g, as described in the previous paragraph. Significant decreases of the values of both the inter-storey drift and the Demand-to-Capacity Ratio were registered in the strengthened frame, in particular at the third storey, (Figures 18 and 19). This confirms the effectiveness of the retrofitting intervention in increasing the strength and stiffness of the structure and in preventing the formation of the soft-storey mechanism at the third floor. The maximum value of the Demand-to-Capacity Ratio for the bare frame was registered in the strong column at the third storey; on the contrary, for the strengthened frame the highest values of the Demand-to-Capacity Ratio were found in the columns of the first two storeys.

Figure 18. Non-linear dynamic analyses: inter-storey drift profiles for the bare and retrofitted frames

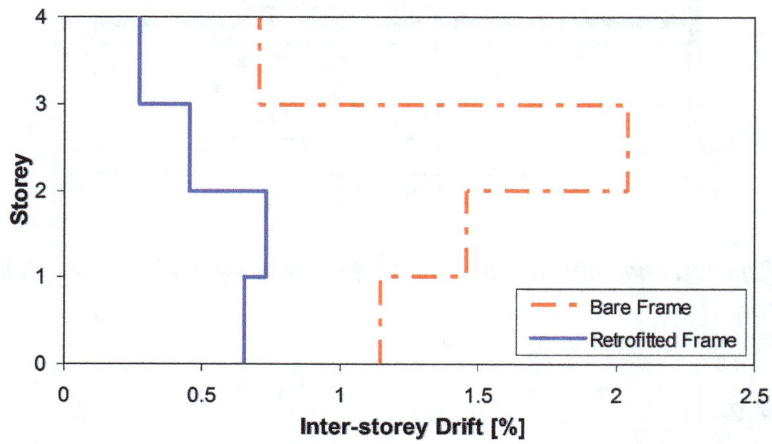

Figure 19. Non-linear dynamic analyses: Demand-to-Capacity Ratio for the bare and retrofitted frames

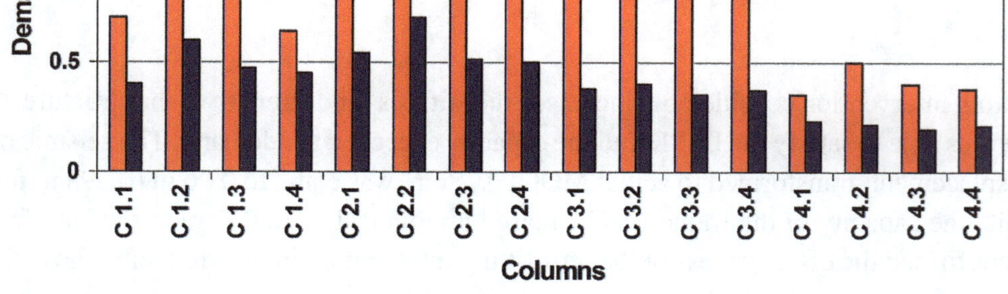

Retrofitting Intervention Using Infill Wall

The third intervention proposed in this study is one of the most common method used for strengthening existing RC structures and consisted in adding a concrete shear wall to the frame. The infill wall was introduced into the short bay of the frame by incorporating existing columns C3 and C4. The existing columns were not adequate to act as boundary elements and therefore additional boundary elements were created with a 30x30 cm cross-section and eight Φ16 vertical longitudinal bars. As regards the connection between the existing members and the infill wall, dowels should be placed between the wall and the columns in order to assure a monolithic connection. The elevation of the retrofitted frame and the cross-section of the wall with reinforcement detailing are shown in Figure 20. The height of the critical region was equal to the height of the first storey. Two structural systems were identified in the building: the wall and the other vertical elements.

The behavior of the retrofitted frame was significantly influenced by the presence of the wall. Non-linear static analyses were performed on the retrofitted frame and the base shear – top displacement curve was obtained. The calculated response of the frame for the case of adding an infill wall in the short bay is compared with the response of the original frame in Figure 21. The insertion of the infill wall substantially increases the stiffness and strength of the original frame. The base shear of the wall

Figure 20. Elevation view of the retrofitted RC frame and cross-section detailing of the infill wall (dimensions in cm). Boundary elements: vertical bars ϕ16, transversal reinforcement ϕ12/10. Web reinforcement: vertical bars ϕ16/17, horizontal bars ϕ14/20.

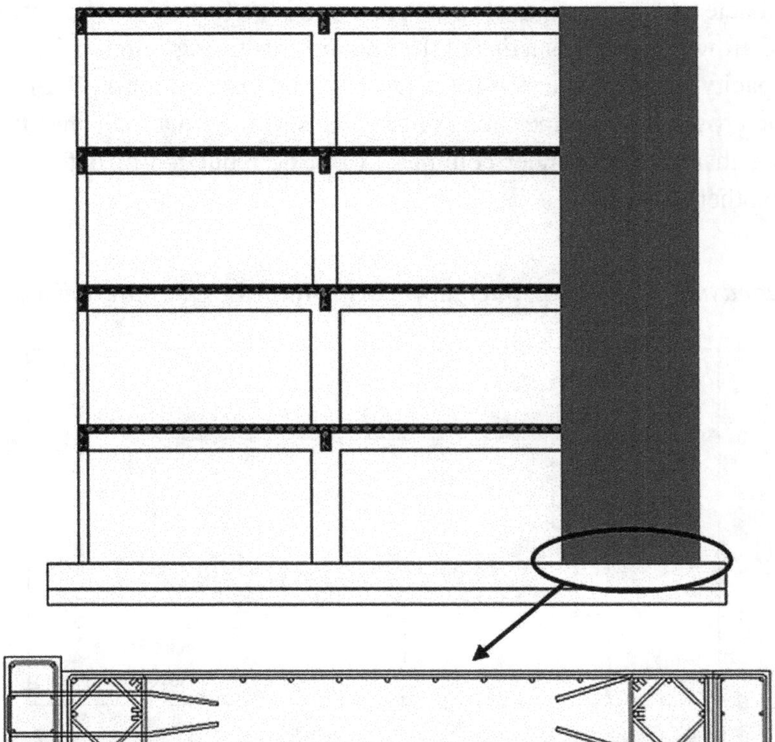

Figure 21. Demand spectra and capacity curve in AD format at the LSSD for the frame retrofitted with infill wall

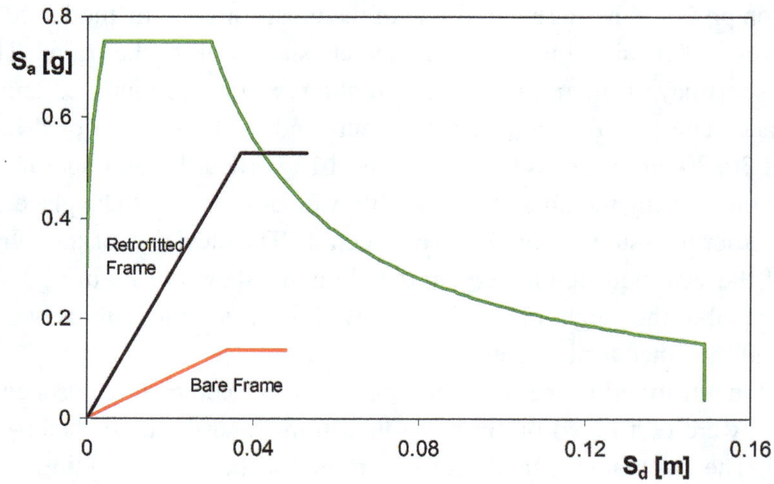

retrofitting solution is about four times greater than the base shear of the bare frame and 1.65 times greater than the base shear of the bracing retrofitting solution.

The effectiveness of the wall retrofit solution was assessed by performing non-linear dynamic analyses with the same records as the previous structural configurations. The inter-storey drift profile imposed by artificial accelerograms with PGA=0.3g is presented and compared with that of the bare frame in Figure 22. A considerable reduction of the inter-storey drifts was registered for all the levels of the frame. The wall acted as a stiff vertical spine preventing the formation of a soft-storey mechanism and the maximum drift was smaller than that of the bracing retrofitting solution. The maximum value of the Demand-to-Capacity Ratio for the retrofitted frame was registered for the wall, as shown in Figure 23. Due to the large cross-section dimensions of the wall, the deformation capacity in terms of chord rotation was smaller than that of slender columns. A drastic reduction of deformation demands was observed for all the other columns.

Figure 22. Non-linear dynamic analyses: inter-storey drift profiles for the bare and wall-retrofitted frames

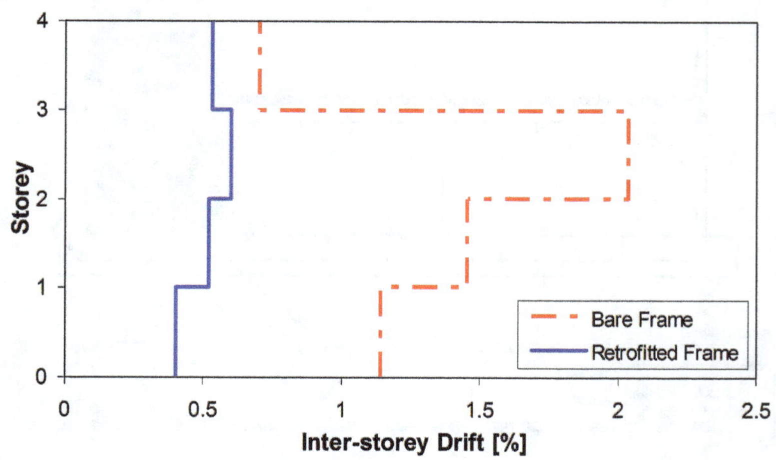

Figure 23. Non-linear dynamic analyses: Demand-to-Capacity Ratio for the bare and wall-retrofitted frames

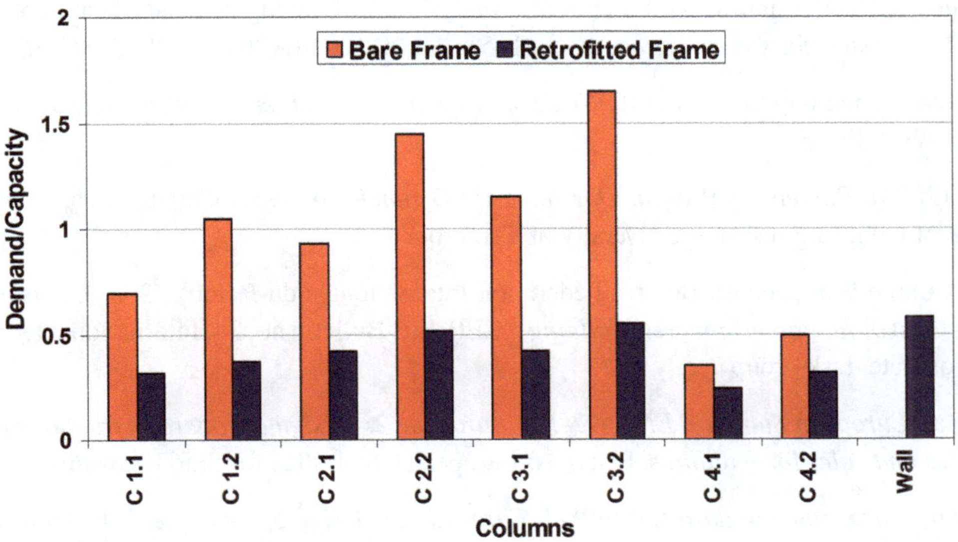

CONCLUSION

The aim of this study was to investigate a simplified displacement based procedure for the seismic assessment and retrofitting of a four-storey RC frame, designed mainly for gravity loads without specific earthquake-resistant provisions and tested at the JRC ELSA laboratory. The effectiveness of three retrofitting interventions with different aims was assessed as well. Detailed numerical models of the RC frame in the bare and retrofitted configurations were developed and static pushover analyses were performed. The theoretical predictions of the simplified procedure in terms of global performance showed that the bare frame was unable to satisfy the demand due to the 0.3g PGA level at the Limit State of Significant Damage. A selective retrofitting intervention was modelled and investigated for the enhancement of the seismic performance of the RC frame. The FRP wrapping increased the deformation capacity of the columns and the RC jacketing was found to be effective for mitigating the abrupt change in the flexural capacity of the strong column at the third floor and for avoiding the soft-storey failure mechanism. The second retrofitting intervention proposed in this study and based on the introduction of eccentric steel bracings reduced the displacement demand on the frame and increased the energy dissipation capacity of the system. The third retrofitting solution based on the addition of a concrete shear wall was extremely efficient in controlling global lateral drift and thus reducing damage in frame members. However, the drawback of this method is related to the strengthening of the existing foundation system to resist the increased overturning moment. The drawbacks and economic losses related to the addition of walls may render the local retrofitting solution carried out by using RC jacketing or FRP wrapping more appealing. Numerical results obtained from the simplified displacement based procedure were confirmed by the non-linear dynamic analyses.

REFERENCES

Antonopoulos, C. P., & Triantafillou, T. C. (2002). Analysis of FRP-strengthened RC beam-column joints. *Journal of Composites for Construction*, *6*(1), 41–51. doi:10.1061/(ASCE)1090-0268(2002)6:1(41)

ATC (Applied Technology Council). (1996). *Seismic evaluation and retrofit of concrete buildings*. ATC (Report No. ATC-40).

Carr, A. J. (2006). *Ruaumoko Program for Inelastic Dynamic Analysis*. Christchurch, New Zealand: Department of Civil Engineering, University of Canterbury.

CEB-FIB (Comité Européen du Béton - Fédération Internationale du Béton). (2003). *Seismic assessment and retrofit of reinforced concrete buildings*. CEB-FIB Bulletin no. 24. International federation for structural concrete, task group 7.1.

CEN. (2004a). *European Standard EN 1992-1-1. Eurocode 2: Design of concrete structures - Part 1-1: General rules and rules for buildings*. Brussels: European Committee for Standardization.

CEN. (2004b). *European Standard EN 1998-1. Eurocode 8: Design of structures for earthquake resistance. Part 1: General rules, seismic action and rules for buildings*. Brussels: European Committee for Standardization.

CEN. (2005). *European Standard EN 1998-3. Eurocode 8: Design of structures for earthquake resistance. Part 3: Assessment and retrofitting of buildings*. Brussels: European Committee for Standardization.

Di Sarno, L., & Elnashai, A. S. (2009). Bracing systems for seismic retrofitting of steel frames. *Journal of Constructional Steel Research*, *65*(2), 452–465. doi:10.1016/j.jcsr.2008.02.013

Fajfar, P. (1999). Capacity spectrum method based on inelastic demand spectra. *Earthquake Engineering & Structural Dynamics*, *28*(9), 979–993. doi:10.1002/(SICI)1096-9845(199909)28:9<979::AID-EQE850>3.0.CO;2-1

Fajfar, P. (2000). A nonlinear analysis method for performance-based seismic design. *Earthquake Spectra*, *16*(3), 573–592. doi:10.1193/1.1586128

Fajfar, P., Marušic, D., & Peruš, I. (2005). Torsional effects in the pushover-based seismic analysis of buildings. *Journal of Earthquake Engineering*, *9*(6), 831–854. doi:10.1080/13632460509350568

Fardis, M. N., Schetakis, A., & Strepelias, E. (2013). RC buildings retrofitted by converting frame bays into RC walls. *Bulletin of Earthquake Engineering*, *11*(5), 1541–1561. doi:10.1007/s10518-013-9435-6

FEMA (Federal Emergency Management Agency). (2000). *FEMA 356: Pre-Standard and Commentary for the Seismic Rehabilitation of Buildings*. Washington, DC: FEMA.

Giannuzzi, D., Ballarini, R., Hucklebridge, A., Pollino, M., & Valente, M. (2014). Braced ductile shear panel: New seismic-resistant framing system. *Journal of Structural Engineering*, *140*(2), 1–11. doi:10.1061/(ASCE)ST.1943-541X.0000814

Kasai, K., & Popov, E. P. (1986). *A study of seismically resistant eccentrically braced steel framesystems. Rep. No. UCB/EERC-86/01*. Berkeley, CA: Earthquake Engineering Research Center, University of California.

Mander, J. B., Priestley, M. J. N., & Park, R. (1988). Theoretical Stress-Strain Model for Confined Concrete. *Journal of Structural Engineering*, *114*(8), 1804–1826. doi:10.1061/(ASCE)0733-9445(1988)114:8(1804)

Martinez-Rueda, J. E., & Elnashai, A. S. (1997). Confined concrete model under cyclic load. *Materials and Structures*, *30*(197), 139–147. doi:10.1007/BF02486385

Masi, A. (2003). Seismic vulnerability assessment of gravity load designed RC frames. *Bulletin of Earthquake Engineering*, *1*(3), 371–395. doi:10.1023/B:BEEE.0000021426.31223.60

Mazzolani, F. (2008). Innovative metal systems for seismic upgrading of RC structures. *Journal of Constructional Steel Research*, *64*(7-8), 882–895. doi:10.1016/j.jcsr.2007.12.017

Menegotto, M., & Pinto, P. E. (1973). Method of analysis for cyclically loaded R.C. plane frames including changes in geometry and non-elastic behavior of elements under combined normal force and bending. Preliminary Report. *IABSE, Zurich*, *13*, 15–22.

Milani, G., & Valente, M. (2015a). Comparative pushover and limit analyses on seven masonry churches damaged by the 2012 Emilia-Romagna (Italy) seismic events: Possibilities of non-linear Finite Elements compared with pre-assigned failure mechanisms. *Engineering Failure Analysis*, *47*, 129–161. doi:10.1016/j.engfailanal.2014.09.016

Milani, G., & Valente, M. (2015b). Failure analysis of seven masonry churches severely damaged during the 2012 Emilia-Romagna (Italy) earthquake: Non-linear dynamic analyses vs conventional static approaches. *Engineering Failure Analysis*, *54*, 13–56. doi:10.1016/j.engfailanal.2015.03.016

Otani, S. (1981). Hysteretic models of reinforced concrete for earthquake response analysis. *Journal of the Faculty of Engineering*, *36*(2), 125–159.

Pinto, A., Verzeletti, G., Molina, J., Varum, H., Pinho, R., & Coelho, E. (2002). *Pseudo-dynamic tests on non-seismic resisting RC frames (bare and selective retrofit frames). EUR Report 20244 EN, ELSA*. Ispra, Italy: Joint Research Centre.

Ricles, J. M., & Popov, E. P. (1987). *Dynamic analysis of seismically resistant eccentrically braced frames. Rep. No. UCB/EERC-87/07*. Berkeley, CA: Earthquake Engineering Research Center, University of California.

Rozman, M., & Fajfar, P. (2009). Seismic response of a RC frame building designed according to old and modern practices. *Bulletin of Earthquake Engineering*, *7*(3), 779–799. doi:10.1007/s10518-009-9119-4

SeismoSoft (2007). *SeismoStruct – A Computer Program for Static and Dynamic Nonlinear Analysis of Framed Structures*.

Spoelstra, M., & Monti, G. (1999). FRP-confined concrete model. *Journal of Composites for Construction*, *3*(3), 143–150. doi:10.1061/(ASCE)1090-0268(1999)3:3(143)

Valente, M. (2011). Eccentric steel bracing for seismic retrofitting of non-ductile reinforced concrete frames.*The Thirteenth International Conference on Civil, Structural and Environmental Engineering Computing CC2011*, Crete, Greece. doi:10.4203/ccp.96.196

Valente, M. (2012). Seismic rehabilitation of a three-storey R/C flat-slab prototype structure using different techniques. *Applied Mechanics and Materials*, *193-194*, 1346–1351. doi:10.4028/www.scientific.net/AMM.193-194.1346

Valente, M. (2013a). Seismic upgrading strategies for non-ductile plan-wise irregular R/C structures. *Procedia Engineering*, *54*, 539–553. doi:10.1016/j.proeng.2013.03.049

Valente, M. (2013b). Seismic strengthening of non-ductile R/C structures using infill wall or ductile steel bracing. *Advanced Materials Research*, *602-604*, 1583–1587. doi:10.4028/www.scientific.net/AMR.602-604.1583

Valente, M., & Milani, G. (2016a). Seismic assessment of historical masonry towers by means of simplified approaches and standard FEM. *Construction & Building Materials*, *108*, 74–104. doi:10.1016/j.conbuildmat.2016.01.025

Valente, M., & Milani, G. (2016b). Non-linear dynamic and static analyses on eight historical masonry towers in the North-East of Italy. *Engineering Structures*, *114*, 241–270. doi:10.1016/j.engstruct.2016.02.004

Vanmarcke, E. H., Cornell, C. A., Gasparini, D. A., & Hou, S. (1976). *SIMQKE: A program for artificial motion generation*. Civil Engineering Department, Massachusetts Institute of Technology.

Yankelevsky, D. Z., & Reinhardt, H. W. (1989). Uniaxial behavior of concrete in cyclic tension. *Journal of Structural Engineering*, *115*(1), 166–182. doi:10.1061/(ASCE)0733-9445(1989)115:1(166)

Youssefa, M. A., Ghaffarzadehb, H., & Nehdia, M. (2007). Seismic performance of RC frames with concentric internal steel bracing. *Engineering Structures*, *29*(7), 1561–1568. doi:10.1016/j.engstruct.2006.08.027

Chapter 3
Single-Run Adaptive Pushover Procedure for Shear Wall Structures

Malik Atik
University of Lille 1, France

Marwan Sadek
University of Lille 1, France & Lebanese University, Lebanon

Isam Shahrour
University of Lille 1, France

ABSTRACT

This chapter proposes a new single-run adaptive pushover method for the seismic assessment of shear wall structures. This method offers two main advantages: it does not require decomposing the structure in nonlinear domain and it avoids the pitfall of previous single-run adaptive pushover analyses in utilizing the modal combination in the determination of the applied loads instead of combining the response quantities induced by those loads in individual modes. After a brief review of the main adaptive pushover procedures, the proposed method is presented as well as its numerical implementation. The predictions of this method are compared to those of other recent adaptive pushover methods and as well as to the rigorous non-linear time history analysis. Analyses show the efficiency of the proposed method.

INTRODUCTION

Nonlinear static procedures or pushover analyses constitute an efficient tool to assess the seismic demand of structures. They constitute a reliable alternative of nonlinear time-history analysis of structures. For tall buildings, the effect of higher modes is not negligible, that's why ignoring their effect is one of the main limitations of pushover analyses. Furthermore, the modes of vibration of the structure can significantly change during strong seismic motion.

DOI: 10.4018/978-1-5225-2089-4.ch003

In recent years, several techniques have been proposed to integrate the effect of higher modes in pushover analyses and to incorporate the variation in dynamic properties associated to structural damages. Gupta and Kunnath (2000) updated the applied load at each increment. The eigenvalue analysis is carried out at each load increment, then a static analysis is carried out for each mode independently. The calculated effects are combined with SRSS and added to the corresponding values from the previous step. Similarly, Aydinoğlu (2003, 2004, 2007) developed and extended this method to estimate the peak demand quantities. These adaptive procedures provide good estimates of seismic demands, however:

1. They are computationally complicated (Chopra and Goel 2002; Baros and Anagnastopoulos 2008). This is mainly due to the absence of a structural equilibrium at the end of each step as the result of using SRSS to combine the responses (Antoniou and Pinho 2004a), so a routine application has to be made to impose the stiffness of the structure at the beginning of each step.
2. In the inelastic domain, the structural system could not be decomposed into several independent systems (corresponding to the desirable number of modes), consequently the application of the modal combination rule in the inelastic domain is no longer valid. To overcome this difficulty, small steps should be taken where the system can be considered linear (Gupta and Kunnath 2000) or the modal response increments in each mode must be scaled in such a way that the response spectrum analysis (RSA) is implemented in a piecewise linear fashion at each pushover step (Aydinoğlu 2003), but it increases the computational demands.

In an attempt to avoid the previous computational complexity (the absence of structural equilibrium) and based on the work of Chopra and Goel (2002), Kalkan and Kunnath (2006) developed an adaptive modal combination procedure that accounts for higher mode effects. They combine the response of individual modal pushover analyses and incorporate the effects of progressive variation in dynamic characteristics during the inelastic response via its adaptive feature. The lateral load distribution used in the progressive pushover analysis is based on instantaneous inertia force distribution across the height of the building for each mode. However, these multi-run methods do not reflect the yielding effect of one mode on other modes and on the interaction between modes in the nonlinear range. On the other hand, this method, as the method of Chopra and Goel (2002), is not applicable to estimating member forces because forces computed by this procedure may exceed the specified member capacity. Therefore, there is a need to recompute the member force from the member deformation(s) determined by this procedure to have member forces consistent with their specified capacity (Goel and Chopra 2005). That needs additional computational effort. A variant of modal pushover analysis (VMPA) has been presented by Surmeli and Yuksel (2015) in order to evaluate the seismic performance of the structures.

In order to combine the advantages of previous methods, single-run adaptive pushover procedures (SRAP) have been proposed. The SRAP procedure does not require decomposing of the structural system into several independent systems corresponding to the desirable number of modes, which is practiced in the multiple-run pushover procedures. In SRAP method, the equivalent seismic load is defined on the base of one of the modal components (floor forces, shear forces, displacement, etc.) which is deduced from different modes using a modal combination rule (SRSS, CQC, etc.). The main critic of these methods consists in the use of the modal combination in defining the applied loads instead of combining the response quantities induced by those loads in individual modes (Chopra, 2007, p. 569; Aydinoğlu, 2003, 2007).

Although in the literature there are applications of pushover methods on shear wall structures (Reyes and Chopra, 2011), it could be noted the absence of the application of single-run adaptive pushover procedures (SRAP) on shear wall structures. After a brief description of the single run adaptive pushover procedures, an innovative new single-run adaptive pushover method for shear wall structures is developed in this chapter. The results of the proposed method are compared to those obtained with the SRAP procedures available in the literature. The results illustrate the superiority of the proposed method in comparison to those available in the literature and may explain why these methods were mainly applied for frame structures.

OVERVIEW AND PRINCIPLE OF SINGLE-RUN ADAPTIVE PUSHOVER PROCEDURES

Antoniou and Pinho (2004a) explored the accuracy of force-based adaptive pushover analysis in predicting the horizontal capacity of reinforced concrete buildings. They proposed a force-based adaptive pushover (FAP) which is an extended version of the fully adaptive pushover algorithm proposed by Elnashai (2001). The lateral load distribution is continuously updated during the process, according to modal shapes and participation factors derived by eigenvalue analysis carried out at each analysis step. The modal floor forces for the desirable modes are evaluated at each step according to the instantaneous stiffness matrix and the corresponding elastic spectral accelerations. Then the lateral load pattern is determined by combining the floor forces of each vibration mode. The loads from all modes are combined using the SRSS rule. It was concluded that, despite its apparent conceptual superiority, current force-based adaptive pushover shows a relatively minor advantage over its traditional non-adaptive counterpart, mainly for the estimation of deformation patterns of buildings, which are poorly predicted by both types of analysis.

Another variant of the method proposed by Antoniou and Pinho (2004b) is the displacement-based adaptive pushover procedure (DAP), whereby a set of laterally applied displacements, rather than forces, is monotonically applied to the structure. In their paper, the authors proposed again the interstory drift as a base instead of the displacement and it has been adopted as the standard DAP variant. The DAP procedure improved the response predictions, throughout the entire deformation range, in comparison to those obtained by force-based methods. Contrary, Casarotti and Pinho (2007) re-adopted the displacement as a base instead of the interstory drift for estimating seismic demands on bridges.

In order to adjust the drawbacks of the FAP procedure, Shakeri et al. (2010) proposed a story shear-based adaptive pushover method (SSAP), where the load pattern is derived from the modal story shear profile. They referred the superiority of the SSAP method over the FAP because it takes into account the change in the sign of the story components along the structure height for higher modes.

From the above, it can be noted that all of these methods have made the previously mentioned pitfall in computing the combined peak value of one response quantity from the combined peak values of other response quantities. However, one does not deny that the results were sometimes satisfactory; the present chapter shows that these methods are not valid for shear wall structures. An innovative new single-run adaptive pushover method is proposed. It is based on the modal overturning moment story, which avoids the previous pitfall and it is valid for shear wall structures.

DESCRIPTION OF THE OVERTURNING MOMENT-BASED ADAPTIVE PUSHOVER PROCEDURE (OMAP)

Theoretical Base of OMAP Method

The idea behind the single-run pushover procedures is that in the inelastic domain, the structural system could not be decomposed into several independent systems. Instead of decomposing the structure into several independent structures, combined equivalent seismic loads are applied to the structure, where one of the modal components (Inertial force, displacement, drift story, shear forçe,...) is chosen to be combined by one of the modal combination rules and to be as a base to define the equivalent seismic loads. It is clear that each base gives different equivalent lateral loads, this leads to the question which base should be chosen in pushover analysis?

In shear wall structures, the moments and the axial force are generally responsible of the plasticity. Where, it is supposed that the standard recommendations concerning the reinforcement details are adopted to avoid shear failure and guarantee a predominantly flexural behavior at plastic hinge zones. But under horizontal earthquake, the axial forces in the shear wall structure remain constant (induced by vertical loads), so the prediction of plastic hinges requires only the determination of the bending moments in the shear wall. Therefore in shear wall structures, the equivalent lateral forces are derived from the combined modal flexural moments, which allow the analysis of the plastic behavior of the structure. In other words, the equivalent lateral forces in the OMAP are utilized to modify the stiffness of the structure instead of using a routine application in order to impose this stiffness as in the adaptive response spectrum analysis (Gupta and Kunnath, 2000; Aydinoğlu, 2003, 2004, 2007). Note that imposing the stiffness of the structure during the analysis is not possible in practical structural engineering software. So the main advantage of the OMAP method consists in its easy implementation maintaining the principle of the adaptive response spectrum analysis for shear wall structures.

On the other hand, it should be emphasized that the equivalent lateral forces are valid for calculating the overturning moments and the corresponding rotations in the structure. These equivalent lateral forces serve to predict the plastic hinge, but they are not valid for estimating other quantities. That's why the other quantities are estimated by combining the peak response quantities in individual modes at each increment. Consequently, it can be noted that if the proposed method (OMAP) is applied to linearly elastic systems, it reduces to the standard response spectrum analysis. This is not the case for the conventional single-run adaptive pushover procedures.

This is the theoretical base of the proposed method (OMAP) and that's why the overturning moments are chosen as the analysis base. In addition, these forces are constantly updated using eigenvalue analysis at each step, which allows consideration of the progressive variation in dynamic properties associated to structural damages.

Algorithm of OMAP Method

The key-elements of the nonlinear static pushover analysis are: the external applied loads and the target displacement. The present chapter is concerned with the first purpose. The capacity spectrum method (CSM) proposed in ATC-40 (1996) could be used to estimate the target displacement or the peak response quantities.

The algorithm of OMAP method offers two improvements:

1. The possibility of choosing different damping values for the modes. It allows incorporating different approaches of modeling damping in nonlinear time history analysis (Charney, 2008, Smyrou et al., 2011).
2. The possibility of calculating the incremental applied loads depending on an incremental roof displacement: at each iteration, the corresponding incremental roof displacement is specified then the incremental applied loads are scaled to give this corresponding incremental roof displacement. Note that specifying an incremental target displacement is more relevant than an incremental base shear as in the SSAP method.

The adaptation at each incremental step in single-run adaptive pushover procedures is just to consider the progressive variation in dynamic properties but in multi-run adaptive procedures where the static analyses are done for each mode separately (Gupta and Kunnath 2000; Aydinoğlu, 2003, 2004), it is not only for this reason but also to avoid overshooting of element yield forces when a modal combination rule is applied. So, a non-adaptive version of single-run adaptive pushover procedures can be performed but this is not the case for the methods proposed by Gupta and Kunnath (2000) and Aydinoğlu (2003, 2004). Consequently, the application of the non-adaptive version of the proposed method allows investigating the effect of neglecting the variation in dynamic properties associated to the structural damage.

The OMAP algorithm includes the following basic steps:

1. Specifying the desirable number of iteration (N) and the corresponding incremental floor displacement (D_{roof}) for each iteration.
2. Defining the elastic response spectrum (Pseudo-accelerations vs. Periods) with the corresponding damping ratio.
3. Performing an eigenvalue analysis of the structure to compute periods (T_j), mode shapes (Φ_{ij}) and modal participation factors (Γ_j) for the (n) desirable modes, where "i" is the story number and "j" is the mode number.
4. Choosing the modal damping ratio of the structure ξ_j, (1≤j≤n).
5. Computing the pseudo-spectral acceleration for each considered mode (Sa_j); if the damping ratio of the jth mode is different from that of the used response spectrum, this latter is adjusted using the following formula (Newmark and Hall, 1982):

$$A_2 = A_1 \frac{\left(2.31 - 0.41^* \ln \beta_2\right)}{\left(2.31 - 0.41^* \ln \beta_1\right)} \tag{1}$$

where:

A_1 = Acceleration corresponding to damping ratio β_1;

A_2 = Acceleration corresponding to damping ratio β_2;

$0 < \beta_1 < 100$ (percentage);

$0 < \beta_2 < 100$ (percentage); and

ln = natural logarithm (base e).

6. Computing the load factor (λ) for this iteration as follows:
   a. Determine the roof displacement before the scaling (D_r) by quadratic combination rule to the peak modal floor displacements

$$D_r = \sqrt{\sum_{j=1}^{n} D_{rj}^2} \tag{2}$$

$$D_{rj} = \Gamma_j \Phi_{rj} \frac{Sa_j}{\omega_j^2} \tag{3}$$

where,

D_{rj} is the peak modal floor displacement at the roof for j^{th} mode before the scaling.

ω_j is the jth natural frequency

   b. Determine the load factor (λ)

$$= \frac{D_{roof}}{D_r} \tag{4}$$

where, D_{roof} the desirable incremental floor displacement for this iteration.

7. Computing the peak modal responses for the (n) modes as follows:

$$F_{ij} = \Gamma_j \Phi_{ij} m_i Sa_j . \tag{5}$$

$$SS_{ij} = \sum_{k=i}^{ns} F_{kj} \tag{6}$$

$$OM_{ij} = \sum_{k=i}^{ns} SS_{kj} * h_k \tag{7}$$

$$D_{ij} = \Gamma_j \Phi_{ij} \frac{Sa_j}{\omega_j^2} \tag{8}$$

$$\Delta_{ij} = D_{ij} - D_{(i-1)j} \tag{9}$$

where:

m_i is the mass of ith story

F_{ij} is the lateral floor force at i[th] floor for j[th] mode

SS_{ij} is the modal story shear at i[th] story for j[th] mode

OM_{ij} is the modal overturning moment at i[th] floor for j[th] mode

h_i is the height of the ith story

ns is the number of stories or floors.

D_{ij} is the floor displacement at i[th] floor for j[th] mode

Δ_{ij} is the story drift at i[th] story for j[th] mode

8. Calculating the desirable combined peak responses by quadratic combination rule for this iteration and add these to the same from the previous iteration:

$$r_i = \sqrt{\sum_{j=1}^{n} r_{ij}^2} \tag{10}$$

For example, the combined overturning moment at i[th] floor is given as follow:

$$OM_i = \sqrt{\sum_{j=1}^{mn} OM_{ij}^2}$$

9. Calculating the equivalent lateral forces which give the combined overturning moment:

$$F_i = \frac{OM_i - OM_{i+1}}{h_i} - \frac{OM_{i+1} - OM_{i+2}}{h_{i+1}}; i = 1, 2, \ldots, (ns - 1)$$

$$F_{ns} = \frac{OM_{ns}}{h_{ns}}; i = ns \quad (11)$$

10. Performing the pushover analysis by using the equivalent lateral forces computed in the step (9) and starting from state at end of the previous iteration.
11. Returning to step 3 and continuing the process N time.

The steps (9) and (10) are the responsible of modifying the structure stiffness, so it can be noted that if the system is elastic, these steps do not modify the eigenvalues. Consequently, for a summation of incremental floor displacements equals to Dr, the method reduces to the standard response spectrum analysis. Noting that the Equation1 is utilized for simplifying the computing of the pseudo-spectral acceleration but that does not prevent using the exact pseudo-spectral acceleration.

Note 1: The square-root-of-sum of squares (SRSS) rule appears to be the obvious choice for modal combination, although complete quadratic combination (CQC) rule may be more appropriate when close modes are present as in the case of coupled lateral-torsional response of three-dimensional systems (Chopra, 2007, 13.7.2 Modal Combination rules).

Note 2: Overturning moments and plastic hinge rotations can be picked up directly from the structure subjected to pushover analysis.

ILLUSTRATIVE EXAMPLES

Introduction

The proposed method (OMAP) as well as the methods proposed by Gupta and Kunnath (2000) and Aydinoğlu (2003, 2004) are based on the principle of the adaptive response spectrum analysis. It was demonstrated that these procedures are able to reasonably estimate the response quantities. The main scope of this chapter is essentially to illustrate the methodology of the use of a single-run adaptive pushover analysis based on the principle of the adaptive response spectrum analysis. The OMAP procedure has been implemented in a Visual Fortran program. The subroutines are linked to the nonlinear version of SAP2000 program in order to calculate and apply equivalent lateral forces at each increment.

Case Study

The case study consists of a shear wall designed for 20 story building and stands for 70 meters tall. The length of the shear wall is 6 m and its thickness is variable with the height. The shear wall properties are summarized in Table 1. The capacity ratio is taken as a result of vertical load of 420 kN/Story. The yield strength of the steel is f_y= 345 MPa and the concrete compressive strength is f'c = 28 MPa.

Ground Motions

The Imperial Valley (1940) is selected in the present study. This record is available in the Pacific Earthquake Engineering Research (PEER) website. Figure 1 shows the seismic record, its response spectrum (5% damping) and frequency content. It has been selected with a frequency content allowing the higher modes to be excited (period less than 1 s in general in tall building structures). The ground motion is scaled by a factor of 2.5 in order to obtain an advanced plasticity state in the structure.

Modeling

The walls are modeled as beam-column elements. The nonlinear behavior of the structure is considered via discrete hinges defined in the wall at the ends of each story. The Yield criterion is based on the interaction of the axial force and the bending moment. Figure 2 shows the load-deformation curve of the hinges. The hinge properties and the modeling parameters a, b, and c (Figure 2) are specified according to FEMA-356. Note that Q and Q_y are the generalized and yield component loads, respectively. However, in recent years, sophisticated modeling for the nonlinear behavior of shear walls became prevalent; simple modeling is adopted in this survey where the objective is essentially to compare the proposed method (OMAP) with the recent adaptive pushover methods available in the literature and the NTHA.

The time history analysis is performed using the numerical implicit Newmark time integration method (SAP 2000). A damping ratio of 5% is considered for the first and third mode of vibrations in order to specify the Rayleigh damping coefficients c_M and c_K:

$$C = c_M \; M + c_K K_t \tag{12}$$

where M is the mass matrix and K_t is the matrix of the tangent stiffness of the structure at each time step.

Table 1. Shear wall properties for 20–story

Story ID	Thickness (cm)	Longitudinal Reinforcement	Capacity Ratio* (Vertical Load)
1→5	28	2φ18/20cm	0.37
6→9	25	2φ14/20cm	0.33
10→11	20	2φ12/20cm	0.35
12→15	18	2φ12/20cm	0.35
16→18	15	2φ10/20cm	0.26
19→20	10	2φ10/20cm	0.33

* The capacity ratio concerns the first story of each group

Figure 1. Ground motion characteristics - Imperial Valley 1940, NS

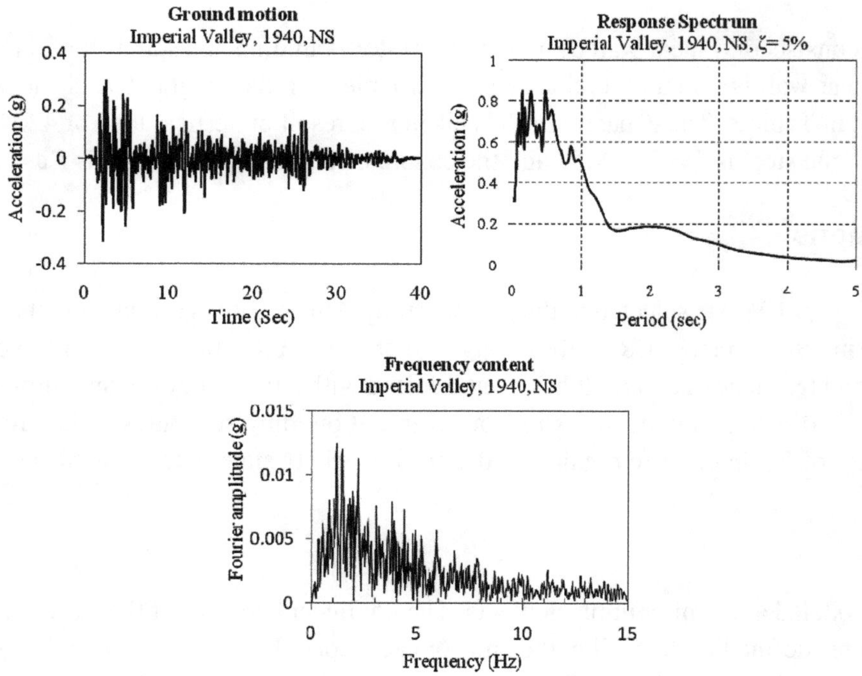

Figure 2. Load-Deformation curve for hinges (FEMA 356)

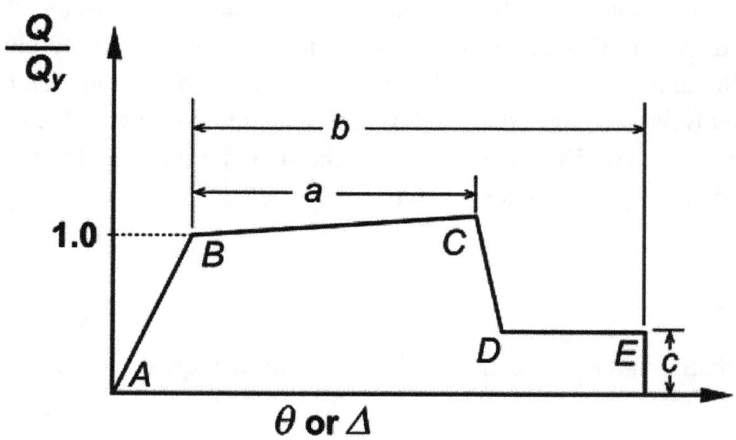

The pushover analyses are performed until obtaining a moment (or a plastic rotation) at a reference point (the base of the building in this example) equals to that resulted by NTHA.

Results and Discussion

The pushover analysis is performed using first six modes. Table 2 shows the natural period T, the modal participation factor Γ, the damping ratio ξ and the spectral acceleration Sa of each mode. Note that the values of Γ are related to the mode shapes which are normalized with respect to the mass matrix such that: $\sum_{i=1}^{ns} \Phi_{ij}^{T} m_i \Phi_{ij} = 1$. On the other hand, the zero value of Γ corresponds to vertical mode, which does not play any role in the computation of the lateral load vector. The proposed "OMAP" procedure is compared with the conventional pushover approach "Mode1", and the above-mentioned FAP, DAP and SSAP procedures.

As previously mentioned, the main difference between the adaptive methods consists in the manner used to build the shape of the applied load. In order to distinguish the effect of the base from that of the adaptation process, results of two analyses are presented: without adaptation and with adaptation. In the first analysis, the comparison with the DAP method is avoided where the non-adaptive displacement-base pushover could hide important structural characteristics and leads to misleading results (Antoniou and Pinho, 2004b).

Without Adaptation

- **Floor Forces:** Figure 3 shows the normalized load patterns obtained from different procedures. Note that the shape is more important than the value itself since the pushover analysis is performed as displacement – controlled load pattern. It can be seen that both SSAP and OMAP methods take into account the sign changes of the floor forces relative to higher modes. That's why Shakeri et al. (2010) have referred the superiority of the SSAP method over the FAP. In fact this explanation is questionable because, any base in which the floor force is one of the derivatives of this base can result in a reverse in the sign of the floor forces.
- **Plastic Hinge Locations:** The locations of plastic hinges are depicted for each method and for the first mode (Mode 1) in the Figure 4. It can be noted that only the "OMAP" procedure predicted some of the plastic hinges due to the higher modes.
- **Flexure Moments:** Flexure hinge properties involve axial force-bending moment interaction as failure envelope. In the present case of shear wall structure, the axial forces remain constant; that's why the difference in the plastic hinge locations between the different methods is referred to the bending moments. Figure 5 shows the bending moment diagrams obtained by different methods compared to that of NTHA where a significant difference can be seen for different methods.

Table 2. Modal properties for 20-story shear wall

Mode	T (sec)	Γ.		Sa (g)
1	2.98	23.00	5	0.10
2	0.54	12.87	2.6	0.87
3	0.20	7.86	5	0.64
4	0.17	0.00	5.9	0.70
5	0.11	5.69	8.8	0.51
6	0.07	4.44	13.5	0.44

In order to interpret this result, elastic structure is used. The resulting elastic moments are presented in term of scaled values where the maximum moment at the base is equal to that obtained by the linear time history analysis (LTHA), since the prediction of the hinges overall the shear wall is governed by the shape of the moment diagram. The obtained result is illustrated in Figure 6. The large difference between the FAP, SSAP and LTHA is due to the previous mentioned pitfall in computing the combined peak value of one response quantity from the combined peak values of other response quantities, while the little difference between the OMAP and LTHA is due to using the SRSS combination in predicting the bending moment.

Figure 7 shows the scaled moment diagrams for LTHA, NTHA and OMAP analyses. Despite the presence of plasticity, the change in moment shape does not occur in the OMAP analysis since the adaptation has not been applied. This result emphasizes the importance of integrating the adaptive feature to the proposed method in order to incorporate the variation in modal properties. This issue is developed in the next section.

With Adaptation

As previously mentioned, the incremental applied loads depend on the specified incremental target displacements. Table 3 shows the target displacements for each adaptation (iteration). The first iteration target displacements have been specified in order to initiate the plasticity in the structure. The incremental progressive variation in dynamic properties (Modal shapes, Period, Damping ratio, Modal participating mass ratios and Spectral accelerations) is detailed in Appendix in Figure 18, Table 7, Table 8, Table 9 and Table 10.

- **Floor Forces:** Figure 8 shows the variation in the incremental applied load pattern and the corresponding plastic hinge locations at different steps during the OMAP analysis.

Figure 3. Floor forces patterns

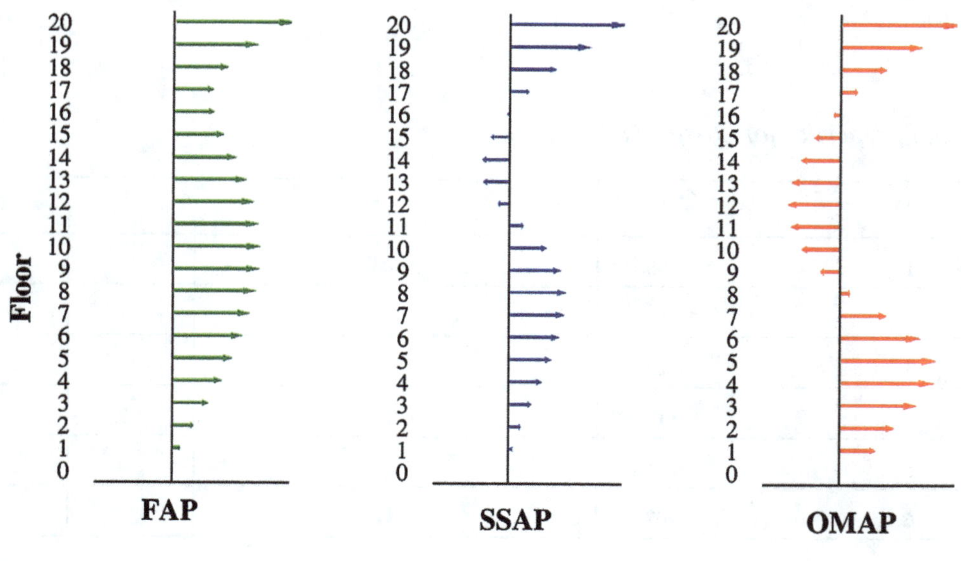

Figure 4. Plastic hinge locations (Non-adaptive pushover analyses vs NTHA)

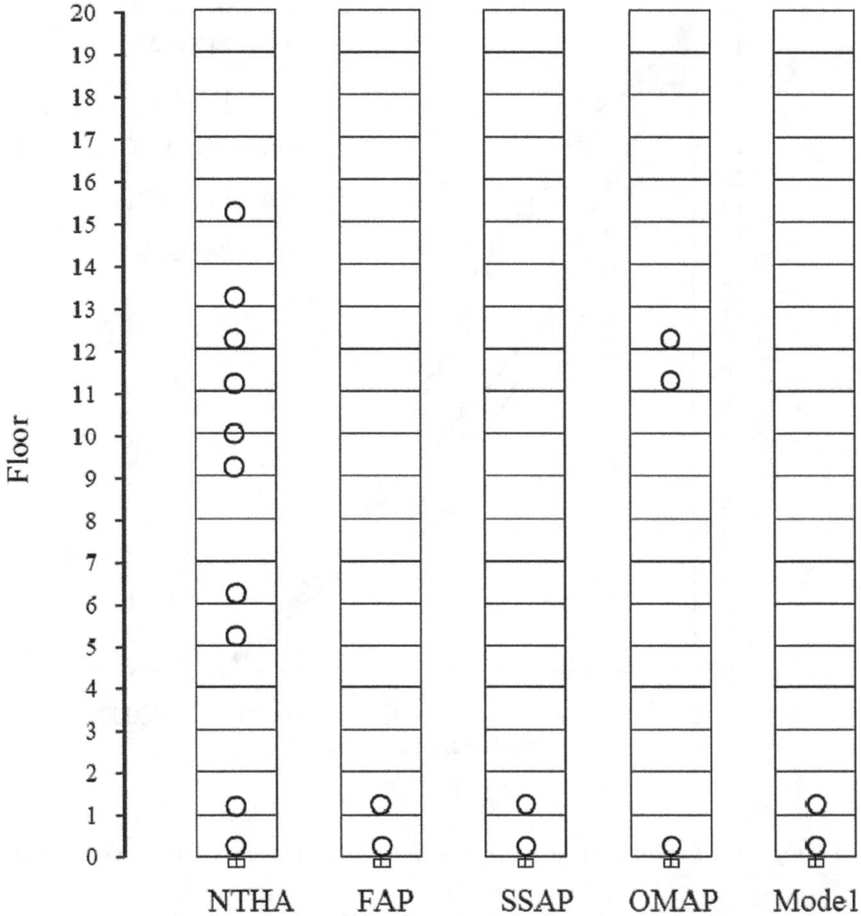

- **Plastic Hinge Locations and Corresponding Moments:** Figure 9 shows the plastic hinges resulting from the adaptive form of the previous analyses. In comparison with the non-adaptive form (Figure 4), it can be noted that the adaptation only improves the results of the OMAP method. On the other hand, the DAP method succeeded in predicting a plastic hinge resulting from the higher modes because of the shape of the corresponding moment diagram (Figure 10).
- **Other Response Quantities:** Figure 11 indicates that OMAP procedure estimates the shear forces, displacement and story drift with a reasonable accuracy.

In order to give an idea about the deformation of the structure in the inelastic range, the obtained pushover curve is depicted in Figure 12 that reveals an advanced yield state.

OTHER CASE STUDIES

The OMAP procedure is applied to other case studies (W20/Erz, W30/Imp and W30/Erz). Another shear wall of a 30-story building is selected, with 105 meters tall, the length of the wall is 6 m and the

Figure 5. Moment diagrams (Non-adaptive pushover analyses vs NTHA)

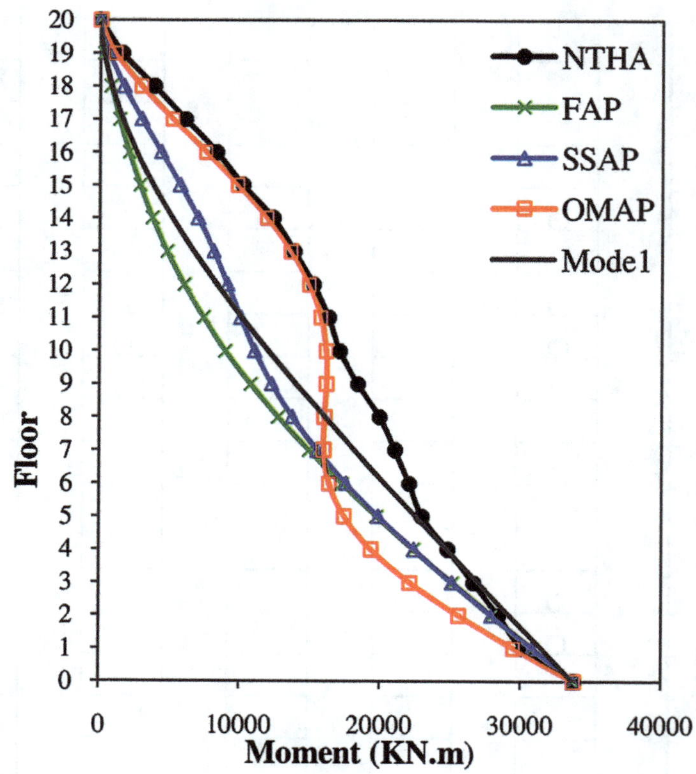

Figure 6. Moment diagrams for an elastic structure (Non-adaptive pushover analyses vs LTHA)

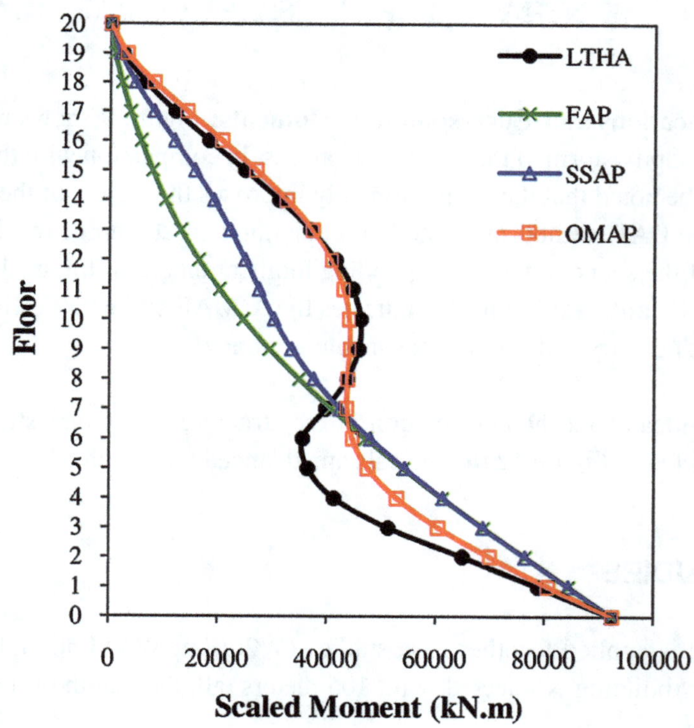

Figure 7. Moment diagrams - Linear vs Nonlinear Analyses

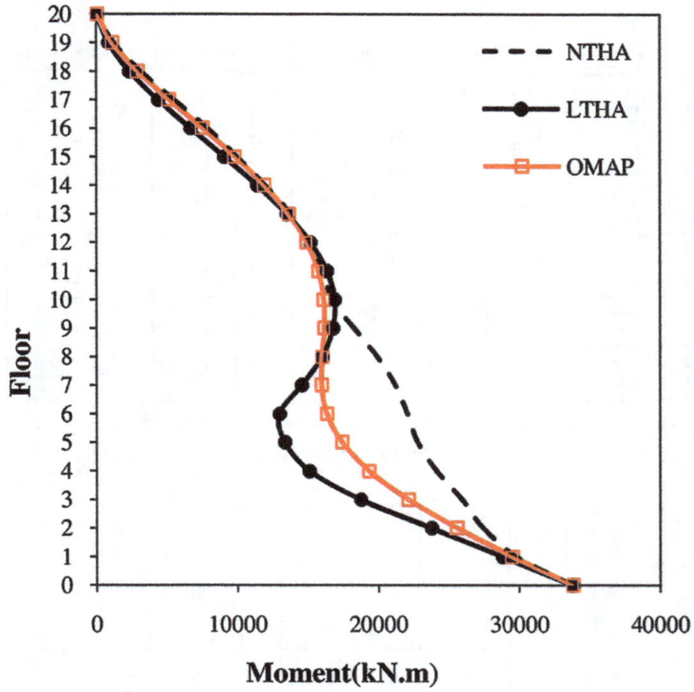

Table 3. Target displacements for each iteration (m)

Number of iteration	FAP	SSAP	DAP	OMAP	Mode 1
1	0.4	0.4	0.35	0.30	0.4
2	0.1	0.1	0.05	0.05	0.1
3 → End	0.1	0.1	0.1	0.05	0.1

Table 4. Shear wall properties for 30–story

Story ID	Thickness (cm)	Longitudinal Reinforcement	Capacity Ratio* (Vertical Load)
1→5	35	2φ20/20cm	0.29
6→10	30	2φ18/20cm	0.29
11→15	28	2φ18/20cm	0.25
16→19	25	2φ14/20cm	0.22
20→21	20	2φ12/20cm	0.22
22→25	18	2φ12/20cm	0.21
26→28	15	2φ10/20cm	0.15
29→30	10	2φ10/20cm	0.14

* The capacity ratio concerns the first story of each group

Table 5. Modal properties for 30-story shear wall

Mode	T (sec)	Γ.	
1	4.91	22.86	5
2	0.86	12.71	2.6
3	0.33	7.86	5
4	0.19	0.00	8.4
5	0.17	5.64	9.1
6	0.11	4.38	14.3
7	0.08	3.67	20.5
8	0.07	0.00	22.4

Figure 8. Variation of the incremental applied load and the corresponding plastic hinges during the OMAP procedure

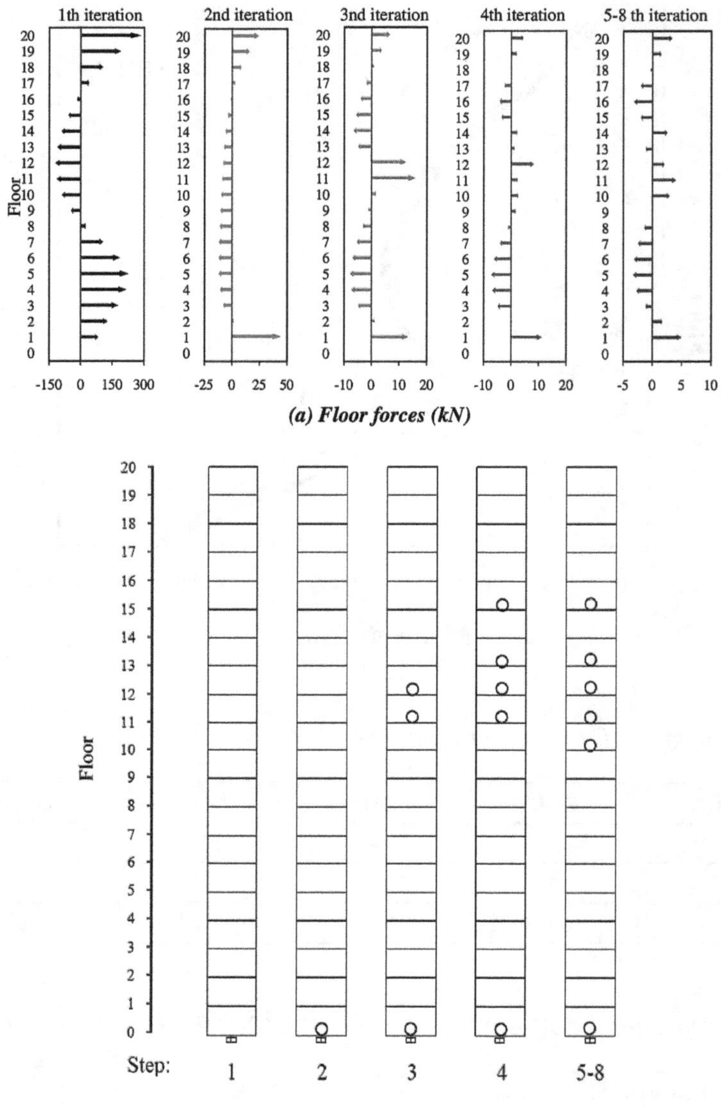

thickness is variable with the height. Tables 4 and 5 summarize the shear wall properties and the modal characteristics respectively. The earthquake of Erzincan 1992 (Figure 13) is applied on the two selected cases of 20 and 30 story shear walls. Table 6 shows the structure references and the correspondent scale factors of the earthquake.

The adaptation effect is summarized in Figure 14, Figure 15, Figure 16, Figure 17. Results confirm the general trend observed for the first case study W20/Imp and point out the great accuracy of the OMAP method with regard to the non-linear time history analysis.

Figure 9. Plastic hinge locations (Adaptive pushover analyses vs NTHA)

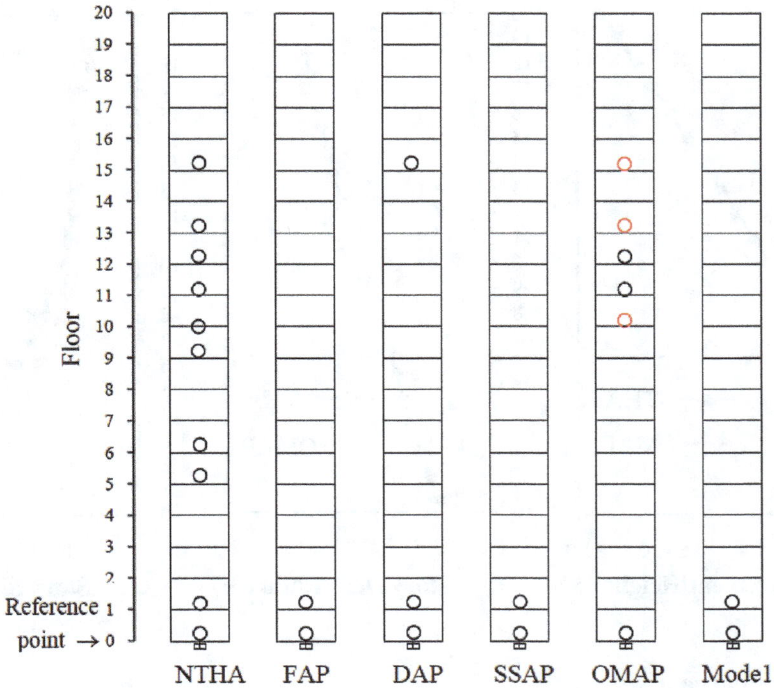

Figure 10. Moment diagrams (Adaptive pushover analyses vs NTHA)

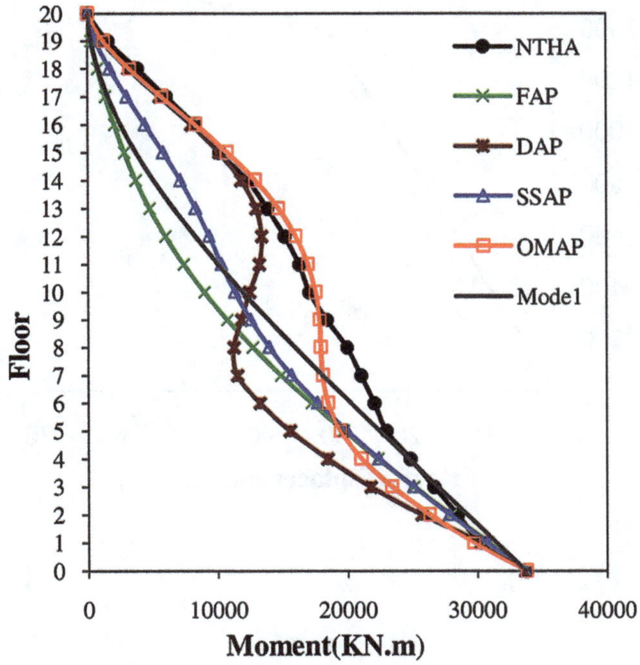

Figure 11. Displacement, Story drift, and Story shear (OMAP vs NTHA)

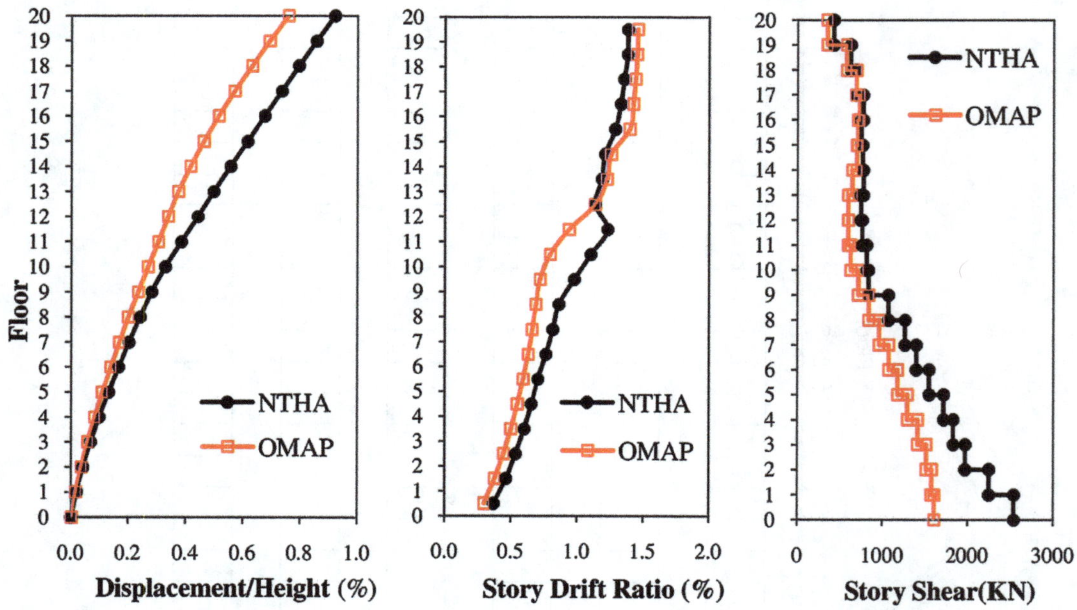

Figure 12. Pushover curve (OMAP)

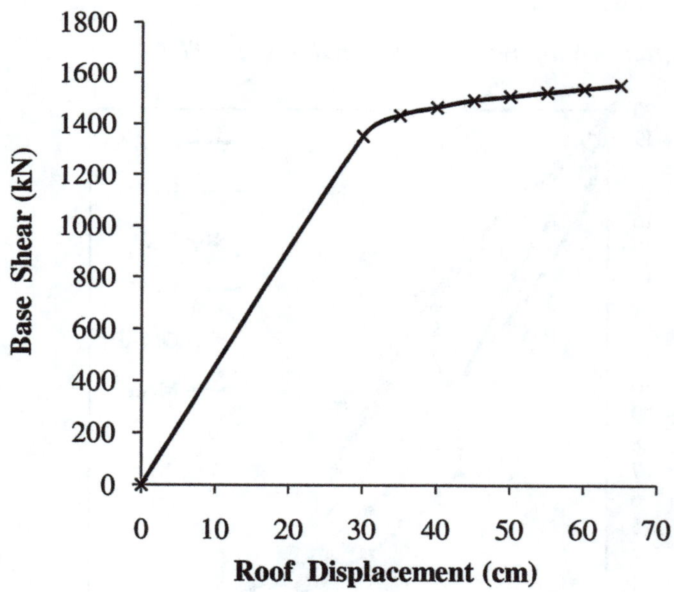

Figure 13. Ground motions characteristics - Erzinkan, 1992, NS

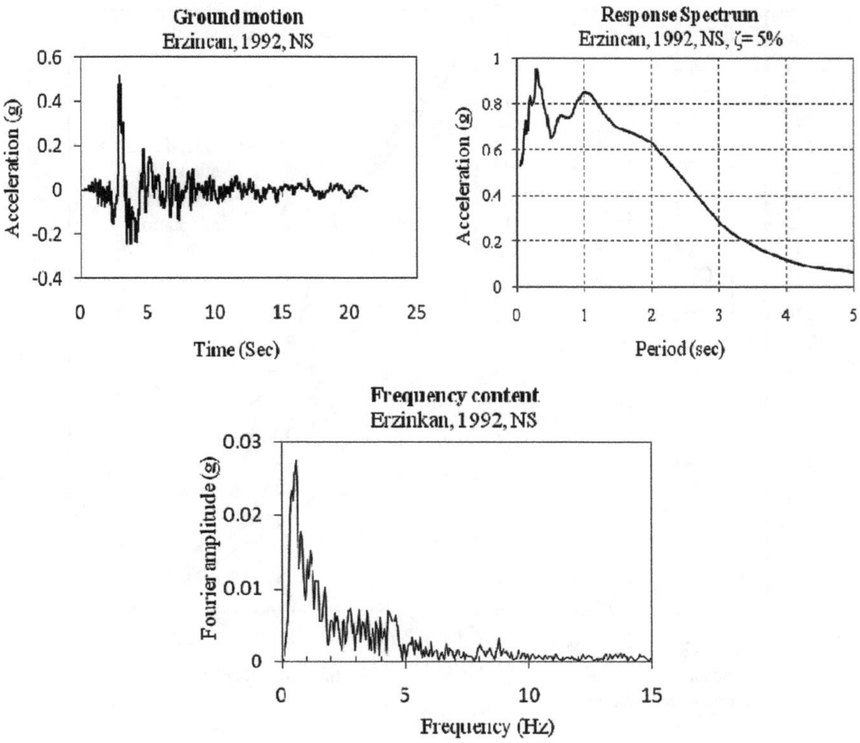

Table 6. Structure references and correspondent scale factors of ground motions

Example	Structure Reference	No. of Stories	Earthquake	Scale Factor
1	W20/Imp	20	Imperial Valley (1940)	2.5
2	W20/Erz	20	Erzincan (1992)	1.0
3	W30/Imp	30	Imperial Valley (1940)	1.5
4	W30/Erz	30	Erzincan (1992)	1.4

CONCLUSION

A new single-run adaptive pushover method "OMAP" is proposed to estimate the seismic response of shear wall structures. The load pattern is derived on the base of the overturning moment as recognition of the evidence that plasticity in the shear wall is mainly governed by this parameter. This method maintains the superiority of not decomposing the mode shapes of the structure. At the same time, it avoids the pitfall which occurred in all previous single-run adaptive pushover analyses in utilizing the modal combination for the applied loads instead of combining the response quantities induced by those loads in individual modes. The OMAP takes into account the progressive changes in the dynamic properties of the structure.

Figure 14. Adaptation effect - Plastic hinge locations and corresponding moments (W20/Imp)

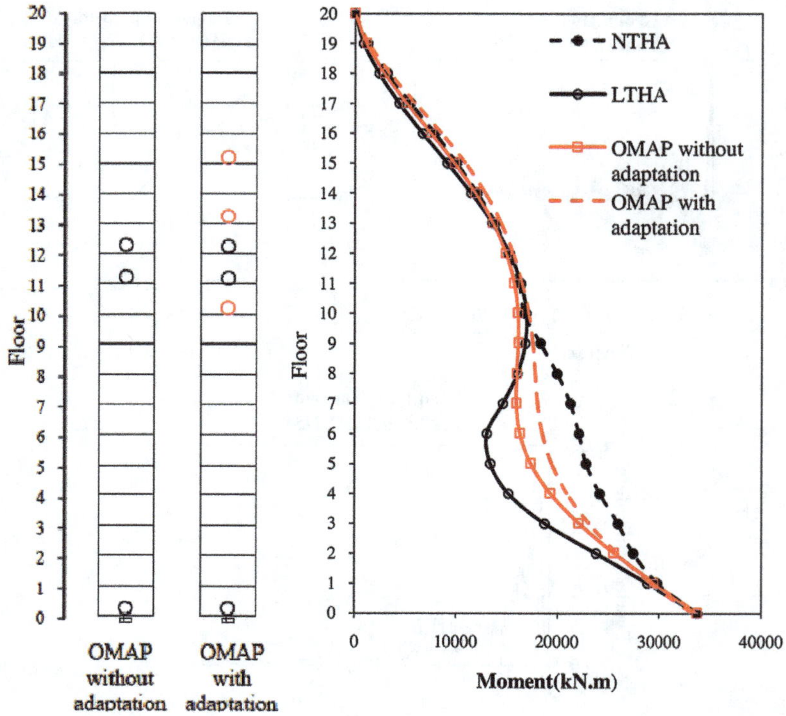

Figure 15. Adaptation effect - Plastic hinge locations and corresponding moments (W20/Erz)

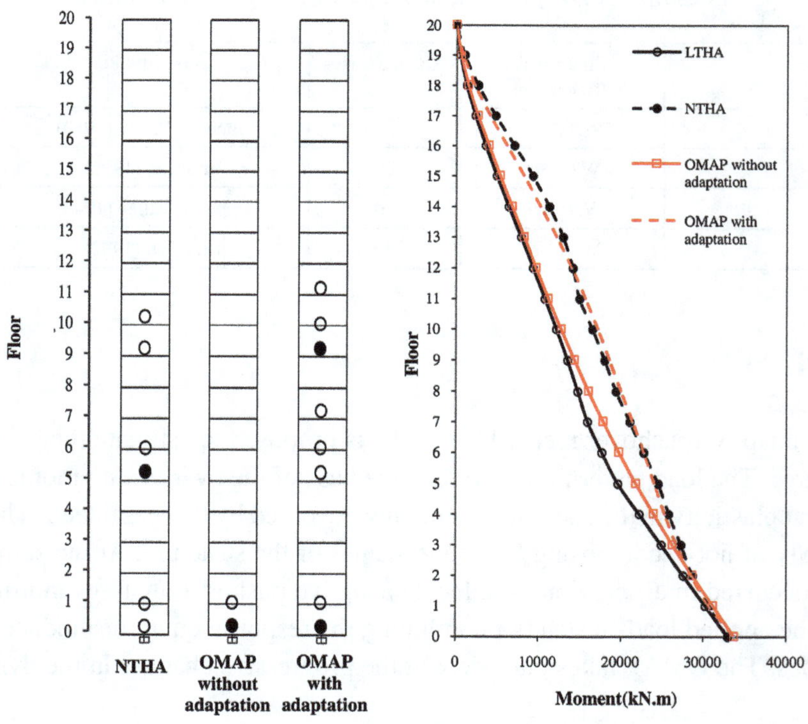

Figure 16. Adaptation effect - Plastic hinge locations and corresponding moments (W30/Imp)

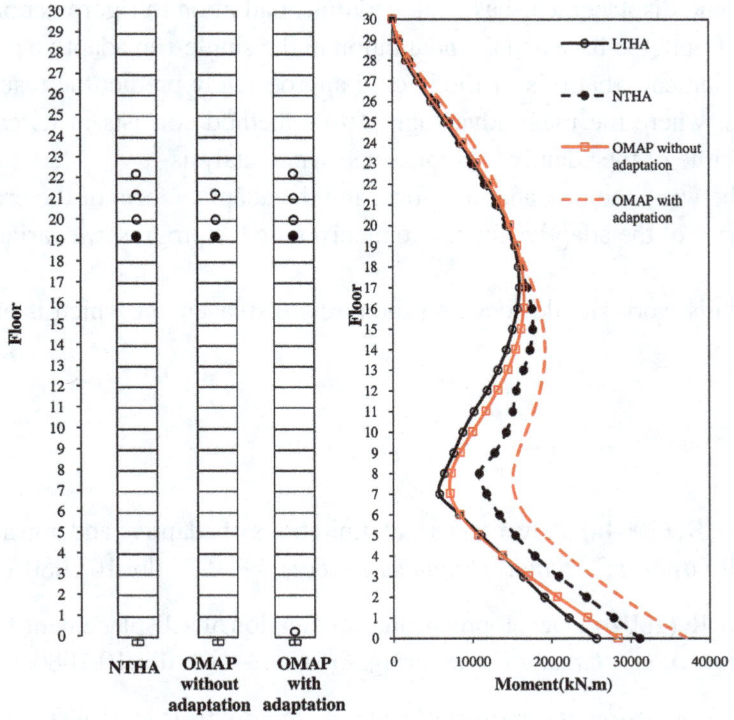

Figure 17. Adaptation effect - Plastic hinge locations and corresponding moments (W30/Erz)

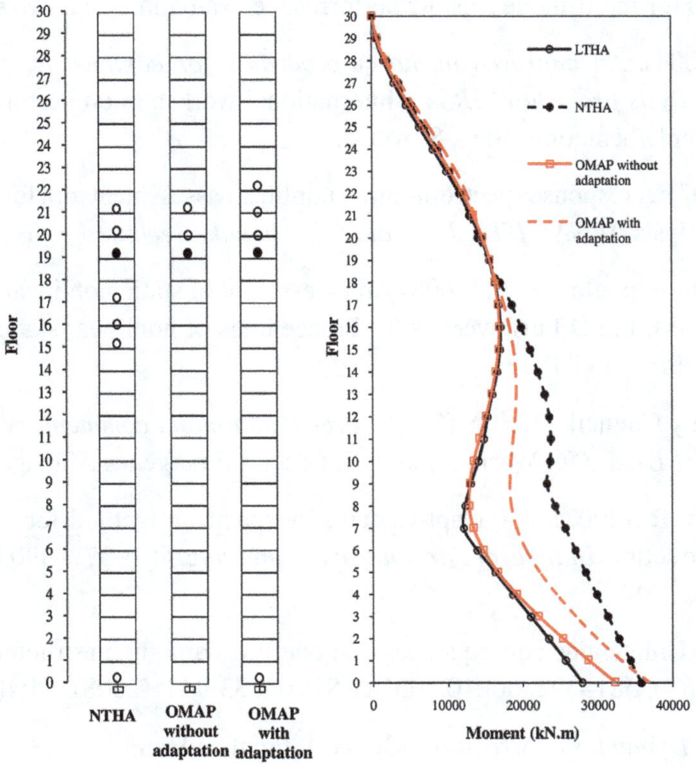

In order to illustrate its potential advantages, the results of the OMAP procedure have been compared to force-based and displacement-based procedures in addition to rigorous nonlinear time history analysis. The effect of both the base and the adaptation in the single-run adaptive pushover analysis are investigated. Results indicate that this method could appropriately predict the results of the nonlinear time history analysis, where the main advantage of this method consists in its easy implementation maintaining the principle of the adaptive response spectrum analysis.

The comparison between the non-adaptive form and the adaptive form of the proposed method emphasizes the importance of the adaptive feature to incorporate the progressive variation in dynamic and modal properties.

The next step of this work should focus on the target displacement which is an important task in pushover Analysis.

REFERENCES

Antoniou, S., & Pinho, R. (2004a). Advantages and limitations of adaptive and non adaptive force-based pushover Procedures. *Journal of Earthquake Engineering*, *8*(4), 497–522. doi:10.1080/13632460409350498

Antoniou, S., & Pinho, R. (2004b). Development and verification of a displacement-based adaptive pushover Procedure. *Journal of Earthquake Engineering*, *8*(5), 643–661. doi:10.1080/13632460409350504

ATC. (1996). *Seismic evaluation and retrofit of concrete buildings*. Report ATC-40. Applied Technology Council.

Aydinoğlu, M. N. (2003). An incremental response spectrum analysis procedure based on inelastic spectral displacements for multi-mode seismic performance evaluation. *Bull Earthquake Eng, 1*, 3-36.

Aydinoğlu, M. N. (2004). *An improved pushover procedure for engineering practice: Incremental response spectrum analysis procedure (IRSA)*. International workshop on performance-based seismic design: concept and implementation, Bled, Slovenia.

Aydinoğlu, M. N. (2007). A response spectrum-based nonlinear assessment tool for practice: incremental response spectrum analysis (IRSA). *ISET Journal of Earthquake Technology, 44*(1), 169–192.

Baros, D. K., & Anagnastopoulos, S. A. (2008) An assessment of static nonlinear pushover analyses in 2-D and 3-D applications, In 3D Pushover 2008 - Proceedings of nonlinear static methods for design/assessment of 3D structures. IST Press.

Building Seismic Safety Council (BSSC). (2000). *Pre-standard and commentary for the seismic rehabilitation of buildings. FEMA-356*. Washington, DC: Federal Emergency Management Agency.

Casarotti, C., & Pinho, R. (2007). An adaptive capacity spectrum method for assessment of bridges subjected to earthquake action. *Bulletin of Earthquake Engineering*, *5*(3), 377–390. doi:10.1007/s10518-007-9031-8

Charney, F. A. (2008). Unintended consequences of modelling damping in structures. *Journal of Structural Engineering, 134*(4), 581–592. doi:10.1061/(ASCE)0733-9445(2008)134:4(581)

Chopra, A. K. (2007). *Dynamics of Structures* (3rd ed.). Prentice Hall.

Chopra, A. K., & Goel, R. K. (2001). *A modal pushover analysis procedure to estimate seismic demands for buildings: Theory and Preliminary Evaluation. PEER Report 2001/03*. Pacific Earthquake Engineering Center, University of California Berkeley.

Chopra, A. K., & Goel, R. K. (2002). A modal pushover analysis procedure for estimating seismic demands for buildings. *Earthquake Engineering & Structural Dynamics*, *31*(3), 561–582. doi:10.1002/eqe.144

Computers, Structures Incorporated (CSI). (2007). *SAP 2000 NL*. Author.

Elnashai, A. S. (2001). Advanced inelastic static (pushover) analysis for earthquake applications. *Structural Engineering & Mechanics*, *12*(1), 51–69. doi:10.12989/sem.2001.12.1.051

Goel, R. K., & Chopra, A. K. (2005). Extension of Modal Pushover Analysis to Compute Member Forces. *Earthquake Spectra*, *21*(1), 125–139. doi:10.1193/1.1851545

Gupta, B., & Kunnath, S. K. (2000). Adaptive spectra-based pushover procedure for seismic evaluation of structures. *Earthquake Spectra*, *16*(2), 367–391. doi:10.1193/1.1586117

Kalkan, E., & Kunnath, S. K. (2006). Adaptive modal combination procedure for nonlinear static analysis of building structures. *Journal of Structural Engineering*, *132*(11), 1721–1731. doi:10.1061/(ASCE)0733-9445(2006)132:11(1721)

Newmark, N. M., & Hall, W. J. (1982). *Earthquake Spectra and Design, monograph. Earthquake Engineering Research Institute*. Berkeley, CA: EERI. Print

Reyes, J. C., & Chopra, A. K. (2011). Three-Dimensional Modal Pushover Analysis of Buildings Subjected to Two Components of Ground Motion, including its Evaluation for Tall Buildings. *Earthquake Engineering & Structural Dynamics*, *40*(7), 789–806. doi:10.1002/eqe.1060

Shakeri, K., Shayanfar, M. A., & Kabeyasawa, T. (2010). A story shear-based adaptive pushover for estimating seismic demands of buildings. *Engineering Structures*, *32*(1), 174–183. doi:10.1016/j.engstruct.2009.09.004

Smyrou, E., Priestley, M. J. N., & Carr, A. J. (2011). Modelling of Elastic Damping in Nonlinear Time-history Analyses of Cantilever RC Walls. *Bulletin of Earthquake Engineering*, *9*(5), 1559–1578. doi:10.1007/s10518-011-9286-y

Surmeli, M., & Yuksel, E. (2015). A Variation of Modal Pushover Analyses (VMPA) Based on a Non-Incremental Procedure. *Bulletin of Earthquake Engineering*, *13*(11), 3353–3379. doi:10.1007/s10518-015-9785-3

APPENDIX: INCREMENTAL PROGRESSIVE VARIATION IN DYNAMIC PROPERTIES DURING THE OMAP PROCEDURE

Figure 18. Variation of the Modal shapes during the OMAP procedure

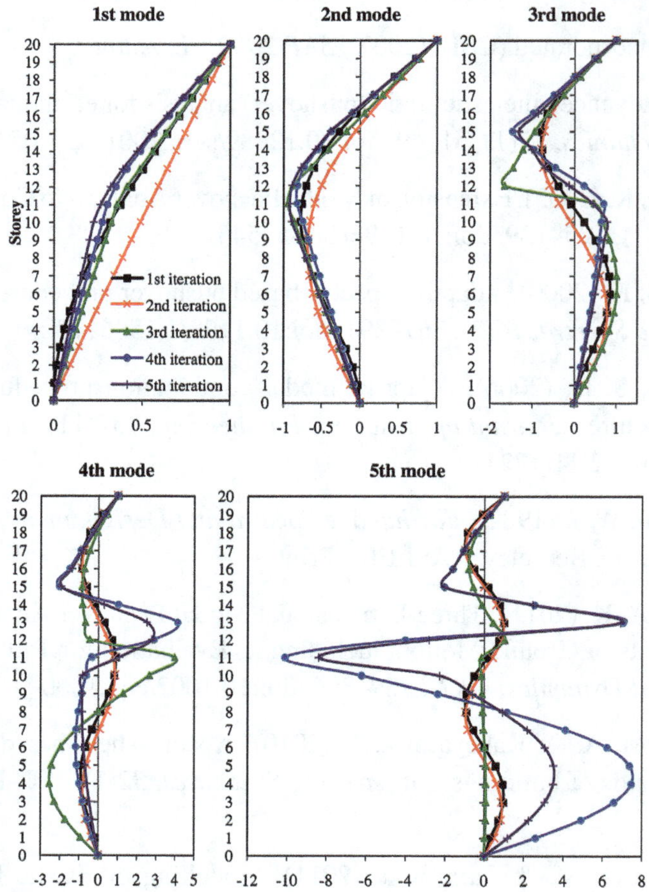

Table 7. Period

T (sec)	Iteration				
Mode	1	2	3	4	5-8
1	2.98	13.65	15.46	16.24	17.59
2	0.54	0.74	3.88	4.75	5.03
3	0.20	0.24	0.39	1.06	1.28
4	0.11	0.12	0.19	0.36	0.48
5	0.07	0.08	0.12	0.21	0.22

Table 8. Damping ratio

ξ%	Iteration				
Mode	1	2	3	4	5-8
1	5.01	21.52	24.35	25.58	27.70
2	2.61	2.44	6.34	7.67	8.09
3	4.99	4.28	3.05	2.56	2.75
4	8.75	7.86	5.24	3.17	2.73
5	13.49	12.52	8.20	4.86	4.69

Table 9. Modal participating mass ratios

%	Iteration				
Mode	1	2	3	4	5-8
1	0.62	0.76	0.67	0.64	0.60
2	0.19	0.13	0.18	0.21	0.25
3	0.072	0.043	0.043	0.024	0.028
4	0.038	0.020	0.052	0.022	0.013
5	0.023	0.012	0.001	0.059	0.031

Table 10. Spectral accelerations

Sa (m/sec^2)	Iteration				
Mode	1	2	3	4	5-8
1	0.862	0.007	0.005	0.005	0.004
2	7.352	0.635	0.033	0.015	0.013
3	5.450	0.906	0.586	0.352	0.200
4	4.352	0.711	0.497	0.531	0.579
5	3.735	0.515	0.484	0.476	0.442

Chapter 4
Influence of the Shear-Bending Interaction on the Global Capacity of Reinforced Concrete Frames:
A Brief Overview of the New Perspectives

Francesco Clementi
Polytechnic University of Marche, Italy

Giovanni Di Sciascio
DI SCIASCIO s.r.l., Italy

Sergio Di Sciascio
DI SCIASCIO s.r.l, Italy

Stefano Lenci
Polytechnic University of Marche, Italy

ABSTRACT

In many seismic countries in the world (e.g. Europe, Northern USA, Japan, Turkey, etc.), the assessment of existing structures is a priority, since the majority of the building heritage was designed according to out-of-date or even non-seismic codes. The uncertainties about the nonlinear behaviour of the structures are, therefore, important and the nonlinear response should be treated directly, with a correspondingly strong increase in complexity of the assessment procedure. The assessment of regular reinforced concrete frame buildings has been performed, according to the Italian Seismic Code, Eurocode 8 and the CNR DT-212 guideline. A lumped plasticity model has been used with the aim of quantifying the differences between a fixed and a continuously updated shear span and between the use of inelastic springs located at the member ends or continuously along the beam elements, and with the purpose of considering the influence of axial-bending-shear interaction on the global capacity of the buildings.

DOI: 10.4018/978-1-5225-2089-4.ch004

INTRODUCTION

The recent seismic events and the importance of seismic prevention, increasingly growing in the last few years, have highlighted the necessity of assessing the capability of the existing building heritage to sustain earthquakes, in order to improve the average safety level of the population. The adequate modelling of existing Reinforced Concrete (RC) frames is a crucial issue (Elnashai & Di Sarno, 2008; Plevris, Mitropoulou, & Lagaros, 2012), related as well to the maintenance and to the structural upgrading possibility. The evaluation of the seismic vulnerability of existing buildings has a key role in determining and reducing the impact of an earthquake (Elnashai & Di Sarno, 2008).

This high vulnerability of RC buildings is due to many aspects, mostly related to the age of the buildings, the low standards of construction execution and maintenance, and the legislation in force at the time, which did not effectively address the seismic problem, even allowing the design for gravity loads in some earthquake-prone areas (Asteris & Plevris, 2015), erroneously considered as non-seismic zones. For this reason, old existing RC framed buildings are today characterized by:

- Poor quality concrete (Fiore, Porco, Uva, & Mezzina, 2013; A. Masi & Chiauzzi, 2013),
- Inefficient construction details (Angelo Masi & Vona, 2012),
- A lack of the fundamental principle of the capacity design, and
- Low column ductility mainly due to the inadequate use of stirrups (Verderame, De Luca, Ricci, & Manfredi, 2011).

Moreover, high shear forces, usually determined by global torsional effects, often result in brittle collapse susceptibility (Francesco Clementi, Quagliarini, Maracchini, & Lenci, 2015).

In most of these structures, the uncertainties about the nonlinear behavior are significant (Mpampatsikos, Nascimbene, & Petrini, 2008): generally, the presence and location of potential inelastic zones, as well as their ductility capacity, are not known. It is, therefore, very hard to define a direct correlation between the nonlinear internal forces that develop in the system during the real seismic excitation and those experienced by an equivalent indefinitely elastic structure. For this reason, a force-based assessment obtained by using an elastic analysis and reducing the internal forces by the behavior factor "*q*", does not achieve, in general, satisfactory results (Mpampatsikos et al., 2008). Hence, the nonlinear behavior of the structure should be faced directly, with a corresponding considerable increase in complexity of the assessment procedure.

In this work, the assessment of RC frame buildings has been performed according to:

- The Italian Seismic Code (NTC 2008, 2008),
- The European code (CEN (Comité Européen de Normalisation), 2005b), and
- The new Italian Guideline (CNR, Consiglio Nazionale delle Ricerche, 2013).

All the Codes consider the nonlinear methods of analysis as the normal way to evaluate the seismic demand. Concerning the assessment of the response, all Codes require a force- (strength-) based procedure for the brittle mechanisms (shear in beams, columns, walls, and joints) and a displacement-based approach for the ductile ones (flexure in beams, columns, and walls): the evaluation of both deformation and shear capacities of the structural members of a building subjected to a ground motion of a certain intensity requires, in general, lengthy and not simple calculations.

Usually, the use of nonlinear static (pushover) analysis leads to the choice of nonlinear constitutive laws. The classical parabola-rectangle diagram for the concrete under compression has been commonly used along with an elastic-hardening diagram for the reinforcement steel. Furthermore, to describe the nonlinear behaviour of columns and beams, a lumped plasticity approach is typically adopted. Usually, default plastic hinges are assigned to the elements:

- Axial force and biaxial bending for columns, and
- Pure bending for beams.

The nonlinear constitutive law of hinges subject to bending is suggested by international provisions, e.g. (CEN (Comité Européen de Normalisation), 2005b; FEMA 356, 2000), in terms of deformation. The shear failure of members is taken into account through the introduction of "shear hinges" with brittle behavior, once defined the shear resistance with the national/international standards.

The major problem with this approach is the absence of interaction between the axial, bending and shear forces during the transversal loading increment. For the most common approach, at present, the interaction is purely numeric and is delegated to the Finite Element solver. Furthermore, these quantities are a function of the shear span, which is always constant during the loading increment: this assumption can lead to under- or over-estimate the capacity of the element. This means that the capacity curve is not sufficient to evaluate the capacity of the structure and this explains the need for local checks.

Nowadays, it is well known (Haselton, Liel, Deierlein, Dean, & Chou, 2011; Liel, Haselton, & Deierlein, 2011) that shear failure can lead to a reduction in lateral building strength, change in inelastic deformation mechanism, loss of axial load-carrying capacity, and, ultimately, building collapse. It is well recognized that the relation between plastic shear demand and shear strength provides useful information in the determination of element (and in particular in the column) failure modes.

Engineering models for shear strength that include uncertainty measures are needed so that safe yet efficient evaluations and designs can be achieved. To assess older existing reinforced concrete buildings, models are needed that are sensitive to the details that are prevalent in those buildings (Haselton, Liel, & Lange, 2008; Ibarra, Medina, & Krawinkler, 2005).

On the basis of these considerations, the aims of this work are:

1. Checking the consistency of the results obtained with "classical" Codes;
2. Checking the error induced by using a constant shear span L_V during the transversal load;
3. Checking the difference between inelastic springs located at the member ends or continuously along the beam elements;
4. Checking the importance of considering the interaction between shear and axial-bending action;
5. Quantifying the errors induced in the estimation of the Seismic Risk of an existing RC structure with new and old formulations.

In order to give answers to the above goals, two different RC frame buildings have been examined varying -in these models- the position and the number of the embedded steel bars, of the stirrups, of the presence of absence of nonlinear geometrical effects, etc. Different Finite Element programs have been used to check all the data.

ANALYSED BUILDINGS

The reference buildings analysed herein have simple and geometrically regular structural schemes (Figure 1), which is typical for RC frame structures. These buildings are located in the city of L'Aquila in the centre of Italy. In a sense, they are representative in terms of typology and dimensions, of the RC buildings present in the Italian territory between the 1950s and the 1990s. In particular, taking into account the last 60 years, the Italian building stock could be split into three periods:

- From 1939 to 1987 (poor general rules for RC structures),
- From 1987 to 2002 (some provisions for RC structures) and
- 2003 to present (appropriate design code).

To obtain a realistic evaluation of the vulnerability, a great deal has been carried out to determine the most important structural characteristics. For these reasons, the frame in Figure 1a has been split into two cases: one where the columns have a sufficient stirrup spacing and the other without sufficient stirrup spacing.

The structures have a square geometry that covers an area of about ~240 m^2 in the plant, with a side equal to 15 m, evaluated with respect to the spacing of the columns (Figure 1). They are characterized by square nets of columns of 5x5 m^2 with an inter-story height equal to 3 m: the columns have a square cross-section of 0.4x0.4 m^2 on the first floor and 0.3x0.3 m^2 on the last three floors (Figure 1a), and 0.6x0.3 m^2 on all the storeys for (Figure 1b).

The frames are bi-directional and all floors are rigid in their planes: in this manner, the influence on pushover analyses is minimized. The foundation system is made up of foundation beams in both the main directions. These characteristics allow us to consider the columns fixed at zero level (Figure 1) and to neglect any effect due to seismic input asynchrony at the base of the various columns. Therefore, fixed foundations have been assumed for these structures.

Figure 1. Geometry of the main frames of the RC buildings

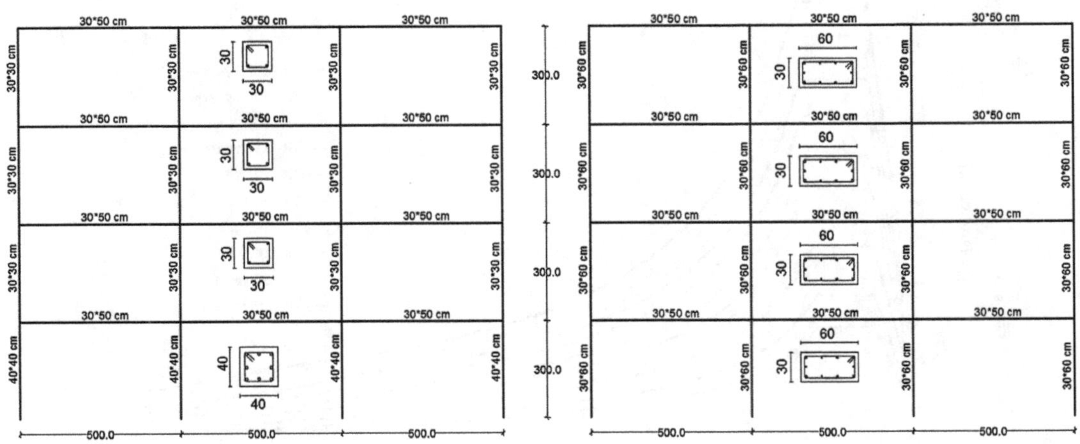

a) b)

The mechanical parameters considered in the modelling of the structure are: f_{cm} = 25 MPa, obtained as the mean peak resistance of concrete, and f_{ym} = 430 MPa, assumed as the mean yielding resistance of both longitudinal and transversal steel (Feb44k).

The considered building belongs to "Class IV" in the Italian seismic code (NTC 2008, 2008). This implies that the Limit State of Significant Damage (LSSD, or SLV in Italian) is associated with a demand recurrence period ($T_{R,D}$) of 1898 years -in black in Figure 2-, which corresponds to an expected peak ground acceleration (P.G.A.) equal to 0.453g ($a_{g,D}$). The other parameters that characterize the elastic response spectrum are (soil type T1 and category of subsoil C are considered): S = 1.092; T_B = 0.183 s; T_C = 0.548 s; T_D = 3.259 s.

METHODS OF ANALYSIS AND ASSESSMENT PROCEDURE

The nonlinear static (pushover) analysis has developed over the last two decades, and has become the most suitable procedure for design and seismic performance evaluation, as it is relatively simple and it considers the post-elastic behaviour. The construction of the capacity (pushover) curves for RC structures requires a certain computational effort. It also requires a careful – and demanding – check of the input data, since the results are very sensitive to the geometric and material modelling. The nonlinear static analysis is performed following the N2 method, originally proposed by Fajfar & Gašperšič (1996), with two distributions, one proportional to the fundamental modes and the other to the mass. In the following, the two distributions will be identified with the labels:

- "PushMode" (proportional to the modes) and
- "PushMass" (proportional to the mass).

A lumped plasticity model is used (Magliulo, Maddaloni, & Cosenza, 2007).

Figure 2. Elastic spectrum for the case study for different return periods T_R

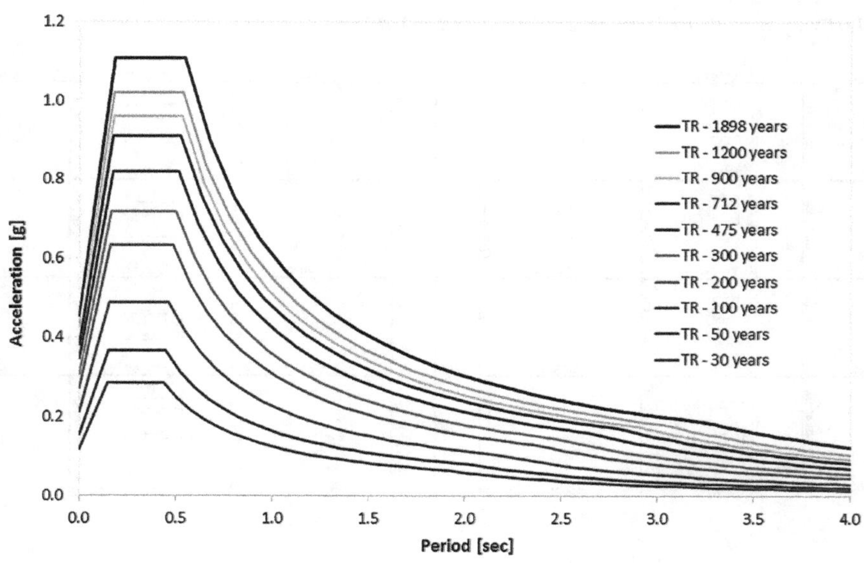

These procedures are displacement-based when considering the ductile mechanisms, and force- (strength-) based when considering the brittle mechanisms. In particular, the ductile modes should be checked in terms of the so-called "chord rotation", while the brittle ones should be assessed in terms of shear strength.

Regarding frame buildings, the structural members (beams and columns) are, in most of the cases, slender elements; the shear forces are, therefore, low compared to the bending moments and, consequently, flexural deformations dominate the behavior. The controlling factor is the so-called "shear span ratio" (Mpampatsikos et al., 2008):

$$\frac{M}{Vh} = \frac{L_V}{h} \tag{1}$$

where L_V (the shear span, Figure 3) may be defined as

$$\frac{M}{V} = L_V \tag{2}$$

In the previous equations, M and V are the bending moment and the shear demand, respectively, and h stands for the height of the element. The lower the shear span ratio is, the more important the shear stresses will be in comparison with the normal stresses.

Ductile Mechanisms

Deformation Capacity: Flexural Rupture

The ductile mechanisms are assessed at the member level, through the evaluation of the chord rotation demand, i.e. θ_D, and the corresponding capacity, i.e. θ_u, at the ends of each structural element (beams and columns).

Figure 3. The shear span definition for the column a) and the beam b)

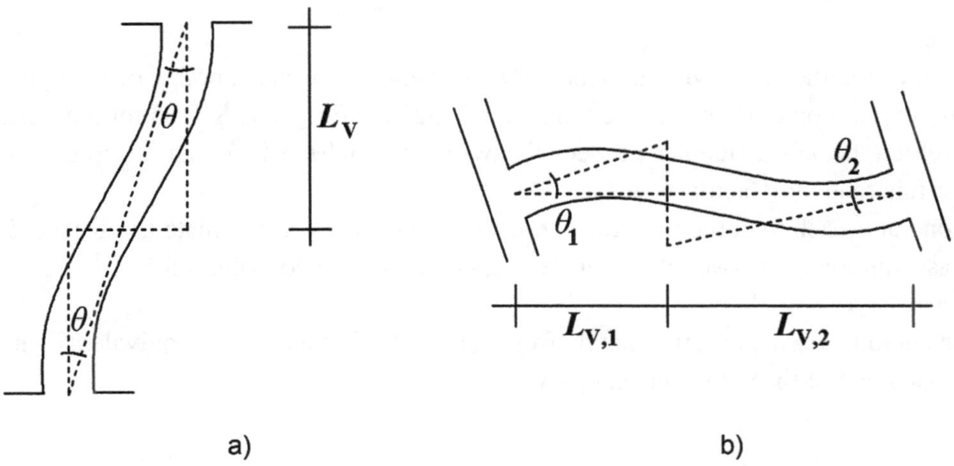

The demand in terms of chord rotation is defined as the angle between the chord connecting the considered end section of the member to the section at which $M = 0$ and the tangent to the member axis at the end section. Since M is proportional to the curvature ϕ, the section where $M = 0$ corresponds to the point of contra-flexure of the member. Therefore, each structural member is considered to be formed by two cantilevers, fixed at the member ends and characterized by a length equal to the shear span L_V, defined in Equation (2).

The chord rotation capacity depends on both the geometrical and mechanical properties of the considered member and on the seismic input (the shear span length is defined as the ratio of bending moment demand to shear demand and the curvature capacity is influenced by the amount of axial load). The chord rotation capacity, hence, may not be defined as an intrinsic property of a member, since the same member may develop different values of capacity as the seismic action changes. Two different approaches to detect the chord rotation capacity, one based on theoretical assumptions and the other based on experimental results, are proposed both in Eurocode 8 and in the Italian Seismic Code.

The empirical expression for chord rotation capacity at flexural failure is based on cycling load results and developed on the basis of statistical methods. According to both Eurocode 8 and Italian Seismic Code:

$$\underset{\approx u}{\prec} = \frac{1}{\gamma_{el}} 0.016 \cdot \left(0.3^{½}\right) \left[\frac{\max\left(0.01; \omega'\right)}{\max\left(0.01; \omega\right)} f_{cm} \right]^{0,225} \left(\frac{L_V}{h}\right)^{0,35} 25^{\left(\alpha \rho_{sx} \frac{f_{yw}}{f_{cm}}\right)} \left(1.25^{100\rho_d}\right) \tag{3}$$

where $\gamma_{el} = 1.5$ as prescribed by Eurocode 8 for primary elements; h is the cross-section depth; $\nu = N/(A_c f_{cm})$ is the axial load normalized with respect to the cross section; ω and ω' are the mechanical reinforcement ratios of the longitudinal reinforcement in tension and in compression zone respectively; f_{cm}, f_{ym} and f_{ywm} are, respectively, the average concrete compressive strength according to (CEN (Comité Européen de Normalisation), 2005a) and the steel yield strength obtained as already specified; $\rho_{sx} = A_{sx}/b_w s_h$, where A_{sx} is the area of transverse steel parallel to the X-direction loading, s_h is the stirrups spacing and b_w is the web width; $\rho_d = 0$ is the ratio of diagonal reinforcement; α is the confinement effectiveness factor equal to

$$\alpha = \left(1 - \frac{s_h}{2b_o}\right)\left(1 - \frac{s_h}{2h_o}\right)\left(1 - \frac{\sum b_i^2}{6h_o b_o}\right) \tag{4}$$

In the case of RC members without details for earthquake resistance, both Codes require multiplying the above expression of θ_u by a reduction factor (0.825 for Eurocode 8, 0.85 for the Italian Seismic Code). Therefore, the only difference between the two Codes in the evaluation of θ_u consists in the value proposed for this reduction factor.

The second approach to computing the chord rotation capacity at Ultimate Limit State is based on theoretical assumptions. The overall curvature is divided into two contributions:

1. Elastic curvature, which grows linearly from zero at the free end of the equivalent cantilever to ϕ_y at the fixed end of the equivalent cantilever;

Plastic curvature (ϕ_u-ϕ_y), constant over a length L_{pl}, called "plastic hinge length". L_{pl} is not a physical quantity, but a conventional one, defined in such a way as to account for any other effects besides the flexural one, like those due to shear and bond-slip.

$$\theta_u = \frac{1}{\gamma_{el}}\left(\theta_y + \left(\varphi_u - \varphi_y\right) L_{pl}\left(1 - \frac{0.5 L_{pl}}{L_V}\right)\right), \tag{5}$$

$$L_{pl} = 0.1 L_V + 0.17 h + 0.24 \frac{d_{bL} f_{ym}}{\sqrt{f_{cm}}} \tag{6}$$

where γ_{el} is assumed equal to 1.5 in the Italian Seismic Code and to 2 in Eurocode 8, for primary seismic elements: the values obtained following the Eurocode 8 procedure are, therefore, more conservative.

According to both considered Codes, no reduction factor should be applied to Equation (5) to deal with the lack of seismic details, since ϕ_y and ϕ_u take already into account the mechanical properties of the cross-sections. Therefore, the ultimate chord rotation demand is evaluated as the sum of an elastic part, due to the elastic curvature, and a plastic part, due to the constant plastic curvature developed along the plastic hinge length. The elastic part may be considered equal to the yielding chord rotation, θ_y, defined as for beams and columns:

$$\theta_y = \varphi_y \frac{L_V}{3} + 0.0013\left(1 + 1.5\frac{h}{L_V}\right) + 0.13\varphi_y \frac{d_{bL} f_{ym}}{\sqrt{f_{cm}}} \tag{7}$$

while the plastic part of the chord rotation, θ_{pl}, is the remaining part of Equation (5).

The yielding curvature ϕ_y may be computed theoretically. The yielding of a section of an RC frame member is commonly considered to correspond to the yielding of the tensile reinforcement, assuming the concrete is indefinitely linearly elastic in compression. Differently, some simplified expressions for evaluating the yielding curvature, which do not require a section calculation, useful in practical applications, are available, see e.g. (Mpampatsikos et al., 2008).

Two aspects of Equations (3), (5), (7) need a discussion, namely

- The way to compute the shear span L_V and
- The choice of the axial load N to be considered.

Concerning the shear span L_V, the application of its correct definition (Equation (2)) will not be trivial. Usually, L_V is used as a constant value during the transversal increment of nonlinear analysis. It can be evaluated by the linear dynamic analysis with two main problems (Mpampatsikos et al., 2008):

1. The values of both moment and shear obtained through an elastic linear analysis grow indefinitely, proportionally to the external forces, and this is true only if the elements do not reach the yield strength;
2. M and V are given as envelope values.

The procedure to compute L_V applying its definition (Equation (2)) may be, hence, complex and long if a dynamic linear analysis is performed. Moreover, L_V should be recomputed whenever the analysis is carried out. The most frequent simplification is to assume L_V equal to half the member length ($L_V = L/2$) and it is what is done in every software with lumped plasticity (F. Clementi, Scalbi, & Lenci, 2016); the length L usually considers the presence or absence of the rigid joint near the node. In this work, the reliability of this simplification has been tested for pushover analysis, by comparing the results obtained considering Equation (2) to those found by assuming $L_V = L/2$.

Regarding the axial load, the correct choice should be the value obtained from the analysis under the seismic load combination. Furthermore, the possibility to consider the axial load due only to the gravity loads of the seismic combination (which is roughly the mean axial value that the columns undergo during a seismic input) is frequently used. In this work, a continuous updating of the axial force is used.

The beam and column elements are characterized by the nonlinear properties, that for lumped plasticity models, are usually elastoplastic curves for bending moments with nonlinear constitutive laws suggested by the Eurocode 8 provisions (CEN (Comité Européen de Normalisation), 2005b) and limited ductility behaviour (Figure 4a). For ductile members, in order to consider the complex interaction between the bending moments about two axes and the axial force, during the transversal load increment a biaxial bending with axial interaction (PMM, that is axial force P and biaxial bending moments M_x, M_y) is usually adopted in the FEM programs.

According to Eurocode 8 and Italian Seismic Code the beam and column verifications for the LSSD (Limit State of Significant Damage) consist in checking that the displacement demand could be achieved by the structure without the elements reaching their ultimate deformation (ductile mechanism). The check is satisfied if $\theta_D \leq \theta_C$, where θ_C is the chord rotation capacity computed with θ_u, which for LSSD is equal to $\frac{3}{4}\theta_u$ (Circolare Ministeriale n. 617, 2009).

The literature provides other models, as reported by (CNR, Consiglio Nazionale delle Ricerche, 2013), which have a smaller experimental data set but more homogeneous data in comparison with Eurocode. They have the advantage to directly refer to the elements without seismically adequate details (Zhu, El-

Figure 4. A dimensional force–deformation relationship adopted for a) bending and b) shear hinges

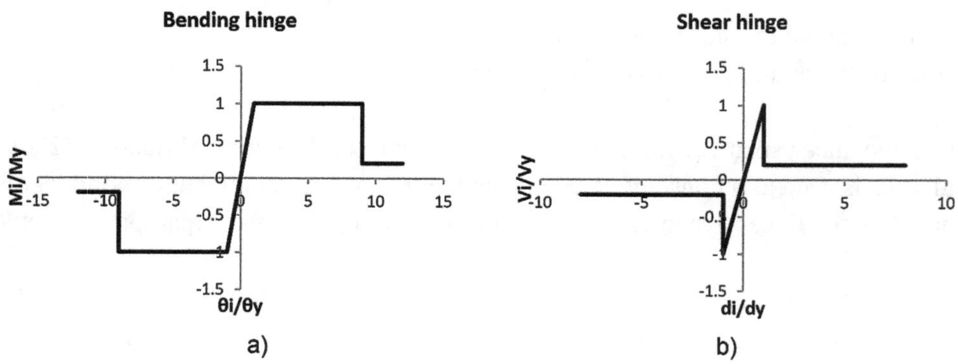

wood, & Haukaas, 2007) or to cover the behaviour range for both "new" and "old" elements (Haselton et al., 2011; Liel et al., 2011). Here, only the model of (Zhu et al., 2007) is reported because it is usable for models without degrading for flexure-dominated failure:

$$\theta_u = 0.049 + 0.716\rho_l + 0.120\frac{\rho_{sx} f_{yw}}{f_c} - 0.042\frac{s}{h} - 0.07\nu \tag{8}$$

where the symbols are the same used in Equation (3), and s is the hoop spacing and ρ_l is the longitudinal reinforcement ratio.

Deformation Capacity: Shear Rupture

In literature, there are many works for shear-dominated elements for which it is possible to establish a law in terms of chord rotation (CNR, Consiglio Nazionale delle Ricerche, 2013). Also, in this case, the model of Zhu et al. (2007) is reported, because it is compatible with the other quantity defined in the previous paragraph and it is usable in models without degradation. The drift capacity models at 20% reduction in lateral strength is:

$$\theta_V = 2.02\rho_{sx} - 0.025\frac{s}{h} + 0.013\frac{L_V}{h} - 0.031\nu \tag{9}$$

and it is obtained with elements without seismic details (e.g., hoops without confinement effect, with reduced anchorage, etc.).

Deformation Capacity: Axial Load Failure

Based on the classical shear-friction model (Bischof, 2002), (Elwood & Moehle, 2005) proposed a model for the drift at axial failure for shear-damaged columns. The effective coefficient of friction along the critical shear failure plane can be calculated by using equilibrium equations, and subsequently related to the drift ratio at axial failure. Elwood and Moehle (2005) found that the effective friction coefficient (calculated by ignoring the contribution of longitudinal reinforcement) providing good agreement with the experimental data, is

$$\mu = \frac{\dfrac{N}{A_{sw} f_{yw} d_c / s} - 1}{\dfrac{N}{A_{sw} f_{yw} d_c / s}\cot\alpha + \tan\alpha} \quad \text{with } \alpha\text{=65°} \tag{10}$$

where d_c=h-c is the depth of the column core (centreline to centreline of transverse reinforcement).

The drift capacity at axial failure can be computed with Zhu et al. (2007) and it can be approximately expressed as an exponential function of the effective coefficient of friction, μ. The median prediction of the drift at axial failure reads

$$\theta_a = 0.184\exp\left(-1.45\mu\right) \tag{11}$$

Limited data are available to develop this capacity model for the drift at axial failure. In Zhu et al. (2007), a database consisting of 28 column specimens is compiled, all for an element without seismic details and it is usable for models without degrading. Similar capacity equations, usable for degrading models, are obtained by Haselton et al. (2011) and Liel et al. (2011), but they are not reported for brevity issue.

Fragile Mechanisms

The brittle mechanisms are assessed at the section level, through the evaluation of shear demand and corresponding capacity at the two ends of each structural member. If a nonlinear analysis is carried out, the values of the internal forces at each step will correctly represent the actual distribution of the demand in structural members: the shear demand is, hence, assumed to be equal to the values obtained directly from the analysis.

The procedure to assess the shear capacity proposed in the Italian Seismic Code is the same as Eurocode 2, and it differs from that recommended in Eurocode 8 and in CNR-DT 212. In the following sections, all the procedures are shown. Considering the more accurate approach suggested by Eurocode 8 and CNR-DT 212 as correct, an aim of this work is to check if the procedure proposed in the Italian Seismic Code may furnish reliable results, although a number of phenomena that characterize the nonlinear response are neglected.

The resisting model is the classical Moërsch-Ritter truss, with 45% inclined compressed concrete struts. The resistance V_R is, hence, the minimum between the value of shear that causes the transverse reinforcement to yield in tension and the one that leads to the diagonal compression failure of the concrete web:

$$V_R = \min\left(V_{R,s} + V_{Rc,t}; V_{Rc,\max}\right) \tag{12}$$

where the shear that causes the transverse reinforcement to yield in tension is calculated as $V_{Rc,t} + V_{R,s}$.

V_{Rc} is the concrete contribution to shear resistance:

$$V_{R,ct} = \left\{0.18 \cdot k \cdot \left(100 \cdot \rho_1 \cdot f_{cm}\right)^{1/3} / \gamma_c + 0.15 \cdot \sigma_{cp}\right\} \cdot b_w \cdot d \geq \left(v_{\min} + 0.15 \cdot \sigma_{cp}\right) \cdot b_w \cdot d \tag{13}$$

where $k = 1+(200/d)^{1/2} \leq 2$, $v_{\min} = 0.035k^{3/2} f_{ck}^{1/2}$, d is the effective depth of the section (in mm); $\rho_1 = A_{sl}/(b_w \cdot d)$ is the geometrical ratio of longitudinal reinforcement (≤ 0.02); $\sigma_{cp} = N_{Sd}/A_c$ is the average compressive stress in the section ($\leq 0.2 f_{cd}$); b_w is the section minimum width (in mm).

$V_{R,s}$ is the resisting force due to transverse reinforcement:

$$V_{R,s} = 0.9 \cdot d \cdot f_{yw} \frac{A_{sw}}{s} \cdot \sin\pm\left(\cot\theta + \cot\alpha\right) \tag{14}$$

The shear that causes diagonal compression failure of the concrete web is computed, according to the Moërsch-Ritter truss model, as:

$$V_{Rc,max} = 0.9 \cdot d \cdot \alpha_c f'_{cd} \cdot b_w \cdot \left(\cot\text{¸} + \cot\alpha\right)\sin^2\text{¸} \tag{15}$$

where f_{cd} is the reduced compressive strength of the concrete core (f_{cd}=0.5•f_{cm}); α_c is a coefficient that takes into account the compression.

It is important to stress the fact that it is common for an existing building to be under-reinforced, in particular concerning the amount of stirrups close to the structural joints: $V_{RC,t}$ becomes, hence, the principal source of the total shear resistance.

The Italian Seismic Code gives the possibility to use the second approach with variable inclined compressed concrete struts for which the V_R is valuable with Equation (12), assuming $V_{Rc,t}$=0 and θ must be inside the range 22°≤θ≤45° (see §4.1.2.1.3.2 (NTC 2008, 2008)) for *Class B*, i.e. "low ductility structures", by imposing the simultaneous rupture $V_{R,s}$=$V_{Rc,max}$. Differently, for "high ductility structures", i.e. *Class A* for the Italian seismic code, cotθ=1 that implicitly assumes a high degradation of the plastic hinges (see §7.4.4.1.2.2 (NTC 2008, 2008)). This approach is typically used for new structures and it gives lower values of resistance if compared with Equation (12).

Typically, the shear failures have been taken into account by the introduction of a shear behaviour with elastic-brittle with limited ductility (Figure 4b), without interaction with axial forces and bending moments, by using Equations (12)÷(15). Otherwise, beams and columns are subject to concomitant bending, shear, and normal force. The interaction between these stresses is one of the relevant phenomena in the inelastic response near collapse of RC structures not designed according to the capacity design. Shear crisis is in most of the cases the main cause of the collapse, whether it precedes the damage for the flexural mechanism, or it occurs following a reduction in shear resistance to cyclic bending deformation (CNR, Consiglio Nazionale delle Ricerche, 2013). The shear strength reduction due to the inelastic response to bending and the resulting ductility demand is a common feature of various models available in the technical literature for shear strength, normally in the form of a linear reductive factor $k(\mu)$ with the (maximum) ductility μ.

Eurocode 8 takes into account the effects of both cyclic loading and inelastic response in the assessment of shear capacity, by decreasing the shear strength with an increase in cyclic inelastic deformation. Since the chord rotation is considered the most meaningful deformation quantity at a member level, the shear resistance in diagonal tension, V_R, is:

$$V_R = \frac{1}{\gamma_{el}} \left\{ \begin{aligned} & \frac{h - x}{2L_V} \min\left(N; 0.55 A_c f_c\right) + \left(1 - 0.05 \min\left(5; \mu_{,,pl}\right)\right) \cdot \\ & \left[0.16 \max\left(0.5; 100\rho_{tot}\right)\left(1 - 0.16 \min\left(5; \frac{L_V}{h}\right)\right)\sqrt{f_c} A_c + V_w \right] \end{aligned} \right\} \tag{16}$$

and it is assumed to be a function of the plastic part of the chord rotation ductility demand. In the previous, x is the neutral axis depth of the section, corresponding to the axial load and bending moment that act on the section when it experiences the considered value of shear demand; A_c is the cross section area, f_c is the compressive strength; γ_{el} = 1.15 as prescribed by Eurocode 8 for primary elements; V_w is the contribution of transverse reinforcement to the shear resistance that for cross-sections with rectangular web of width (thickness) b_w is

$$V_w = \rho_{tot} b_w z f_{yw}, \tag{17}$$

where ρ_{tot} is the total reinforcement ratio; z is the length of the internal lever arm, f_{yw} is the yield stress of transverse reinforcement. The term $\mu_{\Delta,pl} = (\mu_\Delta - 1) = (\theta/\theta_y - 1)$ in Equation (16) represents the plastic part of the ductility demand of the element. Equation (16) can be applied to beams and slender columns and refers only to diagonal tension failure. It is important to stress the fact that the axial load demand affects only the first part of the V_c term of Equation (16); therefore, V_c will not vanish if the member is subject to tensile N but will decrease linearly until $V_{R,min}$ is reached when N passes from compression to tension.

According to Equation (16) it is possible to define the shear resistances corresponding to the flexural yielding $V_y = M_y/L_V$, and the elastic behaviour $V_{R,0}$ when in Equation (16)$\mu_{\Delta,pl}$=0 and using the neutral axis corresponding to the yielding moment M_y, i.e. $x = x_y$. In Equation (11) beyond a limit value of $\mu_{\Delta,pl}$=5, assuming x constant (i.e., when the plastic moment is significantly developed x does not change), V_R is constant at its lowest value $V_{R,min}$. For beams ($N = 0$), when $\mu_{\Delta,pl} \geq 5$, $V_{R,min} = 0.75 V_{R,0}$. Hence, with reference to Figure 5 it is possible to distinguish three different ruptures (CNR, Consiglio Nazionale delle Ricerche, 2013):

Figure 5. Modes of the collapse of a reinforced concrete element subject to bending, shear, and axial forces (CNR, Consiglio Nazionale delle Ricerche, 2013): a) "brittle shear", b) "ductile shear", c) flexural failure

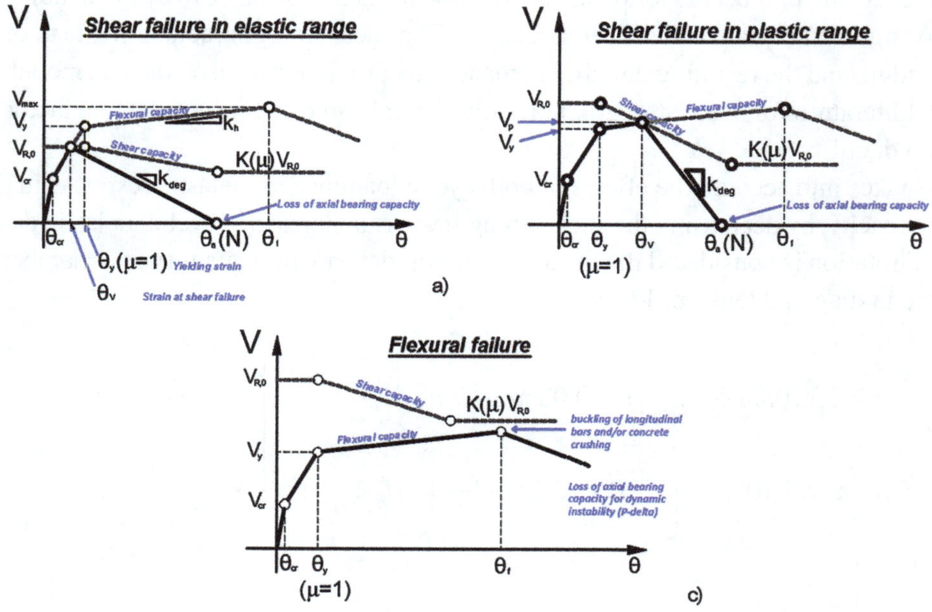

- If $V_{R,0} < V_y$, the element will undergo a "*brittle shear*" failure. The degradation in the lateral load capacity occurs *before* the yielding of the longitudinal reinforcement due to shear distress (i.e., diagonal cracking) in the element for a chord rotation θ_V. The subsequent behaviour is degrading until the formation of a crossing shear cracking and a subsequent sliding, a condition in which the shear force is reduced to negligible values and the element loses almost completely its axial strength, in correspondence to a chord rotation θ_a (Figure 5a).
- If $V_{R,min} < V_y < V_{R,0}$, the element will undergo a "*ductile shear*" failure. The degradation in the lateral load capacity occurs *after* the yielding of the longitudinal reinforcement but results from shear distress in the element. If the value of the residual shear strength $kV_{R,0}$ is less than the maximum shear, V_{max}, the shear failure occurs after the bending yielding to an intermediate value between the initial shear resistance $V_{R,0}$ and the residual, to an angular deformation θ_V. The subsequent behaviour is again degrading until the formation of a crossing shear cracking and the crisis in correspondence to an angular deformation θ_a (Figure 5b).
- If $V_y < V_{R,min}$, the element will fail in flexure and not in shear. The degradation in the lateral load capacity occurs *after* the yielding of the longitudinal reinforcement due to damage related to flexural deformations (i.e., spalling of concrete, buckling of longitudinal bars, concrete crushing, etc.), if the value of the residual shear strength $kV_{R,0}$ is bigger than the maximum shear, V_{max} (Figure 5c). In this case, a degradation of the flexural resistance occurs in correspondence to an angular deformation θ_u. The slope of the post-peak branch is associated with the ultimate strain for which the shear goes to zero, and it depends on the concrete degradation process (slope post-peak of σ-ε bond) and on the axial stress level at which the element is loaded, considering the geometric stiffness loss for large deformations. The degradation is sudden in the case in which the bars are subject to buckling.

While various existing shear strength models incorporate some of these variables, it was of interest to introduce another model (Sezen & Moehle, 2004) because it is used in the criterion of discrimination of failure between shear (*ductile* or *brittle*) and flexure of the model of Zhu et al. (2007):

$$V_R = k\left(\mu_\Delta\right)\left[\frac{0.5\sqrt{f_c}}{L_V / h}\sqrt{1+\frac{N}{A_g 0.5\sqrt{f_c}}}0.8A_g + V_w\right] \tag{18}$$

where A_g is the total area of the section, and the factor $k(\mu_\Delta)$ is defined to be equal to 1.0 for displacement ductility μ_Δ less than 2, to be equal to 0.7 for displacement ductility exceeding 6, and to vary linearly for intermediate displacement ductility. In this model, the influence of the normal stress does not result in a distinct contribution to the resistance but in a modification of the strength of the concrete term. Moreover, the reductive factor $k(\mu_\Delta)$ is applied to the sum of concrete and steel contributions, rather than only to the first term.

SEISMIC VULNERABILITY ASSESSMENT

In this section, a summary of the major results of the seismic vulnerability assessment of the building illustrated in the previous section is reported. All the structural models have been developed with the same commercial Finite Element (FE) programs STRAUS7© - STRAND7© outside Italy - (STRAUS7, 2016) and EasyOver© (EasyOver, 2016). To have a better perception of the variability of the numerical results, other comparisons are made with the FE program Midas GEN© (Midas GEN, 2015), for both the linear and nonlinear analyses. It is worth remarking that, in general, for displacements attained at the end of the pushover curves, second order effects could influence the results. To have a better understanding of this aspect, all the analyses have been conducted with the P-delta effects.

The programs Straus7©/Strand7© and EasyOver© use a Moment-Curvature relationship with a continuous update of the shear span L_V and of the axial force N during the transversal loads increment for ductile mechanisms, and the plastic behaviour can rise in any point of the beam axis. They also have the possibility of considering the shear interaction with axial forces and bending moments described above and choosing the capacity equations for safety verification of shear mechanisms, i.e. with Equation (12) with 45% inclined compressed concrete struts or Equation (16). Differently, Midas Gen© has a classic phenomenological approach with lumped plasticity model for ductile mechanisms, i.e. Moment-rotation (Figure 4), with $L_V=L/2$, where the plastic behavior is associated only to the extremities of the beam element; the "shear hinges" are governed by Equation (12) with 45% inclined compressed concrete struts.

The FE models are reported in Figure 6, and they are analysed in order to discuss all the cases reported in Figure 5. In particular, all the models have the same geometry and number of rebars for the main and secondary beams of the frames. The model in Figure 6a has been divided into two different cases. They have the same number of rebars in the columns but different stirrup spacing:

- The first has stirrups ϕ6/20 (in order to capture the "*brittle shear*" mechanism),
- The second has the columns of the first elevation with stirrups ϕ8/10 and in the remaining elevations ϕ8/20 (in order to capture the "*ductile shear*" mechanism).

Figure 6. Finite Element models: a) brittle frame, b) ductile frame

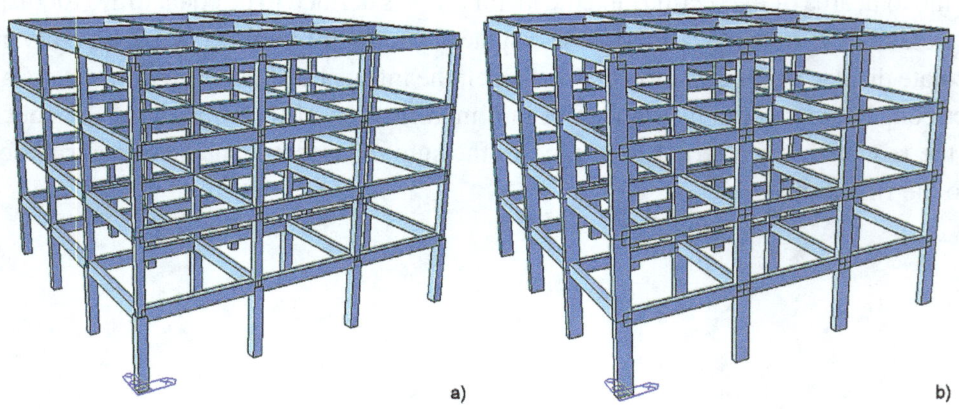

The FE model in Figure 6b has strong columns and ductile beams in order to have a ductile global behaviour.

To understand the actual vulnerability of the structures the seismic risk index I_R is evaluated for each of the models. The I_R is computed both in term of PGA or T_R by means of:

$$I_R = \left(\frac{a_{g,C}}{a_{g,D}}\right) I_R = \left(\frac{a_{g,C}}{a_{g,D}}\right) \text{ and } I_R = \left(\frac{T_{R,C}}{T_{R,D}}\right)^{0.41}. \tag{19}$$

It might be worth keeping in mind that $I_R \geq 1$ corresponds to a safe structure, $I_R < 1$ corresponds to an unsafe building with respect to the standard of the new buildings. The values $a_{g,C,}$ and $T_{R,C}$, in (19) for the LSSD correspond to the seismic action that produces the breakage of a number (two/three) of neighbor elements so as to trigger a ductile (i.e., chord rotation) and/or brittle (shear resistance) mechanism, and that comes from the nonlinear analysis (Figure 7). The exponent 0.41 in (19)b allows the comparison between the two I_R. The use of the Seismic Index Risk I_R gives the possibility to work, indifferently, with PGA (International country) or T_R (Italian country) once the value is known.

Modal Analysis

In order to assess the dynamic characteristics of the structure, modal analyses have been initially performed. The spatial models have the mean values of resistance for concrete and steel. The natural periods with the corresponding participation masses are reported in Table 1 for the brittle frame and in Table 2 for the ductile frame. As expected (due to the rigidity of the floor), 12 modes are necessary to get a total amount of effective masses equal to 100%.

Figure 7. Flow chart of the technique followed in the analysis to determine global and local failure mechanisms

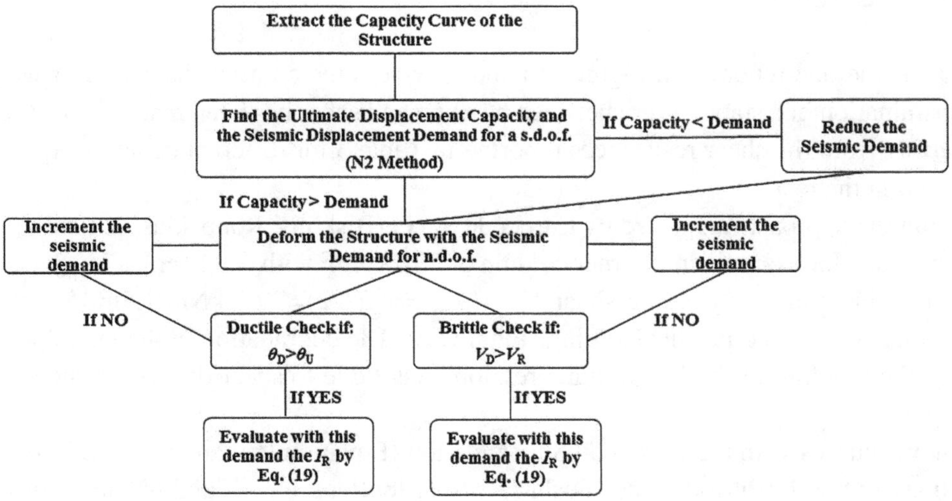

Table 1. Periods and modal effective masses for transversal, longitudinal and vertical direction for the principal modes of Finite Element model in Figure 6a

Mode n°	Period	TRAN-X		TRAN-Y		ROTN-Z	
	[sec]	Mass(%)	Sum(%)	Mass(%)	Sum(%)	Mass(%)	Sum(%)
1	0.680	0.000	0.000	85.936	85.936	0.000	0.000
2	0.680	82.409	82.409	0.000	85.936	0.000	0.000
3	0.588	0.000	82.409	0.000	85.936	83.669	83.669
4	0.374	0.000	82.409	9.986	95.921	0.000	83.669
5	0.283	0.000	82.409	0.000	95.921	10.924	94.593
6	0.283	11.632	94.041	0.000	95.921	0.000	94.593

Table 2. Periods and modal effective masses for transversal, longitudinal and vertical directions for the principal modes of Finite Element model in Figure 6b

Mode n°	Period	TRAN-X		TRAN-Y		ROTN-Z	
	[sec]	Mass(%)	Sum(%)	Mass(%)	Sum(%)	Mass(%)	Sum(%)
1	0.613	0.000	0.000	78.345	78.345	0.000	0.000
2	0.443	78.345	78.345	0.000	78.345	0.000	0.000
3	0.441	0.000	78.345	0.000	78.345	78.559	78.559
4	0.201	0.000	78.345	0.000	78.345	0.000	78.559
5	0.138	0.000	78.345	0.310	78.655	0.000	78.559
6	0.134	0.310	78.656	0.000	78.656	0.000	78.559

From both the FE programs, the same principal modes are detected. The participation masses of the principal modes are higher than 75% in both directions, and this means that the pushover analysis is applicable due to the high regularity of the buildings.

"Brittle Shear" Building

The first case is the one reported in Figures 1a and 6a where the columns have a big spacing of stirrups. A discrimination of a static failure between bending and shear has been made before the nonlinear analysis, starting from the shear resistances reported in Table 3 for different capacity equations previously explained in the text.

Considering the approach suggested by Eurocode 8, i.e. (Biskinis, Roupakias, & Fardis, 2004), it is possible to evaluate, for example in a corner column of 30x30 cm² with ϕ6/20 cm, a $V_{R,0}(\mu_{\Delta,pl}=0)=46.68$ kN. Comparing this with the yielding shear $V_y=M_y/L_V=M_y/(L/2)=79.5$ kNm/1.5m=53 kN a "*brittle shear*" behaviour is expected for this building (Figure 5a). The degradation in the lateral load capacity occurs *before* the yielding of the longitudinal reinforcement due to shear distress in the element for a chord rotation θ_V.

The pushover curves with the Eurocode 8 formulation (Equation (6)) are reported in Figure 8 with continuous lines for the "PushMode" and "PushMass" distributions, for +X and +Y directions, evaluated

Table 3. Values of shear resistance evaluated with different capacity equations for a corner column of 30x30 cm² with ϕ6/20 cm

	EUROCODE 8 Equation (16)		NTC2008 Equation (12)	NTC2008 Equation (12) with variable θ and $V_{Rc,t}$=0	NTC2008 Equation (12) with cotθ=1 and $V_{Rc,t}$=0
	$\mu_{\Delta,pl}$=0	$\mu_{\Delta,pl}$=5			
V_R	46.68 kN	35.94 kN	92.25 kN	76.55 kN	30.6 kN

with EasyOver©. A brittle global behaviour due to the absence of a minimum ductility of the columns is observable. In the same Figure 8 are reported with dotted lines the analogous four pushover curves obtained with the same software but with a shear resistance evaluated with Equation (12) contained in the Italian Seismic Code. A more ductile global behaviour is observed from these capacity curves, and it is confirmed from Table 3 where now the V_y=53 kN is compared with V_R= 92.25 kN, and then a "*ductile shear*" behaviour can be observed.

These results are then compared with another FE program, i.e. Midas GEN©, and the curves are those reported in Figure 8 with dashed lines. Also, in this case, a clear "*ductile shear*" behavior is observable. Comparing the dotted (EasyOver©) with the dashed curves (Midas GEN©) both obtained with Equation (12) a different initial stiffness is visible. This is due to a different shear span L_V used in the programs: the former has a continuously updated L_V, the latter a fixed value of L_V=*L*/2. Furthermore, this difference is also due to a different plastic capacity: the Midas GEN© approach is a phenomenological approach where the plastic hinge can rise only at the end sides of the beams and columns; for EasyOver© a more diffuse damage is possible, not only concentrated at the ends but also in the middle of the elements. For this issue, other analyses have been made excluding a priori the rise of plastic deformation in the middle of the elements: very similar curves are obtained compared with those of Midas GEN©.

Figure 8. Pushover curves for the "brittle shear" building

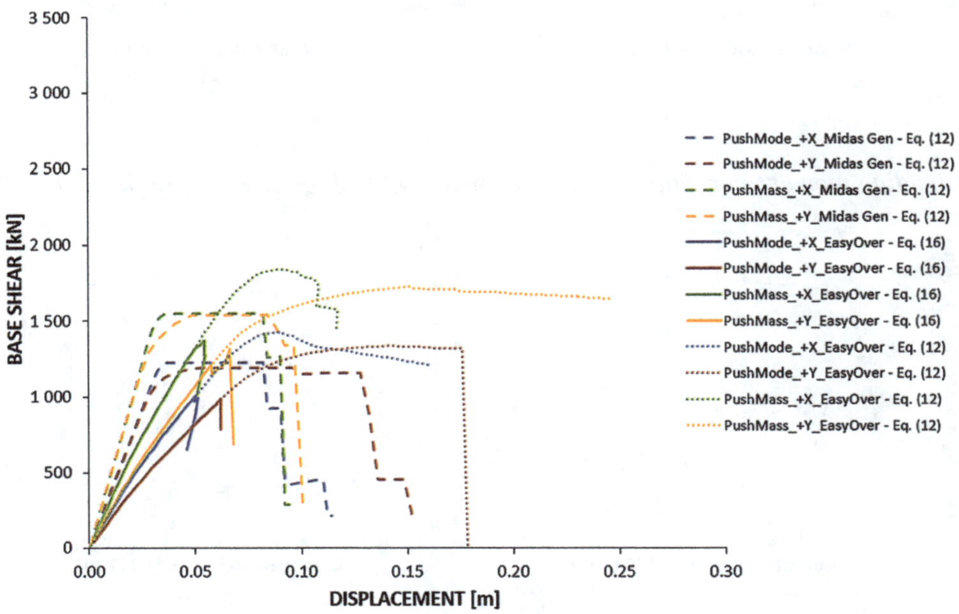

Differently, the ultimate displacement (i.e., when the vertical drop in the capacity curve is clear) of the Eurocode 8 and CNR-DT212 is always smaller than that of the Italian Seismic Code.

From a sectional (bending moment-curvature diagram) point of view, comparing the results obtained with the two different formulations in EasyOver©, it is possible to observe, for the same column analyzed in Table 3, a ductile behavior in the two axes in all the loading steps, when the Italian Seismic Code is used (Figure 9). This is no longer true when the Bending-Shear interaction according to the Eurocode 8 formulation is taken into account. This confirms the interaction curve reported in Figure 5a and plotted in terms of moment-curvature in Figure 10a: the shear crisis occurs in the linear range when the yielding of the first bar has not yet occurred. This brittle sectional behavior corresponds to a more brittle global behavior in X-direction (see Figure 8). Analyzing Figure 10b, where the sectional behavior in the other column axis is reported, a more ductile behavior is observable, which corresponds to a higher dissipation capacity in the Y-direction and thus to a bigger ultimate displacement than the one in the X-direction.

From a global point of view, it is possible to evaluate the T_R of the seismic action that corresponds to a clear drop of the Base Shear on the capacity (pushover) curves or, alternatively, when the shear decreases of about 15% with respect to the maximum value. For Midas GEN© the action producing a

Figure 9. Bending moment-curvature for the column of Table 3 for the Italian Seismic Code formulation

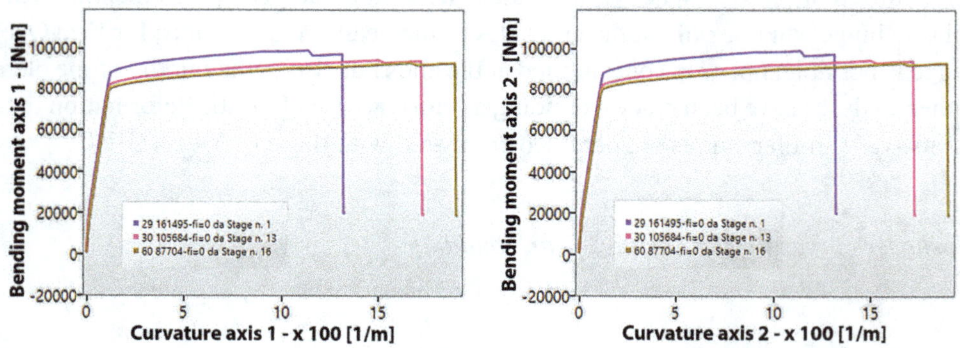

a) b)

Figure 10. Bending moment-curvature for the column of Table 3 for Eurocode 8 formulation

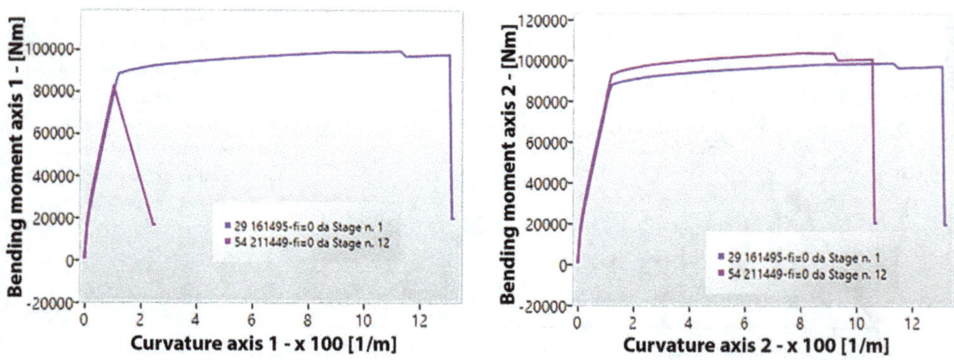

a) b)

global mechanism is for T_R=330 years (I_R=0.49), for EasyOver© with Equation (12) a T_R=900 years (I_R=0.74) and with Equation (16)T_R=60 years (I_R=0.24).

Now it is possible to compare the previous results with those of the local checks, evaluated as explained in the introduction of this section (Figure 7), in terms of I_R with Equation (19b). For ductile mechanisms, with Midas GEN© with Equation (12) a minimum T_R=40 years (I_R=0.21) is observable, for EasyOver© with Equation (16) a T_R=285 years (I_R=0.46) and with Equation (12) the global mechanism is gained before the local rupture. For brittle mechanism, with both Midas GEN and EasyOver© with Equation (12), it is impossible to see a shear rupture before the global mechanism; differently, for EasyOver© with Equation (16), a T_R=25 years (I_R=0.17) is observed.

When comparing the safety verifications of ductile members of EasyOver© and Midas GEN©, both with Equation (12), a difference of about 55% in terms of I_R is noticeable. This means that a continuous update of the shear span L_V and a not univocal determination of the position of the bending hinges can lead to a better description of the nonlinear behavior of the building, which brings to big values of I_R. This difference is lesser comparing EasyOver© with Equation (16) and Midas GEN©: an increment of about 15% in terms of I_R is noticeable. This means that even if the shear resistances of the elements are different (see Table 3), with Eurocode 8 or the Italian Seismic Code similar values of I_R are obtainable, and this means that the methods are sufficiently aligned.

"Ductile Shear" Building

The second case has the same geometry as the previous one, but the columns have a better spacing of stirrups. A discrimination of a static failure between bending and shear is made as in the previous case, starting from the shear resistances reported in Table 4 for different capacity equations.

Considering the approach suggested by Eurocode 8, i.e. (Biskinis et al., 2004), it is possible to evaluate in the same column of the previous case, i.e. 30x30 cm² with ϕ8/20 cm a $V_{R,0}(\mu_{\Delta,pl}=0)$=67.4 kN and a $V_{R,min}(\mu_{\Delta,pl}=5)$=51.48 kN. Comparing this with the yielding shear $V_y=M_y/L_V=M_y/(L/2)$= 79.5 kNm/1.5m=53 kN a "*ductile shear*" behaviour is expected for this building ($V_{R,min}<V_y<V_{R,0}$, Figure 5b). The degradation in the lateral load capacity occurs *after* the yielding of the longitudinal reinforcement, and it corresponds to a degradation of the flexural resistance in correspondence of an angular deformation θ_a.

The EasyOver© pushover curves with the Eurocode 8 formulation (Equation (6)) are reported in Figure 11 with continuous lines. It now evident the quite ductile global behaviour that is due to the presence of a minimum ductility of the columns. In the same Figure 11 with dotted lines are reported the analogous EasyOver© pushover curves obtained with shear resistance evaluated with the Italian Seismic Code (Equation (12)). A similar ductile global behaviour is observed from these capacity curves because the influence of the shear capacity is minimal and the nonlinear behaviour is associated, at least in the first part of the curves, to the axial-bending interaction.

The results obtained with Midas GEN© are reported in Figure 9 with dashed lines. Also, in this case, a clear "*ductile shear*" behavior is observable. Comparing the dotted (EasyOver©) with the dashed curves (Midas GEN©) both obtained with Equation (12) a different initial stiffness is visible. This is always due to a different shear span L_V used in the programs and to a difference in the plastic capacity as explained in the previous case.

From a sectional (bending moment-curvature diagram) point of view, comparing the results obtained with the two different formulations in EasyOver©, it is possible to observe, for the same column analyzed in Table 4, a ductile behavior in the two axes in all the loading steps, when the Italian Seismic Code is

Table 4. Values of shear resistance evaluated with different capacity equations for a corner column of 30x30 cm² with ϕ8/20 cm

	EUROCODE 8 Equation (16)		NTC2008 Equation (12)	NTC2008 Equation (12) with variable θ and $V_{Rc,t}$=0	NTC2008 Equation (12) with cotθ=1 and $V_{Rc,t}$=0
	$\mu_{\Delta,pl}$=0	$\mu_{\Delta,pl}$=5			
V_R	67.4 kN	51.48 kN	116.1 kN	136.1 kN	54.44 kN

used (Figure 12). This is no longer true when the Bending-Shear interaction according to the Eurocode 8 formulation is taken into account. This confirms the interaction curve reported in Figure 5b and plotted in terms of moment-curvature in Figure 13a: the shear crisis occurs in the plastic range when the yielding of the bars has occurred. The major effect is observable comparing Figures 13a and 13b, where, with the increasing of load steps, the moment-curvature diagram in axis 1 presents a reduced ductility compared to the other axis (see green lines), as a consequence of the effect of the above mentioned Bending-Shear interaction.

This "ductile shear" sectional behavior corresponds to a not so brittle global behavior in the X-direction compared to the case of "brittle shear". For "*ductile shear*" behavior, the ultimate displacement (i.e., when the vertical drop in the capacity curve is clear) is similar considering both the Eurocode 8 and CNR-DT212 or the Italian Seismic Code. This means that when a mixed rupture mode is observable the two different approaches give a similar global behavior. For the Midas GEN© model, the action producing a global mechanism is for T_R=350 years (I_R=0.50), for EasyOver© with Equation (16) a T_R=1145 years (I_R=0.81) and with Equation (12)T_R=590 years (I_R=0.62).

Now it is possible to compare the previous results with those of the local checks always in terms of I_R. For ductile mechanisms, with Midas GEN© a minimum T_R=40 years (I_R=0.21) is observable, for EasyOver© a T_R=310 years (I_R=0.48) for both models. For brittle mechanism, with Midas GEN© with Equation (12) it is impossible to see a shear rupture before the global mechanism; differently for EasyOver© with Equation (16) a T_R=60 years (I_R=0.24) is obtainable and with Equation (12) the global mechanism is observed before the local shear crisis.

When comparing the safety verifications of the global mechanism evaluated with EasyOver© and with Midas GEN©, both with Equation (12), a difference of about 38% in terms of I_R is observable: this is the effect of a more exact evaluation of the nonlinear behavior of the member. This difference in the activation of the global mechanism is smaller comparing EasyOver© with Equation (16) and Midas GEN©, and it is about 19% in terms of I_R. Finally, comparing the EasyOver© results with Equation (16) and Midas GEN© in the achievement of the local failure, a bigger difference is noticeable (-493%): this is due to a lower value of shear strength obtained with EUROCODE 8 compared with the value obtained with NTC2008 (see Table 4); however, more investigations must be carried out on this issue.

Global Ductile Building

The last case has a different geometry compared to the previous buildings and different rebars and stirrups inside the columns. A discrimination of a static failure between bending and shear has been made as in the previous case, starting from the shear resistances reported in Table 5 for different capacity equations.

Figure 11. Pushover curves for the "ductile shear" building

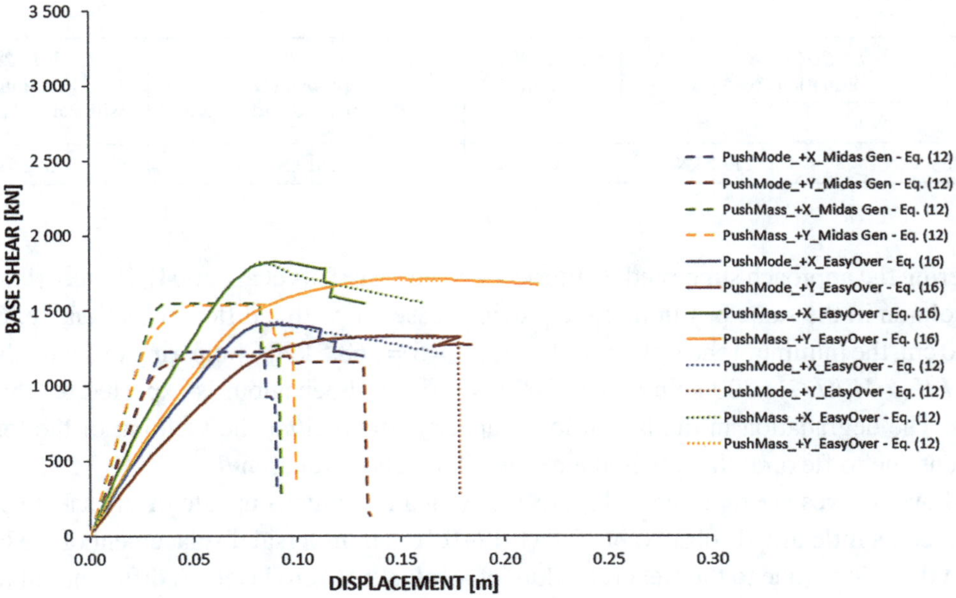

Figure 12. Bending moment-curvature for the column of Table 4 for the Italian Seismic Code formulation

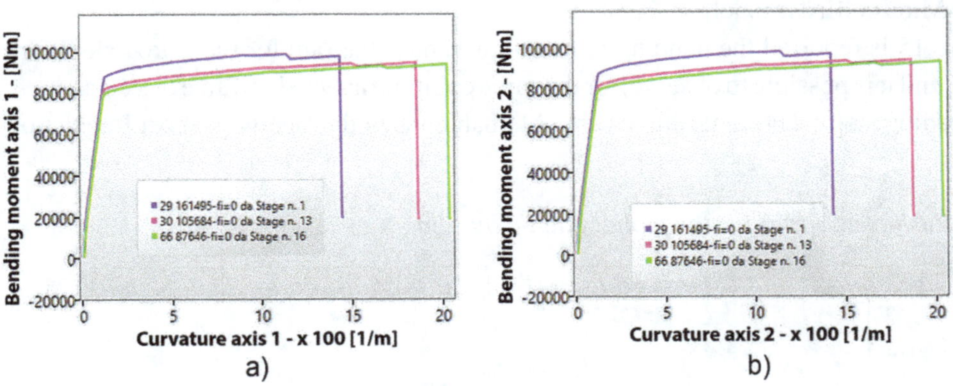

Figure 13. Bending moment-curvature for the column of Table 4 for Eurocode 8 formulation

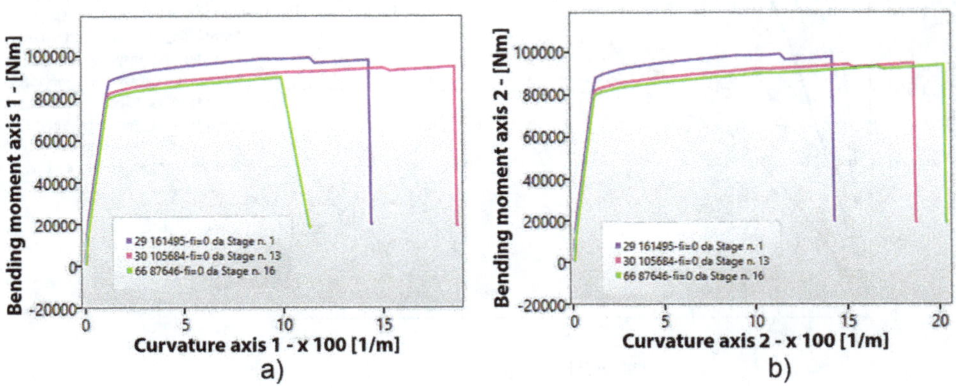

Table 5. Values of shear resistance evaluated with different capacity equations for a corner column of 60x30 cm² with ϕ10/10 cm

	EUROCODE 8 Equation (16)		NTC2008 Equation (12)	NTC2008 Equation (12) with variable θ and $V_{Rc,t}=0$	NTC2008 Equation (12) with cotθ=1 and $V_{Rc,t}=0$
	$\mu_{\Delta,pl}=0$	$\mu_{\Delta,pl}=5$			
V_R	368.5 kN	279.2 kN	434.1 kN	761 kN	352.4 kN

Considering the approach suggested by Eurocode 8, i.e. (Biskinis et al., 2004), it is possible to evaluate in the column in the same position of the previous case but with a different section, i.e. 60x30 cm² with ϕ10/10 cm, the minimum shear strength $V_{R,min}(\mu_{\Delta,pl}=5)$=368.5 kN. Comparing this with the yielding shear $V_y=M_y/L_V=M_y/(L/2)$=300 kNm/1.5m=200 kN a flexural behaviour is expected for this building (Figure 5c). The degradation in the lateral load capacity occurs *after* the yielding of the longitudinal reinforcement due to flexural distress in the element for a chord rotation θ_u.

The pushover curves are reported in Figure 14. With a continuous update of the shear span L_V, i.e. with EasyOver©, a little drop is observable for a 0.12÷0.15m transversal displacement of the building in the different directions (due to the flexural failure of a column at zero level). A different initial stiffness is always noticeable due to the different plasticity models used: a phenomenological and a continuous one respectively for Midas GEN© and EasyOver©. In addition, a different global capacity is observable when considering Midas GEN© and EasyOver©: this is ascribable to a limit number of plastic zones used in the Midas GEN© model.

In Figure 15 is reported the bending moment-curvature diagram for the Eurocode 8 formulation in EasyOver©, and it is possible to observe, for the same column analyzed in Table 5, a clear ductile behavior in all the loading steps. The same curves are obtainable when the Bending-Shear interaction according

Figure 14. Pushover curves for the global ductile building

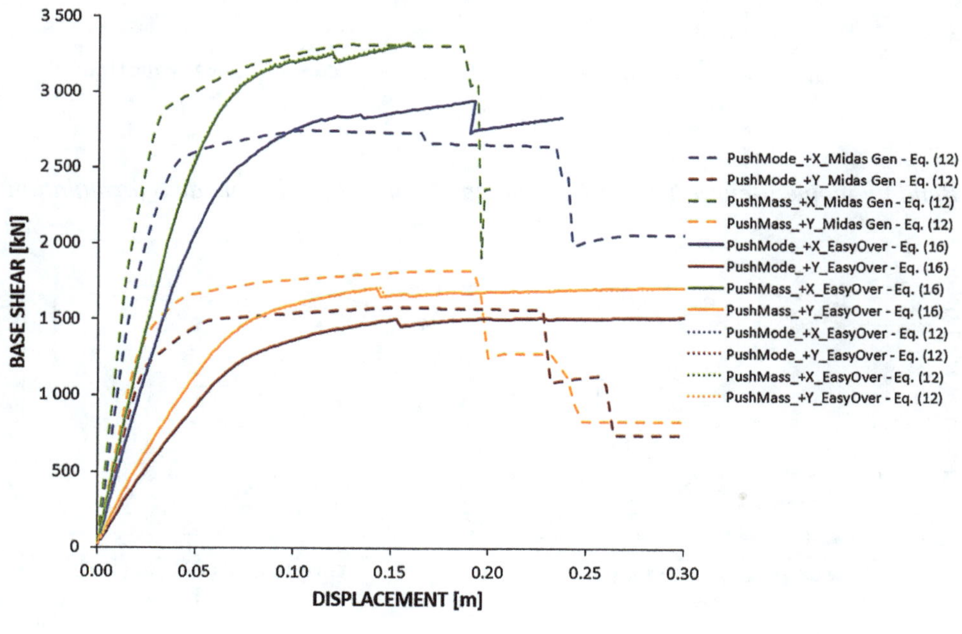

Figure 15. Bending moment-curvature for the column of Table 5 for Eurocode 8 formulation

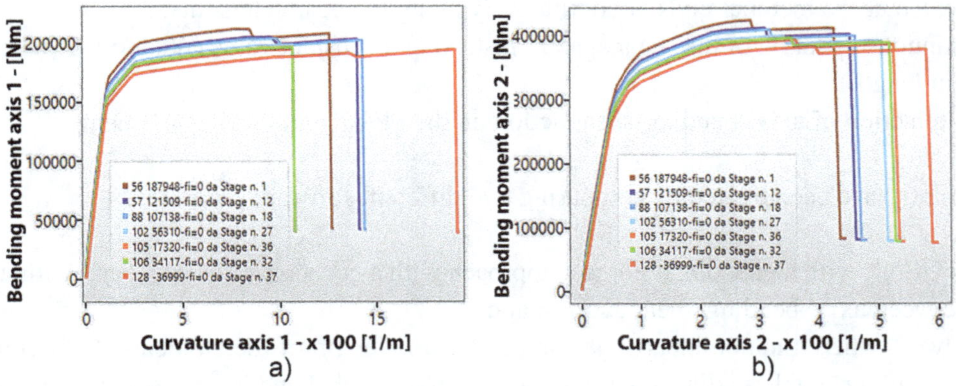

to the Italian Seismic Code formulation is taken into account but they are not reported for brevity issue. This confirms the interaction curve reported in Figure 5c and plotted in terms of moment-curvature in Figure 15, where the shear crisis does not occur during all the transversal loading. The same effects are obtainable from a global point of view where the two formulations give the same ultimate displacement and the same global ductility. Different ultimate displacements are obtainable with the classic lumped plasticity and constant values of the shear span L_V implemented in Midas GEN©. The effects of these parameters in the global ductility need further investigation.

For all the models the action producing a global mechanism is greater than the seismic demand, i.e. T_R>1898 years, and this means that I_R>1.0. Now it is possible to compare the previous results with the local checks always in terms of I_R. For ductile mechanisms, with Midas GEN© a minimum T_R=1100 years (I_R=0.80) is observable, for EasyOver© with Equation (16) a T_R=1140 years (I_R=0.81) and with Equation (12) a T_R=1285 years (I_R=0.85). For brittle mechanism, a T_R>1898 years (I_R>1.0) is obtained for all the models.

In conclusion, though the ultimate global displacement for the various curves is dissimilar, the local safety verifications show minor differences in terms of I_R. Comparing EasyOver© with Equation (16) and Midas GEN© a difference of I_R of about 1.5% is obtained in terms of chord rotations, and about 6% comparing EasyOver© with Equation (12) and Midas GEN©.

CONCLUSION

In this work, RC frame buildings designed or not designed with the capacity design criterion (i.e. with out-of-date seismic codes) have been considered. All the possible approaches for the nonlinear static analysis with lumped plasticity have been checked.

Eurocode 8, the Seismic Italian Code and CNR-DT 212 have been applied in order to assess global, ductile and brittle local behaviors of all considered frame structures. The consistency of the results obtained by applying the different codes has been checked, in particular studying the effect of:

1. The use of a lumped plasticity model with a phenomenological approach where inelastic hinges are located at the member ends or continuously along the beam elements;
2. A continuous updating of the shear span L_v instead of a constant value equal to $L/2$ (where L is the element length);
3. The interaction of axial-bending-shear action on the global and local assessment.

These points have been treated by comparing two different software programs:

- Midas GEN©, with a phenomenological approach with a constant value of L_V and without interaction between axial-bending-shear actions, and
- EasyOver©, where the nonlinearity is spread over the length of the elements, L_V is continuously updated and the axial-bending-shear interaction on the global and local assessment is considered.

The first conclusion concerns the global behavior. In particular, the initial elastic stiffness of the capacity curves is influenced by the choice of the plastic hinge location: a stiffer behavior is observed if the plastic behavior is concentrated at the end of the element. Further investigations with more complex in-plan structures should be carried out in order to have a better perception of the influence of the various parameters of the nonlinear model.

Regarding the assessment of ductile mechanisms, the obtained results show that the theoretical formula of θ_u (Equation (5)) is too sensitive to the value assumed by the shear span L_V; hence, there is the need to consider its correct definition (Equation (2)) instead of $L_V = L/2$. These assumptions lead to an evaluation of the capacity, which is more demand-dependent.

Concerning the assessment of brittle mechanisms, the approach proposed by the Italian Seismic Code (Equation (12)) does not furnish results consistent with those obtained through the most accurate procedure suggested in Eurocode 8, which takes into account the seismic degradation in plastic regions.

FUTURE RESEARCH DIRECTIONS

There are several research directions along which this work can be developed. The need to consider the effect of bi-axial bending in local checking for ductile mechanisms should be tested with bi-dimensional failure curves. The numerical models used to analyze the structures may affect the results. For this reason, the results obtained cannot be extended to all RC frame buildings. The authors are working to modify the characteristics of the numerical model and to increment the number of analyzed buildings, in order to deal with a wide range of structural configurations and to get to more general conclusions. More comparisons should be made with fiber models (Pierdicca, Clementi, Maracci, Isidori, & Lenci, 2015). In addition, the real strength of the constituent material should be the subject of further research.

REFERENCES

Asteris, P. G., & Plevris, V. (Eds.). (2015). *Handbook of Research on Seismic Assessment and Rehabilitation of Historic Structures*. IGI Global; doi:10.4018/978-1-4666-8286-3

Bischof, C. R. (2002). ACI Manual of Concrete Practice, 2010. *Practice, 552*, 2002–2002.

Biskinis, D. E., Roupakias, G. K., & Fardis, M. N. (2004). Degradation Of Shear Strength Of Reinforced Concrete Members With Inelastic Cyclic Displacements. *ACI Structural Journal*, (101): 773–783.

CEN (Comité Européen de Normalisation). (2005a). *Eurocode 2 - Design of concrete structures Part 1-1: General rules and rules for buildings - EN 1992-1-1*. CEN.

CEN (Comité Européen de Normalisation). (2005b). *Eurocode 8 - Design of structures for earthquake resistance - Part 3: Assessment and retrofitting of buildings - EN 1998-3*. CEN.

Circolare Ministeriale n. 617. (2009). *Cons. Sup. LL. PP., 'Istruzioni per l'applicazione delle Nuove Norme Tecniche per le Costruzioni' di cui al decreto ministeriale del 14.01.2008*. G.U. del 26.02.2009 n. 47, supplemento ordinario n. 27. (in Italian).

Clementi, F., Quagliarini, E., Maracchini, G., & Lenci, S. (2015). Post-World War II Italian school buildings: Typical and specific seismic vulnerabilities. *Journal of Building Engineering*, *4*, 152–166. doi:10.1016/j.jobe.2015.09.008

Clementi, F., Scalbi, A., & Lenci, S. (2016). Seismic performance of precast reinforced concrete buildings with dowel pin connections. *Journal of Building Engineering*, *7*, 224–238. doi:10.1016/j.jobe.2016.06.013

Consiglio Nazionale delle Ricerche (CNR). (2013). *CNR-DT 2012/2013 - Istruzioni per la Valutazione Affidabilistica della Sicurezza Sismica di Edifici Esistenti*. Retrieved from http://www.cnr.it/documenti/norme/IstruzioniCNR_DT212_2013.pdf

EasyOver. (2016). *Analysis Manual*. Author.

Elnashai, A. S., & Di Sarno, L. (2008). Fundamentals of Earthquake Engineering. John Wiley & Sons. http://doi.org/ doi:10.1002/9780470024867

Elwood, K. J., & Moehle, J. P. (2005). Axial Capacity Model for Shear-Damaged Columns. *ACI Structural Journal*, *102*(4), 578–587.

Fajfar, P., & Gašperšič, P. (1996). The N2 method for the seismic damage analysis of rc buildings. *Earthquake Engineering & Structural Dynamics*, *25*(1), 31–46. doi:10.1002/(SICI)1096-9845(199601)25:1<31::AID-EQE534>3.0.CO;2-V

FEMA. (2000). *356. Pre-standard and Commentary for the Seismic Rehabilitation of Buildings - Applied Technology Council*. Washington, DC: ATC.

Fiore, A., Porco, F., Uva, G., & Mezzina, M. (2013). On the dispersion of data collected by in situ diagnostic of the existing concrete. *Construction & Building Materials*, *47*, 208–217. doi:10.1016/j.conbuildmat.2013.05.001

Haselton, C. B., Liel, A. B., Deierlein, G. G., Dean, B. S., & Chou, J. H. (2011). Seismic Collapse Safety of Reinforced Concrete Buildings. I: Assessment of Ductile Moment Frames. *Journal of Structural Engineering*, *137*(4), 481–491. doi:10.1061/(ASCE)ST.1943-541X.0000318

Haselton, C. B., Liel, A. B., & Lange, S. T. (2008). Beam-Column Element Model Calibrated for Predicting Flexural Response Leading to Global Collapse of RC Frame Buildings. *Peer 2007, 3.*

Ibarra, L. F., Medina, R. A., & Krawinkler, H. (2005). Hysteretic models that incorporate strength and stiffness deterioration. *Earthquake Engineering & Structural Dynamics*, *34*(12), 1489–1511. doi:10.1002/eqe.495

Liel, A. B., Haselton, C. B., & Deierlein, G. G. (2011). Seismic Collapse Safety of Reinforced Concrete Buildings. II: Comparative Assessment of Nonductile and Ductile Moment Frames. *Journal of Structural Engineering*, *137*(4), 492–502. doi:10.1061/(ASCE)ST.1943-541X.0000275

Magliulo, G., Maddaloni, G., & Cosenza, E. (2007). Comparison between non-linear dynamic analysis performed according to EC8 and elastic and non-linear static analyses. *Engineering Structures*, *29*(11), 2893–2900. doi:10.1016/j.engstruct.2007.01.027

Masi, A., & Chiauzzi, L. (2013). An experimental study on the within-member variability of in situ concrete strength in RC building structures. *Construction & Building Materials*, *47*, 951–961. doi:10.1016/j.conbuildmat.2013.05.102

Masi, A., & Vona, M. (2012). Vulnerability assessment of gravity-load designed RC buildings: Evaluation of seismic capacity through non-linear dynamic analyses. *Engineering Structures*, *45*, 257–269. doi:10.1016/j.engstruct.2012.06.043

Midas GEN. (2015). Analysis Manual.

Mpampatsikos, V., Nascimbene, R., & Petrini, L. (2008). A Critical Review of the R.C. Frame Existing Building Assessment Procedure According to Eurocode 8 and Italian Seismic Code. *Journal of Earthquake Engineering*, *12*(sup1), 52–82. http://doi.org/10.1080/13632460801925020

NTC. (2008). *Decreto Ministeriale 14/1/2008. Norme tecniche per le costruzioni*. Ministry of Infrastructures and Transportations, G.U. S.O. n. 29 on 2/4/2008.

Pierdicca, A., Clementi, F., Maracci, M., Isidori, D., & Lenci, S. (2015). Vibration-Based SHM of Ordinary Buildings: Detection and Quantification of Structural Damage. *ASME 2015 International Design Engineering Technical Conferences and Computers and Information in Engineering Conference*, 8. http://doi.org/ doi:10.1115/DETC2015-46763

Plevris, V., Mitropoulou, C. C., & Lagaros, N. D. (Eds.). (2012). *Structural Seismic Design Optimization and Earthquake Engineering*. IGI Global. doi:10.4018/978-1-4666-1640-0

Sezen, H., & Moehle, J. P. (2004). Shear strength model for lightly reinforced concrete columns. *Journal of Structural Engineering*, *130*(11), 1692–1703. doi:10.1061/(ASCE)0733-9445(2004)130:11(1692)

STRAUS7. (2016). *Theoretical manual-theoretical background to the Strand7 finite element analysis system*. Author.

Verderame, G. M., De Luca, F., Ricci, P., & Manfredi, G. (2011). Preliminary analysis of a soft-storey mechanism after the 2009 LAquila earthquake. *Earthquake Engineering & Structural Dynamics*, *40*(8), 925–944. doi:10.1002/eqe.1069

Zhu, L., Elwood, K. J., & Haukaas, T. (2007). Classification and Seismic Safety Evaluation of Existing Reinforced Concrete Columns. *Journal of Structural Engineering*, *133*(9), 1316–1330. doi:10.1061/(ASCE)0733-9445(2007)133:9(1316)

KEY TERMS AND DEFINITIONS

Case Study: It is a construction taken as a reference to perform surveys, modelling, and analyses.

Italian Seismic Code: It is the set of rules to design new structures or to assess the existing ones, considering the seismic loads.

Lumped Plasticity Model: It normally consists of elastic beam elements with inelastic springs located at the member ends or continuously along the members themselves. Usually, the springs have zero-length and for that reason, they are also called "point hinges". The stiffness matrix of the entire element is computed from the stiffness of the single parts that compose the member. Lumped inelasticity models have been widely used in earthquake engineering applications because their simple formulation allows very fast analyses.

Pushover Analysis: It is a nonlinear analysis to assess the structural capacity under static horizontal loads increasing to the collapse of the structure. The results of the pushover analyses are capacity curves identified by the variation of the base shear in function of the displacement of a control point on the structure.

Reinforced Concrete Building: It is a construction formed by columns and beams made of concrete, with embedded rebars and transversal stirrups.

Return Period: It is the average time span between earthquake occurrences in a fault or in a source zone.

Seismic Risk Index: The seismic risk index is a number lower or higher than one that defines the level of the risk of a construction subject to a definite earthquake. It is obtained by the ratio between the structural capacity and the seismic demand, generally expressed in terms of PGA (peak ground acceleration).

Chapter 5
A Comparative Investigation of Structural Performance of Typical and Non-Ducitle Public RC Buildings Strengthened Using Friction Dampers and RC Walls

Erkan Akpinar
Kocaeli University, Turkey

Seckin Ersin
OTS Engineering and Consultancy Ltd., Turkey

ABSTRACT

Strengthening of non-ductile public buildings is a never-ending issue. Selection of the suitable strengthening method and appropriate analysis type for the assessment of pre- and the post-intervention performances are still open to question. The displacement or drift limitations are crucial as well as demand capacity ratios for determination of such buildings performance under severe ground motion. In this chapter, an investigation of seismic performance focused on displacement criterion of strengthened non-ductile public RC buildings in Turkey is presented. Both the nonlinear static and response history analysis were conducted. Friction dampers which are fairly modern technique and conventional RC wall implementation method were introduced to as-is building. For the simplicity and the easy of the process, 2D frame selected for investigation. Comparison of the aforementioned techniques for non-ductile public RC buildings and performances particularly by means of displacement obtained using different methods for those investigated schemes are carried out and presented in the chapter.

DOI: 10.4018/978-1-5225-2089-4.ch005

INTRODUCTION

Adequate structural performance is the key-point of the earthquake-proof buildings. The arising question is how the engineers define this adequacy and how they supply the sufficient structural performance for the existing buildings. The seismic assessment and the performance evaluation methods are varying from linear to nonlinear manners as well as from equivalent analysis techniques to the most sophisticated state of the art method namely response history analysis. There exists also another balance needs to be considered for the assessment issue which is time and cost effectiveness. Although the linear and the static analysis methods are the easiest ways of the evaluation task, it is a well-known fact that they lead to accumulated errors. On the other hand, there is no doubt for the nonlinear way of thinking and the dynamic analysis methods converge to more realistic results. Recently, nonlinear static pushover method is the most widely preferred way of the seismic assessment and performance evaluation issue. Relative simplicity and fair reliability are the reasons behind the popularity. Despite that existing numerous variation of the method available, force-based single mode pushover analysis is the most applied technique for the regular, low height to mid-height buildings. Time history response analysis is the all-time winner for the structural analysis cluster as it considers the real time equation of the strong ground motion. Unfortunately, its time consuming manner due to the intensive calculations in time domain and related complex and costly nature generally results in that it is not preferred in design practice carried out in regular design offices.

Reinforced concrete (RC) buildings dedicated to public services are generally subjected to intensive superimposed loads and expected to be almost functional and to have limited damage after strong ground motions. Therefore, smaller displacements and drift ratios are expected as a good performance indicator along with the other performance parameters for such buildings as compared to regular residential counterparts. There exist considerable amount of such RC public buildings as well as other aged structures which were constructed before modern building codes. Unfortunately, neither ductile behavior nor adequate structural performance can be obtained from such buildings and immediate measurements need to be taken for them to satisfy the public safety. Once structural performance was assessed using aforementioned analysis methods, most probably the next step will be the action to improve the behavior of the building of which performance level evaluated as not adequate under considered strong ground motion. There exist numerous ways of strengthening techniques for improving the structural performance of RC structures. They classify as two main stream philosophy called member strengthening and system strengthening. If the overall behavior and the drift ratios are in a range of satisfying level, only the member strengthening such as RC jacketing, FRP wrapping, steel encasing or FRP or steel plate implementation to the surfaces could be enough to achieve the desired performance. On the other hand, if the exhibited global behavior may not be as good as at an expected level in terms of either lateral drifts or horizontal displacement, beyond the essential member improvements, then global structural intervention techniques such as RC shear wall addition, diagonal steel bracing inclusion, damper implementation or base isolation utilization need to be used.

Global system interventions are proved as considerably adequate techniques to satisfy the performance objectives for the RC buildings designed according to old seismic codes by not only analytical studies but also through experimental investigations. Conventional RC shear wall addition inside the existing frames is the most popular example for the system strengthening methods (Figure 1a and 1b). However, introducing such rigid RC shear walls to structures as a global intervention leads to increasing the lateral load demand. Consequently, high seismic loads on these additional wall elements always lead

Figure 1. (a) and (b) Structural RC wall addition (OTS), (c) and (d) Friction damper implementation (Pall Dynamics)

to construction of new foundation and enlargement of the existing foundations of the attached columns. Sometimes additional touch to the foundation with micro piles is unavoidable during this intervention. Moreover, special measurements should be taken for the anchorage between the new RC wall and the old RC column along the whole neighboring surfaces. Furthermore, erection of the new RC infill wall along with the construction of its foundation result in enormous and disturbing impact on architecture such as demolition and replacement of partitions, architectural finishes and other interior non-structural components and dramatic and long-lasting disruption of the occupancy (Fardis 2009). Friction dampers are also good choices as a system strengthening method (Constantinou et al., 1998, Soong et al., 1999, Symans et al., 2008, Pall et al., 2004) to reasonably fulfill the performance objectives in point of decreasing lateral displacements and interstory drifts without folding the base/story shear demand (Figure 1c and d1). They have simple behavior, stable and large rectangular hysteresis loops with negligible fade and relatively low costs. Also, they do not require periodic maintenance and they are independent of the frequency and velocity contents of the earthquakes unlike other passive energy dissipation systems. The drawbacks of these devices are the loss of clamping forces between plates over the time, surface distortions as physical conditions and the requirement of nonlinear analysis methods due to nonlinear characteristics of the devices (Pall et al., 2004). Comparatively, damper implementations are thought to be more advantageous way of global strengthening task respect to conventional RC wall addition counterpart in terms of the simplicity and the time consumption of intervention method.

This chapter presents the results of the performance assessment of the non-ductile RC public building that constructed before the modern seismic codes using two different nonlinear analytical methods. Additionally, consequences in terms of displacement and drift regarding to performance objective of the RC shear wall and the friction damper implementations as system strengthening intervention individually were exposed. Typical and non-ductile RC public building having a symmetrical plan was selected to investigate. Conventional nonlinear static analysis namely pushover analysis and time history response analysis were conducted on single representative frame in transverse direction of the selected building. Sap2000 v.16 software is used for the structural analysis. Moreover three different stories (5, 7, and 9 stories) were included to obtain the effect of the intervention profit as well as analysis type along with the altering building height.

ANALYSIS PROCEDURES FOR THE PERFORMANCE EVALUATION OF STRUCTURES

Structural evaluation and performance assessment are the kickoff points of the rehabilitation process and the attempt to achievement of the proper and "collapse proof" behavior for existing structures. Especially since early-'80s, considerable amount of work has been conducted to develop appropriate seismic assessment methodologies by national and international organizations such as CEB (Comite Euro – International du Beton), ATC (Applied Technology Council), Japan Building Disaster Prevention Association and FEMA (Federal Emergency Management Agency (fib. T.G. 7.1, 2003). In modern codes, the performance assessment is done based on either strength or displacement parameter. The force based evaluation considers the ratio of strength capacity to demand for the structural assessment. In contrast, the displacement based evaluation, which is a relatively new methodology, takes into account the displacement related quantities interrelated with damage to assessment rather than the strength. In modern codes, the only displacement related quantity to be checked in force based evaluation procedure is interstory drift ratio whereas displacement and deformation quantities belonging to either individual elements or overall system are considered during the displacement based assessment. The terminology "force based / displacement checked" is used in the literature for the previous method and no attempt is included to determine the damage level of the structure (Priestley et al., 2007). Although the linear analysis can be a tool for force based evaluation, it is not the case and nonlinear analysis methods should be used for the proper displacement based assessment so that the inelastic behavior of the structural members can be included. The nonlinear incremental static analysis called pushover and the nonlinear time history response analysis are the analysis methods for obtaining the inelastic response. While the time history response analysis is an already set method, the pushover analysis method has many variations such as conventional, adaptive, modal etc. The conventional pushover procedure with invariant load distribution as described in Turkish Earthquake Code 2007 (TEC 2007) is considered herein. It should be noted that certain amount of site survey on the investigated structure and cross sectional analyses for the structural members according to findings from this survey must be carried out to establish the reasonable nonlinear behavior. According to TEC 2007 as well as other modern building codes and manuals, the "knowledge factor" about the building depending on the reliability of the information available from the site survey should be determined as a first step for the investigation. Next, the strength of the concrete and reinforcement steel should be obtained from in-situ and laboratory tests. In addition to material tests, reinforcement detailing in the structural members and any deterioration in members or previous damage state should be clearly obtain from the as-built structure. The cross-sectional analyses using all these outcomes gathered from the building put forward the inelastic capabilities of the individual structural member which let the nonlinear structural analysis can be done (TEC 2007, Fardis, 2009).

While the nonlinear incremental static procedure, pushover, is chosen for finding the performance level of a structure; dominating natural frequency and mode shape are determined by linear modal analysis. The strong ground motion load demand of the structure is determined as a displacement or a force value. The structure is to be pushed using monotonically increasing displacements (or forces), as soon as the demanding point of the structure has been reached. Before starting nonlinear pushover procedure, a nonlinear case which contains all vertical loadings such as dead loads from slabs, walls and cladding, live loads; has to be analyzed to find the initial condition of structure before lateral incremental loading. After the pushover analyses has started; variations in internal forces of structural elements, inelastic deformations of structural elements are recorded cumulatively for each step of the procedure. When the demand of the structure has been reached, these cumulative records are used for finding the maximum values of the plastic deformations and forces in structural elements. The structure which is intended to be analyzed by using the incremental equivalent earthquake load method needs to satisfy some requirements. The structure should not have torsional irregularities. The dominating mode shape of the structure has to have 70% mass participation ratio (TEC 2007). It is assumed that, the participation of the equivalent earthquake load occurs independently from the formation of plastic hinges during the pushover analysis. This assumption enables multiplying the mode shape amplitude which has determined by linear modal analysis before starting the lateral incremental analysis with the mass to make the participation of loading values.

The nonlinear time history response analysis is another powerful method used for the performance based assessment by using existing or artificial strong ground motion records as acceleration per time step values. Each acceleration value is used in corresponding time step in general equation of motion and the related displacements and internal forces are calculated for the demand of the earthquake. The calculated values are integrated by using some numerical methods to obtain earthquake performance of the structure. Nonlinear characteristics of materials are also taken into account. One of the most common numerical integration method used in time history analysis is Newmark β method which is going to be mentioned below. All numerical integration methods used in time history response analysis are used to solve basically the general equation of motion (Clough and Penzien, 1995). Average acceleration method is more advantageous method than linear acceleration method since it remains stable independent from time step and period values. For this reason, average acceleration method is more commonly used one than linear acceleration method. While using nonlinear time history method in performance based evaluation, damping value also has a big importance and so ought to be controlled. TEC 2007 as well as other guidelines and codes present some special considerations for selecting the strong ground motion record. Those requirements can be stated as follows:

1. The time duration of the strong ground motion record weather real or artificial has to be longer than 15 seconds,
2. The average value of the spectral acceleration values based on zero period has to be greater than the spectral acceleration of the linear design spectrum and
3. The average spectral acceleration values for the artificial earthquake records generated for 5% damping ratio, has to be greater than 90% of elastic spectral accelerations in the period range of T_1 and 0.2 times T_1 with respect to dominant natural period, (TEC 2007).

Moreover, although filtering the noise in earthquake and scaling are really important issues for the time history response analysis, since they deeply related with e7arthquake engineering field, such efforts does not taken into account herein.

STRENGTHENING OF EXISTING STRUCTURES

The considerable amount of existing structures needs to be either retrofitted or strengthened in Turkey as well as in the other countries. Past earthquakes have revealed this necessity in dramatic ways. The lessons learned from earthquake reconnaissance visits have shown that there exist several reasons behind the fatal structural failures and building collapses (Bruneau & Saatciolu 1993; Mitchell et al., 1995; Saatcioglu et al., 2001). They can be summarized as:

- Deficiency in material properties,
- Poor workmanship,
- Inadequate strength and/or deformation capacities,
- Improper structural configuration and
- Non-suitable member detailing.

Although some of them are regard to site application and manufacturing phase, the rest of the mentioned issues are related to design consideration. Existing RC structures which were built before mid-70's are more or less far from the earthquake resistant design philosophy due to the inadequacy of the old fashion design codes about seismic aspect as well as the other aforementioned deficiencies.

The strengthening is aiming that modifying the seismic demand and/or the member capacities so that satisfying the general safety rule which is providing the member capacities equal to or bigger than the demand quantities. This statement can be achieved by following approaches or even combining them:

1. Reducing the demand quantities on either members or the entire structure due to seismic events,
2. Increasing the individual member capacities.

There exist several strengthening techniques and one or more interventions among them need to be implemented to obtain the successful achievement of either approach. Each strengthening method has its own pros and cons, in addition need to be considered all aspects carefully (Fardis 2009). The selection of strengthening method is complex issue and also related with not only technical considerations but also architectural, financial and social point of view. Some of the related aspects can be summarized as:

3. Strengthening versus reconstruction cost,
4. Strengthening materials, technology and knowledge available,
5. Effects and duration of partial or total evacuation,
6. Restriction related to architectural, functional and circumferential issues,
7. Social and historical significance (fib. T.G. 7.1, 2003).

The selection of the proper strengthening approach should be done according to special consideration of individual case after a detailed performance assessment effort. If the deficiencies are diagnosed only a few structural members of which locations are not gathered together in a single area, then the intervention for only the individual members and joints can be appropriate (fib. T.G. 7.1, 2003). In general, such deficiencies result from:

- Localized improper material usage and construction/workmanship problems,
- Wrong reinforcement detailing,
- Increasing vertical loads,
- Durability related issues.

In other words, increasing the member capacity approach can be enough to achieve an appropriate seismic behavior for such situation. On the contrary, if the deficient members are localized somewhere in the structure either vertically or horizontally as an indication of having a torsional problem or a rigidity/strength diversity among the stories, more comprehensive measures such as enlargement of the columns, implementing new lateral-load-resisting system, adding new shear walls should be applied (fib. T.G. 7.1, 2003). The plan or vertical irregularities as well as lack of lateral rigidity either partially or entirely may result in such extensive interventions generally called system strengthening. Moreover, removal of the mass even entire story, weaken some members as cutting some of the reinforcing bars, taking apart the structure into independent parts can be alternative approaches for achieving appropriate seismic behavior.

If the structure involves already damaged members, repair works should be applied prior to the further strengthening interventions. The repair works can be summarized as:

1. Injection either with grout or epoxy of cracks,
2. Replacement of loose concrete,
3. Renewal of spalled concrete,
4. Replacement of buckled or fractured reinforcing bars.

After the repair work on damaged RC members varying from moderately to heavily if needed, decided strengthening technique can be implemented for the improvement of the structural behavior. The intervention methods for increasing capacity of the RC structural member can be jacketing or adding extra reinforcement using different materials. Although some of the intervention techniques for member strengthening have multiple effects and improve more than one structural parameter at the same time, several techniques can be used for increasing the selected capacity only. Thus, entire strengthening scheme as well as intervention for individual member should be determined carefully. System or structure-level strengthening techniques can be stated as addition of new RC walls, addition of bracing system, implementation of energy dissipation devices, introducing base isolation systems. Addition RC shear wall and friction damper implementation interventions are explained in following paragraphs as they are main concern of the chapter.

Introducing additional RC shear walls to existing structures is one of the most common system-strengthening technique. The additional shear wall is very effective to limiting the interstory drifts and lateral displacements as well as damage state in other RC members and also non-structural elements. Generally, the most convenient and preferred application is that infilling strategic bays with fully or partially manner in existing frame. In the case of fully infilling the existing frame, it incorporates the beams and neighboring two columns and those existing framing members act as boundary elements. The proper anchorage and detailing of reinforcement should be provided to obtain intended plastic behavior and high deformation capability. Moreover, continuity of the longitudinal re-bars should be provided and ensured to achieve the strength increase (Fardis 2009). Besides the aforementioned advantageous, adding new RC shear wall intervention has its own drawbacks that can be stated as:

1. Increasing the shear demand,
2. Need to qualified site work especially for reinforcement detailing and anchorages,
3. Generally lead to extra effort to foundation strengthening,
4. Disruption of occupancy longer than any other member intervention,
5. Messy and noisy application.

Implementation of friction dampers to existing structure is another overall system-strengthening method. Friction dampers dissipate energy when the relative movement causes friction between the touching faces of two solid objects. For example, oval holed - bolted friction dampers have static friction force because of pretention on the bolts. When a bigger static friction force is applied to the damper, plates connected by bolts slip on each other. Last decade, there is a commonly used friction damper type called Pall friction damper. In Pall type friction dampers; friction mechanisms occur at the connection of X shape diagonal elements. Damper is installed to the center of the diagonals via bolts. When the lateral loads such as earthquake loads act onto the system, two of the bracing elements take compression force while the other two in tension. Damper located at the center of the diagonals transforms to a parallelogram shape from a square shape by deformation and dissipates energy. According to the experimental research done by Pall and Marsh (1982), structural behavior of friction dampers can be modeled by using Coulomb formula:

$$P = \mu N \operatorname{sgn}(\dot{u}) \tag{1}$$

where μ, N and $\dot{u}$ stand for friction coefficient, normal force and velocity respectively. It can be seen from the equation, that in Coulomb model, force and friction coefficient are not considered as time dependent values. On the other hand it is almost impossible to be valid in reality. Because of this reason, friction force is needed to be taken as a variable for structural design. Pall type friction dampers are used in this chapter. Pall type friction dampers are the first kind of the dampers which had been erected to structures in North America. For example, the biggest structure in the world regard to its settled area, Boeing Plane Factory was retrofitted using Pall friction dampers. The structure has 37 m height and 107 m opening and covers almost 98000 m^2 areas. 537 pieces of damper had been used with capacities varied between 330 and 880 kN. It was noted that the dampers dissipate 60% of the energy income due to earthquake (Vail et al., 2004). Using friction dampers have certain advantages. First of all, erection of dampers to the structure does not cause a necessity to evacuate the structure entirely. This provides a huge ease when it is comprised with adding shear wall method. Friction dampers are also replaceable unlike permanent elements such as RC shear walls. Analytical researches have shown that implanting friction dampers to the existing structure does not increase of base shear demand, which also means that the probability of strengthening the existing elements of structure is lower. On the other hand, it is really difficult to obtain the optimum design load for a friction damper. If the design load is lower than needed; dampers may activate by simple service loads such wind loads; if the design load is larger than needed, cyclic loadings cause a less number of cycles and both situations cause reduction in energy dissipation.

CHARACTERISTICS OF THE INVESTIGATED BUILDING AND THE INTERVENTION METHODS IMPLEMENTED

The only hidden performance objective of the conventional force-based linear analysis is to supply the life safety without a total collapse under design earthquake having 10 percent probability exceedance in 50 years. However no damage classification is either defined or achieved by performing a force-based linear analysis. It was the main tool to obtain the seismic safety for long time until the performance based design philosophy took place. The construction practice and design codes in Turkey as well as in numerous countries in the world has been led to design the buildings as non-ductile or limited ductile manner unintentionally for long time. Relatively low strength concrete, lack of confinement, non-satisfactory column cross-sections etc. have resulted in unable to deform as much as expected for such buildings. A real representative public building plan designed in a non-ductile manner contradicting with the TEC 2007 was used for analytical investigation. In Figure 2, 3D view, plan and section in longitudinal direction of the building are displayed. The building is accepted to be located in a high seismicity zone which is Level 1 according to the TEC2007 and effective ground motion acceleration, A_0, is 0.4. Site specific periods are $T_A = 0.15$ and $T_B = 0.6$ according to soil type (type C) and the site class (Z3) of the area. Live load participating factor is taken as 0.3 and building importance factor is regarded as 1 for the "as is" case.

The material strength values are selected according to old fashion practice for poor structural behavior. Characteristic concrete strength is 18 MPa and characteristic yield strength of longitudinal and transverse reinforcements are 220 MPa considering them as plain bars. In columns and beams, reinforcement contents as well as rebar details are assumed as suitable to the old fashion design codes. The rebar arrangement for the columns is 12ϕ14 and longitudinal reinforcement ratio is 0.008. The reinforcement detailing for the supports is 4ϕ14 for negative (top) side and 3ϕ14 for the positive (bottom) side in the beams. Stirrup arrangement is ϕ8 with a spacing of 150 mm both for the beams and the columns in the building with inadequate end hook condition (90° bent). Therefore, the concrete is considered in an unconfined condition for nonlinear cross-sectional behavior.

Dimensions of the beams are 60 x 25 cm and the columns are 70 x 30 cm for the entire structure. Although all the columns have unique dimensions, their orientation varies in the building. Plan view of the columns is shown in Figure 2. There exist three bays of which the spans are 5.25 m, 3.75 m, 5.25 m located in the global *X* direction which is transverse direction of the building. The story height is constant and 3 m for each single story.

In spite of the aforementioned numerous unfavorable aspects, such quite old typical public buildings generally have some positive features that built up with successive frames and having regular plans. Regarding to these favorable characteristics of the investigated building come along with the framing style, the analysis are carried out on a single frame spanning in the weak direction of the building. The structural strengthening elements and interventions are distributed in the building taking into account of convenient and appropriate seismic overall behavior. The selected frame analyzed in this study is also containing one vertically continuous intervention. Varying story numbers as 5, 7 and 9 are considered to present the effect of the change in total building height on performance assessment with different selected strengthening schemes.

Two intervention methods are selected namely RC wall addition and friction damper implementation as a conventional technique and a modern method respectively. Analyzed frame and implemented interventions are depicted in Figure 3 with the variation of the height. The RC shear wall added to the frame is designed with high ductility level according to TEC2007 and by using 30 MPa of compressive

Figure 2. Investigated building

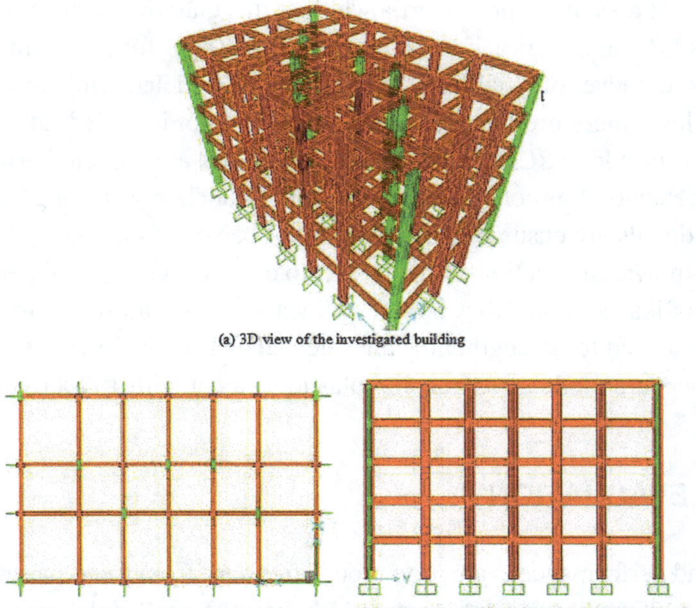

(a) 3D view of the investigated building

(b) Plan view of the investigated building

(c) Section in longitudinal direction of the investigated building

Figure 3. Schematic presentation of the analyzed frame

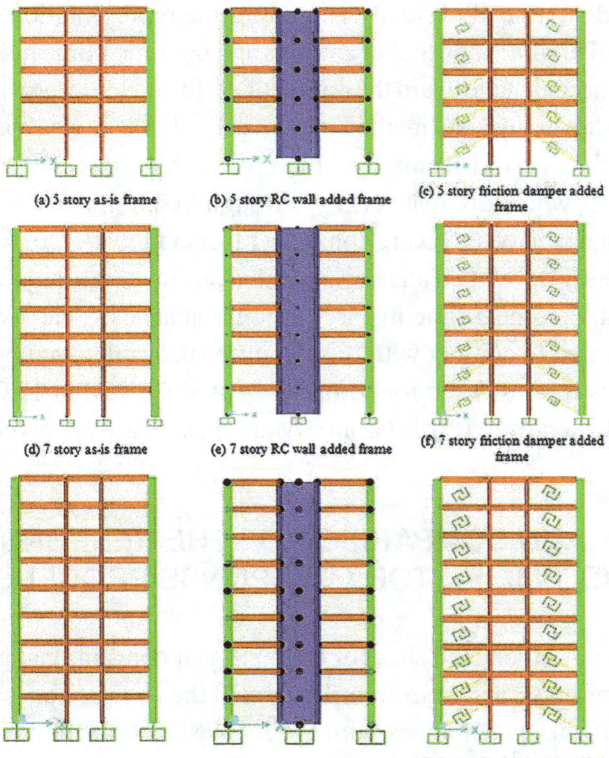

(a) 5 story as-is frame

(b) 5 story RC wall added frame

(c) 5 story friction damper added frame

(d) 7 story as-is frame

(e) 7 story RC wall added frame

(f) 7 story friction damper added frame

(g) 9 story as-is frame

(h) 9 story RC wall added frame

(j) 9 story friction damper added frame

concrete strength. Since the shear walls are introduced to building as new members, their moment of inertia are considered as gross moment of inertia whereas this quantity is taken into account as cracked section property for old RC members according to TEC2007 (0.4 I_g for the beams and 0.4-0.8 I_g for the columns depending on the level of axial load). The walls are modeled with equivalent frame elements and fixed at the base level therefore foundation assessment is not included into the investigation. The nodes in the vicinity of the added RC shear wall at the floor level are modeled using body constraints to achieve suitable wall behavior. It is considered that all related anchorages along the contact surfaces and proper reinforcement details are ensured and fully integral behavior supplied between the old and new RC elements. As it is known, the friction dampers start to dissipate energy and perform plastic behavior after a certain amount of load level and they behave as linear elements until that limit. Optimum slip loads for friction dampers that used for strengthening are selected as 150 kN for first three stories and 100 kN for the rest. The dampers are modeled as uniaxial plastic element with 192000 kN/m pre-slip stiffness.

PERFORMANCE EVALUATION

Seismic assessment and performance evaluation procedures vary from force-based methods to displacement-based techniques either linear or nonlinear. Although the aim of the performance evaluation is damage determination of the structure under considered earthquake, displacement and inter-story drift values are strong indicators for the overall behavior and structural performance. In this study, instead of final damage level, these displacement dependent indicators as well as story shear demands are evaluated and compared. RC infill wall and friction damper interventions individually investigated using nonlinear analytical analysis methods. Lumped plasticity is used for the representation of the nonlinearity of the RC elements. Conventional nonlinear pushover analysis and nonlinear time history response analysis are carried out for the evaluation and put forward the results of performance improvement regarding to "as-is" case. Even though the pushover analysis method can reveal only the overall behavior enough-well, it is thought that this quick and less effort required method is suitable for the preliminarily analysis of such structures including damper which dissipate energy through hysteretic behavior. Since the investigated building has regular plan and behave as flexure dominant manner globally, conventional pushover method is preferred instead of recently developed and sophisticated pushover techniques. More realistic and state of the art analysis method, nonlinear time history response analysis is carried out with three different real earthquake records scaled to comply with the response spectrum given by the code and maximum response among the results is considered for comparison as suggested in TEC2007. The details of the earthquakes which are obtained from PEER Ground Motion Database are given in Table 1 and Figure 4.

ANALYSIS RESULTS AND COMPARISON OF THE RESPONSES OF THE PUSHOVER AND TIME HISTORY RESPONSE PROCEDURES

Outcomes regarding to displacement and shear of the nonlinear conventional pushover and time history response procedures are presented in this section. The aim of the comparison is to find out the excellence of the investigated strengthening techniques with the increasing story number based of displacement and story shear performances.

Table 1. Details of the selected ground motion records

Earthquake	Date	Station	Moment Magnitude	Direction	Distance from Fault (km)	Duration (sec)	Scale Factor
Landers	28/06/1992	Yermo Fire	7.28	FP	23.6	43.9	2
Düzce	12/11/1999	Bolu	7.14	FP	12	55.89	0.8677
Kocaeli	17/08/1999	Duzce	7.51	FN	15.4	27.2	1.4282

Figure 4. Earthquake Records for Nonlinear Response History Analysis Procedure, a) Landers Earthquake Record, b) Duzce Earthquake Record, c) Kocaeli Earthquake Record

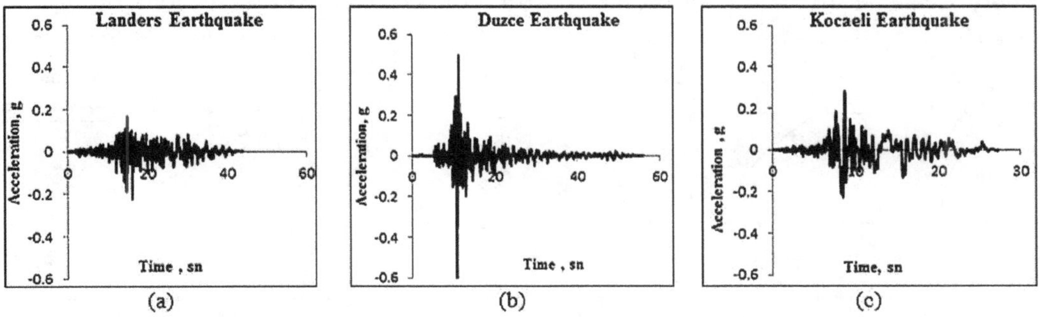

Figure 5 represents the results of the story numbers versus story shears of the investigated cases. The pushover procedure results are placed in the left side column and the outcomes of the time history response procedure are displayed in the right side column. It is clearly shown in the figures that addition of the RC shear wall increase the story shear demand as expected. In contrast, implementation of the friction damper does not change the story shear demand relative to as-is building even though the demand increases slightly at the lower stories with the increasing total number of building story number. The demand of the story shear trend explained in here is almost overlapped for both pushover and time history response procedures results, with different outcome values. It is obvious that enormously increased shear force demand will carried by added shear wall depending on the convenience of the workmanship and the application details of the implementation. In case of insufficient and inaccurate site application of the wall as well as design phase, increasing shear force demand affects the other structural elements and building performance in a dramatically negative manner. It should be noted that the story shear demand of the friction damper added buildings can be well predicted from a preliminary analysis by pushover procedure since the outcomes of the two procedures are similar.

The story displacement responses are shown in Figure 6. The pushover responses are displayed in left side column and the time history response procedure results are depicted in right side column. The results reveal that both of the intervention methods can be classified as effective strengthening methods for the improvement of the displacement performance. It is clearly understand from the figure that the pushover analysis is misleading in the displacement performance for the friction damper implemented buildings with the increasing building story number. However for the case of 5 or less story building application, it is thought that the pushover analysis can be a proper analytical tool for a preliminary analysis since the displacement performance responses are similar for both nonlinear analytical procedures.

Figure 5. Story shear responses, pushover and time history response procedures

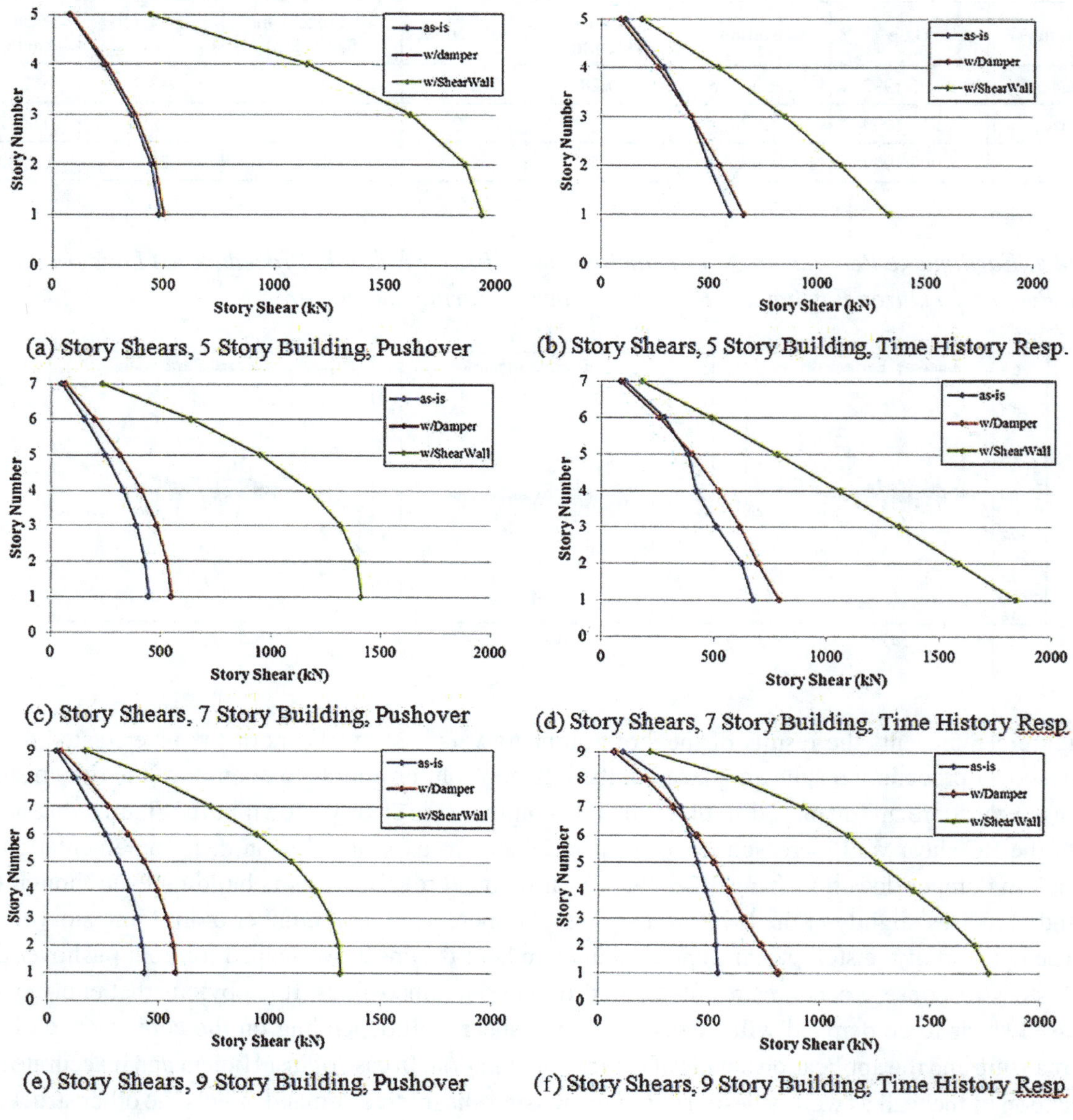

(a) Story Shears, 5 Story Building, Pushover

(b) Story Shears, 5 Story Building, Time History Resp.

(c) Story Shears, 7 Story Building, Pushover

(d) Story Shears, 7 Story Building, Time History Resp

(e) Story Shears, 9 Story Building, Pushover

(f) Story Shears, 9 Story Building, Time History Resp

Figure 7 displays interstory drift responses according to story numbers regard to pushover and time history response procedures. The left sided graphics belong to pushover analysis results whereas right sided graphics refer to time history responses. It is found out that the pushover procedure underestimate the interstory drift responses with the increasing total story numbers even though the behavioral trends are quite similar to the time history procedure outcomes. Both the strengthening methods are found to be useful intervention measures for the investigated type 7 and less story buildings. On the other hand realistic analysis procedure, time history response tool, reveals that the friction damper implementation is not a convenient method for the 9 story building with applied slip initiation load and/or selected stiffness.

Figure 6. Story displacement responses, pushover and time history response procedures

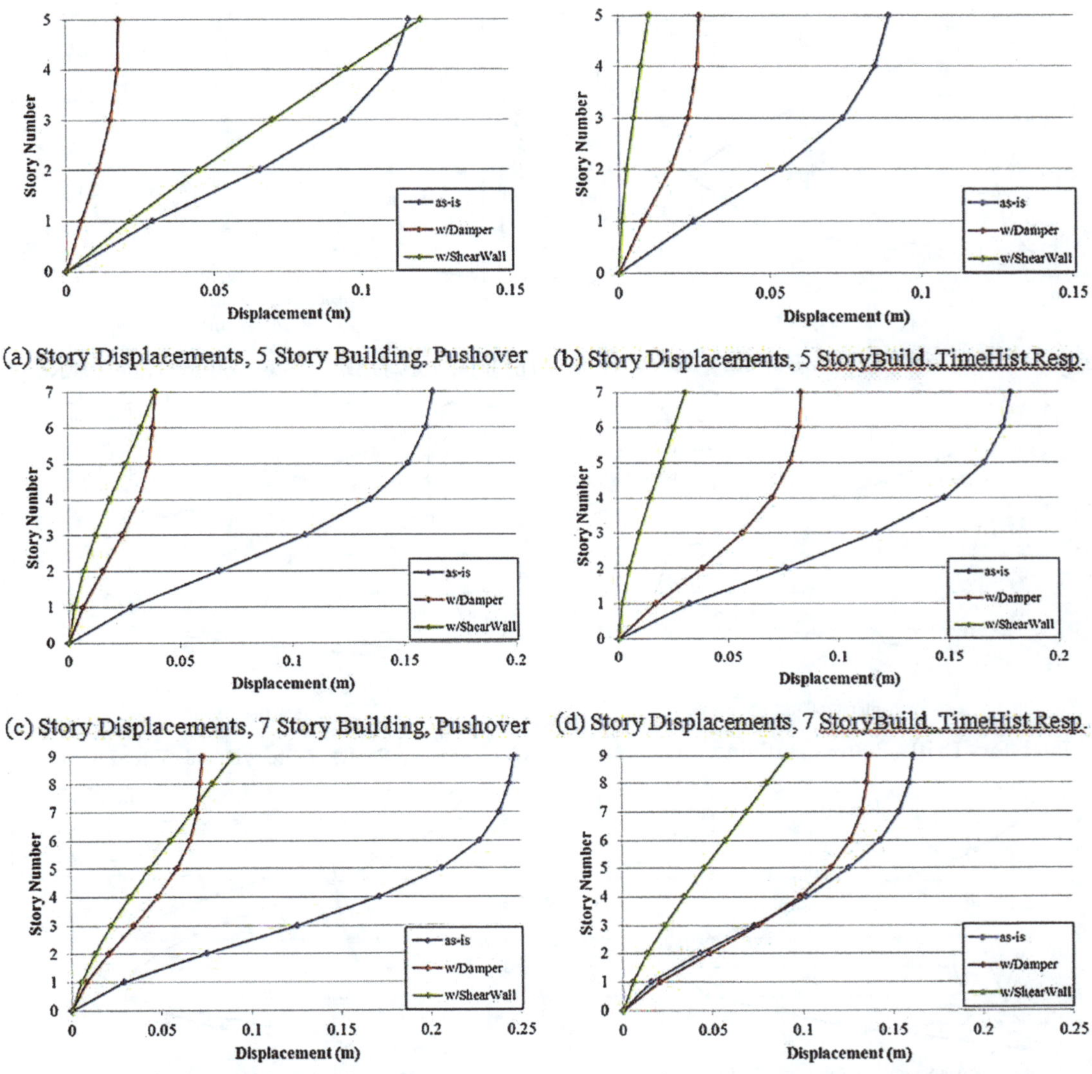

(a) Story Displacements, 5 Story Building, Pushover (b) Story Displacements, 5 StoryBuild. TimeHist.Resp.

(c) Story Displacements, 7 Story Building, Pushover (d) Story Displacements, 7 StoryBuild. TimeHist.Resp.

(e) Story Displacements, 9 Story Building, Pushover (f) Story Displacements, 9 StoryBuild. TimeHist.Resp.

CONCLUSION

Analysis results reveal that the friction damper installation is an effective alternative technique against the conventional methods for strengthening of the non-ductile reinforced concrete buildings. Structural performance indicators such as lateral story displacement and interstory drift for the friction damper inclusion are almost alike with the result of conventional RC wall addition measurement for the 5 story buildings. It is understood that life safety or collapse prevention performance targets in regard to design earthquake can be easily obtained using these devices without abandoning the occupancy and the building related activities. The overall performance and effectiveness of the damper implementation technique decrease with the increasing number of story of the building. It is thought that either the number of dampers or the slip initiation load and stiffness values should be reconsidered for the increasing total

Figure 7. Interstory drift responses, pushover and time history response procedures

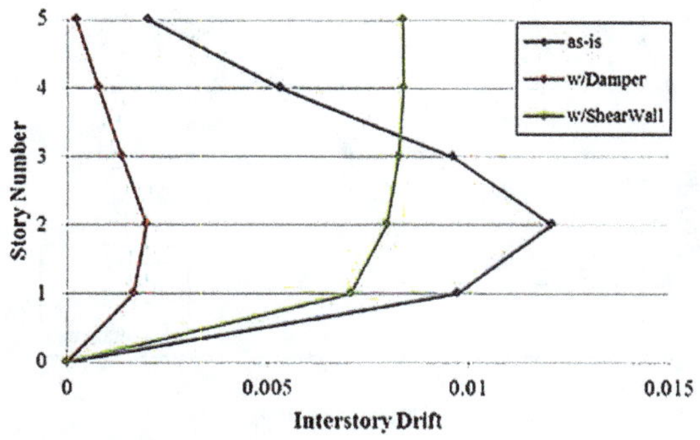

(a) Interstory Drifts, 5 Story Building, Pushover

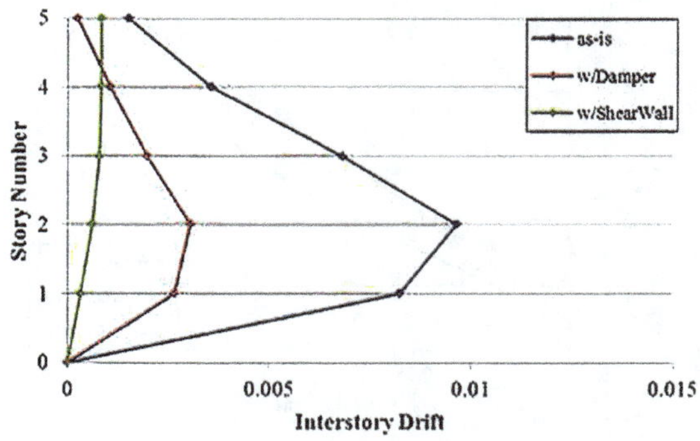

(b) Interstory Drifts, 5 StoryBuild.,TimeHist.Resp.

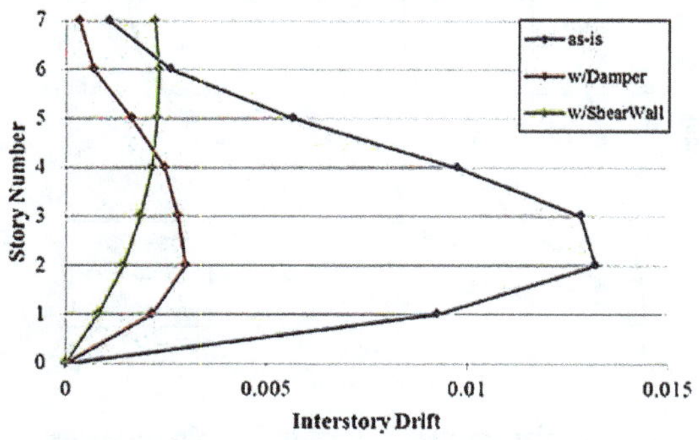

(c) Interstory Drifts, 7 Story Building, Pushover

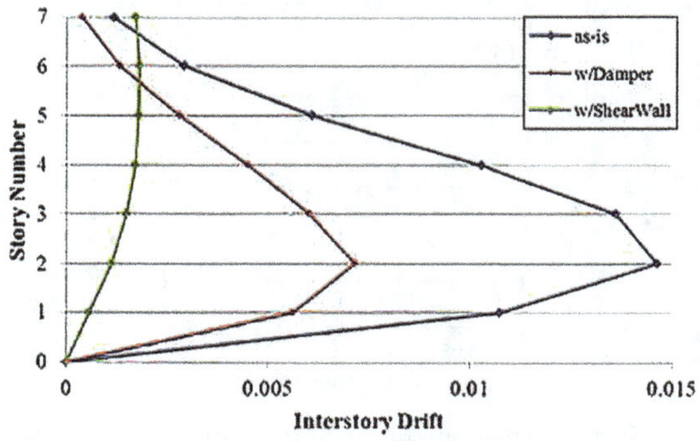

(d) Interstory Drifts, 7 StoryBuild.,TimeHist.Resp.

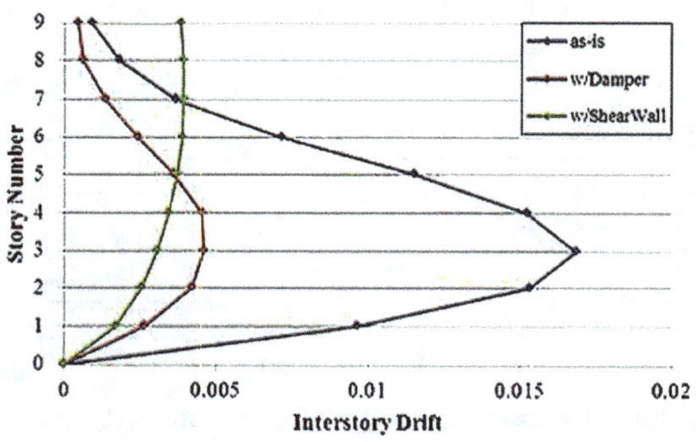

(e) Interstory Drifts, 9 Story Building, Pushover

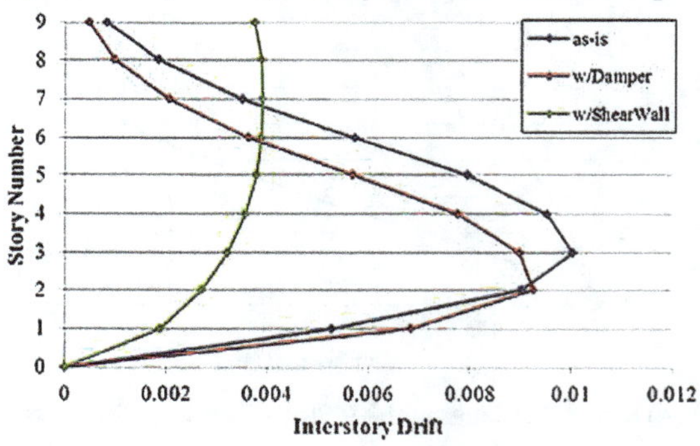

(f) Interstory Drifts, 9 StoryBuild.,TimeHist.Resp.

number of story. It is also revealed that story shear demand of the buildings with the friction damper does not shift as much as the demand responses come along with RC shear wall intervention. It is also figured out that pushover analysis procedure is a proper tool for preliminarily analysis of friction damper strengthening method for buildings with 5 or less stories. On the other hand, the increasing total story number adversely affects this shining capability even it vanishes with the 9 story building in regard to time history response procedure. It should be kept in mind that these outcomes are obtained from single 2D frame analysis and need to be confirmed with 3D models and varying damper specifications.

REFERENCES

Bruneau, M., & Saatciolu, M. (1993). Performance of Structures during the 1992 Erzincan Earthquake. *Canadian Journal of Civil Engineering*, *20*(2).

Clough, R. W., & Penzien, J. (1993). *Dynamics of Structures* (2nd ed.). New York: McGraw-Hill.

Constantinou, M. C., Soong, T. T., & Dargush, G. F. (1998). Passive Energy Dissipation Systems for Structural Design and Retrofit, MCEER Monograph No. 1.

Fardis, M. (2009). *Seismic Design, Assessment and Retrofitting of Concrete Buildings* (3rd ed.). New York: Springer. doi:10.1007/978-1-4020-9842-0

fib. T.G. 7.1. (2003). *Seismic Assessment and Retrofit of Reinforced Concrete Buildings: State-of-the-art report*. Bulletin 24, Sprint-Digital-Druck.

Mitchell, D., DeVall, R. H., Saatcioglu, M., Simpson, R., Tinawi, R., & Tremblay, R. (1995). Damage to concrete structures due to the 1994 Northridge earthquake. *Canadian Journal of Civil Engineering*, *22*(1).

Pall, A., & Pall, R. T. (2004). Performance-Based Design using Pall Friction Dampers – An economical design solution. *13th World Conference on Earthquake Engineering*, Vancouver, Canada.

Pall, A. S., & Marsh, C. (1982). Seismic Response of Friction Damped Braced Frames. *Journal of Structural Division, ASCE, St.*, *9*, 108.

Priestley, M. J. N., Calvi, G. M., & Kowalsky, M. J. (2007). *Displacement-Based Seismic Design of Structures*. Pavia, Italy: IUSS Press.

Saatcioglu, M., Mitchell, D., Tinawi, R., Gardner, N. J., Gillies, A. G., Ghobarah, A., & Lau, D. et al. (2001). The August 17, 1999, Kocaeli (Turkey) Earthquake Damage to Structures. *Canadian Journal of Civil Engineering*, *28*(4).

Soong, T. T., & Dargush, G. F. (1999). *Passive Energy Dissipation and Active Control, Structural Engineering Handbook* (4th ed.). CRC Press.

Symans, M. D., Charney, F. A., Whittaker, A. S., Constantinou, M. C., Kircher, C. A., Johnson, M. W., & McNamara, R. J. (2008). Energy Dissipation Systems for Seismic Applications: Current Practice and Recent Developments. *Journal of Structural Engineering*, *134*(1), 3–21. doi:10.1061/(ASCE)0733-9445(2008)134:1(3)

TEC. (2007). *Specification for Buildings to be Built in Seismic Zones 2007*. Ministry of Public Works and Settlement Government of Republic of Turkey.

Vail, C., Hubbell, J., O'Connor, B., King, J., & Pall, A. (2004). *Seismic upgrade of the Boeing Commercial Airplane Factory at Everett, WA, USA*. The 13th World Conference on Earthquake Engineering, Vancouver, British Columbia, Canada.

Chapter 6
Dynamic Stability and Post-Critical Processes of Slender Auto-Parametric Systems

Jiří Náprstek
Institute of Theoretical and Applied Mechanics AS CR, Czech Republic

Cyril Fischer
Institute of Theoretical and Applied Mechanics AS CR, Czech Republic

ABSTRACT

High-rise structures exposed to a strong vertical component of an earthquake excitation are endangered by auto-parametric resonance effect. While in a sub-critical state, the vertical and horizontal response components are independent. Exceeding a certain limit causes the vertical response to lose stability and induces dominant horizontal response. This effect is presented using two mathematical models: (1) the non-linear lumped mass model; and (2) the one dimensional model with continuously distributed parameters. Analytical and numerical treatment of both leads to three different types of the response: (1) semi-trivial sub-critical state with zero horizontal response component; (2) post-critical state (auto-parametric resonance) with a periodic or attractor type chaotic character; and (3) breaking through a certain limit, the horizontal response exponentially rises and leads to a collapse. Special attention is paid to transition from a semi-trivial to post-critical state in case of time limited excitation period as it concerns the seismic processes.

INTRODUCTION

This chapter is devoted to dynamic analysis of slender structures (towers, masts, chimneys, bridges, etc.) under pure vertical excitation acting in the base. Dynamic behaviour of such structures is a widely discussed topic in earthquake engineering literature. However, authors are dealing predominantly with an influence of horizontal excitation components. On the other hand, a strong vertical excitation component, especially in the earthquake epicentre area, can be decisive. Joint effect of horizontal and vertical components is rarely taken into account because the widely used linear approach usually does not

DOI: 10.4018/978-1-5225-2089-4.ch006

provide any interesting knowledge in such a case. In linear models, the vertical and horizontal response components are independent. If no horizontal excitation is taken into account, no horizontal response component is observed. The mutual interference of horizontal and vertical components of the response becomes visible only if the non-linear models are adopted. The corresponding non-linear model is then of an auto-parametric character. As a consequence, e.g., if the frequency of a vertical excitation in a structure foundation finds in a certain interval and its amplitude exceeds a certain limit, the vertical response component loses dynamic stability and dominant horizontal response component is generated — even if no horizontal excitation is assumed. This post-critical regime (auto-parametric resonance) follows from a non-linear interaction of vertical and horizontal response components that can lead to a failure of the structure. On the other hand, until the amplitude or frequency of excitation remain outside the critical values, the non-linear relations do not manifest themselves. The response remains in sub-critical linear regime and follows the stable semi-trivial (i.e., almost linear) solution.

Qualitatively different post-critical response types can be detected when sweeping throughout the frequency interval beyond the stability limit. Deterministic as well as chaotic response types were observed. In general, they can occur in a steady state. Quasi-periodic and completely undetermined regimes, including a possible energy transflux, occur between degrees of freedom.

Similar effects can be encountered in different branches. For instance, a railway car running on an imperfect track generating periodical motion can result in a stability loss and possible derailment of the car (although other causes of this effect are not negligible) (Náprstek, 2013). Another example is the capsizing of ships on a wavy surface due to a periodic change in the metacentric height of the vessel or due to non-linear coupling between heave-roll or pitch-roll motions (Nabergoj & Tondl, 1994; Nayfeh, Mook, & Marshall 1973; Tondl, Ruijgrok, Verhulst, & Nabergoj, 2000).

The whole problem is solved with two simplified non-linear models. One is the lumped mass model, where the elastic console is modelled by lumping the mass at the top and relating the mass to the basement by a massless spring. The other model assumes uniformly distributed stiffness and mass in order to respect the whole eigen-frequency spectrum. Both models serve their particular purposes. The lumped mass model is simpler and allows better analytical treatment to get a more detailed insight into its internal structure. The continuous model offers greater flexibility with respect to sensitivity to individual eigen-frequencies.

The outline of the chapter is as follows. First, the mathematical models of the structure are defined in the form of Lagrange equations. Although the massless spring and continuous console bending is considered linear, the models comprise non-linear relation between displacement and rotation of the base point and the console bending. The next section presents theoretical formulation of the semi-trivial solution and stability limits for both models. The linear perturbation approach is used. The response and perturbations are assumed in the form of harmonic functions. Existence of the stationary solution is assessed. Influence of system parameters for both models is studied. This is followed by a section discussing various types of the post-critical response. The assessment is mostly based on numerical evaluations of the stability conditions for various parameters. The case when the stability condition is overstepped is examined. It is shown that the system is able to withstand a limited duration of excitation exceeding the bounds defined by the stability conditions. If the excitation is stopped, the system is able to return to a standstill. Conclusions are presented in the last section.

BACKGROUND

Auto-parametric systems have been intensively studied for the last three or four decades. These investigations are motivated by various technical branches and by basic theoretical research in rational mechanics. A few theoretical studies dealing with these effects have been published sooner, namely in the period of 1968-1985 (Haxton & Barr, 1974; Nayfeh et al., 1973; van der Burgh, 1968). Following this pioneering time, many papers contributing to analytical, numerical, and experimental aspects of auto-parametric systems have been published (mostly by Tondl, Nabergoj, and co-authors) (Nabergoj, Tondl, & Virag, 1994; Tondl, 1991, 1997; Tondl et al., 2000). Comprehensive overview of partial results and methods appeared in monographs by Guckenheimer and Holmes (1983) or Hatwal, Mallik, and Ghosh (1983). Basic definitions and results on a level of the rational mechanics can be found in the monograph by Chetayev (1961). For extension onto the stochastic approach, see Bolotin (2013), Pugachev and Sinitsyn (1987), Lin and Cai (1995) or Náprstek (1996,1998).

Some papers dealing with auto-parametric systems close to the system presented in the current work under deterministic and random excitation were recently published by the authors of this chapter (Náprstek & Pirner, 2002; Náprstek & Fischer, 2008, 2009, 2011). The mathematical models used in these studies idealized the vertical structure as one concentrated mass related with the basement by a massless spring. Selected results are included in the current work. However, follow-up research (Náprstek & Fischer, 2012) revealed that such approach is not satisfactory in many particular cases. In principle, easily deformable tall structures are the most sensitive regarding effects of auto-parametric resonance. Therefore, the structure should be modelled as a console with continuously distributed stiffness and mass in order to respect the whole eigen-frequency spectrum. This methodology is referred by the second model used in the presented text. Concerning subsoil, conventional model including internal viscosity can be retained.

The excitation caused by an earthquake was traditionally limited to a pure horizontal movement in engineering literature. With the growth of computational abilities, such limitation gradually abandoned. This process could lead to including all six degrees of freedom into consideration (Zembaty, 2013). The effects of vertical ground acceleration were not taken into account as the peak vertical acceleration was considered definitely lower than its horizontal counterpart (Newmark, Blume, & Kapur, 1973) until the early 1990s even though aggravating effect of the vertical ground acceleration was known before (Shih & Lin, 1982). Investigations by Elnashai and co-workers (Broderick & Elnashai, 1995; Elnashai & Papazoglou, 1996) identified observed failure modes as being caused dominantly by vertical ground motion (Ambraseys & Douglas, 2000). Several authors, including Bozorgnia, Niazi, and Campbell (1995), pointed out that vertical-to-horizontal ratio (V/H) of selected near-field strong motion records have significantly overstepped the widely assumed 2/3 ratio based originally on recommendation by Newmark and Hall (1978). Importance of the V/H ratio is regularly confirmed in papers dealing with its significance to engineering design (Bozorgnia & Campbell, 2004; Salazar & Haldar, 2000; Silva, 1997) or in reviews of particular earthquake effects where the V/H ratio is usually mentioned (Kalkan, Adalier, & Pamuk, 2004). Carydis, Castiglioni, Lekkas, Kostaki, Lebesis and Drei (2012) reported observation of damages caused by purely vertical ground motion. Such kind of vertical motion is created by local conditions, including reflections, interferences, and other phenomena. However, it has to be taken into account for serious assessment of slender structures.

From the mathematical point of view, problem of the vertical stability of slender structures is similar to problem of an inverted pendulum. Whereas the classic inverted pendulum problems studies stability conditions of the unstable upward position, the case of slender structures is set up in the opposite direction: the upward position is stable thanks to non-zero (in fact very significant) stiffness in the base (or pivot) of the structure, which keeps the structure erect. Vertical movement of the pivot of the pendulum or basement of the structure can have a stabilising or destabilising effect, depending on system parameters as stiffness in the basement, excitation frequency and amplitude, etc. Dynamics of similar systems is usually very complicated even when only a few degrees of freedom are involved. This shows that the complex behaviour arises from the non-linearity of the equation of motion. Theoretical and experimental examination of the stability properties of inverted pendulum-like systems is generally well established. The first theoretical work dates back to origin of the twentieth century (Stephenson, 1908), however, the stability of the upward position was first analysed by the method of averaging by Kapitsa (1951). Vast number of theoretical works appeared since that time. Numerous works on experimental investigation can be represented, e.g., by the paper by Smith and Blackburn (1992). A dynamical stability of the inverted pendulum is a worthy topic of various demonstrations, which are available in the internet.

Models presented in the current chapter differ from the usual inverted pendulum by adding extra degrees of freedom representing bending of the console and deformation of the subsoil. These extra variables introduce an additional complexity to the model, together with terms describing damping and stiffness in the base point. The Mathieu's equation describing the classical inverted pendulum does not cover such a complex case. However, the approach used in the present chapter follows procedures applicable for the inverted pendulum analysis and the reader is assumed to be familiar with such techniques, see monographs by Arnold (1978), Nayfeh and Mook (2008) or Tondl, Ruijgrok, Verhulst and Nabergoj (2000).

MATHEMATICAL MODEL

The two simplified theoretical models used in this chapter are outlined in Figure 1. Both are assumed in a vertical plane only, taking into account only one horizontal component. Damping is assumed as proportional and harmonic excitation is assumed in the base point. Both models are described by the vertical displacement and angular rotation of the system in the B point $y(t)$, $\varphi(t)$, and bending deformation of the vertical console. In case of the lumped-mass 3-DOF model (referred to as model A in the following text, Figure *1*a) is the bending deformation denoted as $\xi(t)$, in case of the continuously distributed mass of the console (model B, Figure *1*b) the bending deformation depends on the longitudinal coordinate (height) of the console and is denoted as $u(x,t)$.

Both systems are Hamiltonian so that the governing differential systems can be deduced in the form of Lagrange equations. The kinetic and potential energies of the moving system A are formulated as follows:

$$T = \frac{1}{2}M\left(\dot{y}^2 + r^2\dot{\varphi}^2\right) + \frac{1}{2}m\left(\dot{y}^2 + (l\dot{\varphi} + \dot{\xi})^2 - 2\dot{y}\left(l\dot{\varphi} + \dot{\xi}\right)\sin\varphi\right)$$

$$U = Mg \cdot y + \frac{1}{2}C\left((y - y_0)^2 + r^2\varphi^2\right) + mg\left(y - l\left(1 - \cos\varphi\right) - \xi\sin\varphi\right) + \frac{3EJ}{l^3}\cdot\xi^2, \tag{1}$$

and for model B:

Figure 1. Outline of the auto-parametric 3-DOF models used: a) lumped mass model A and b) model with uniformly distributed mass of the vertical console – model B

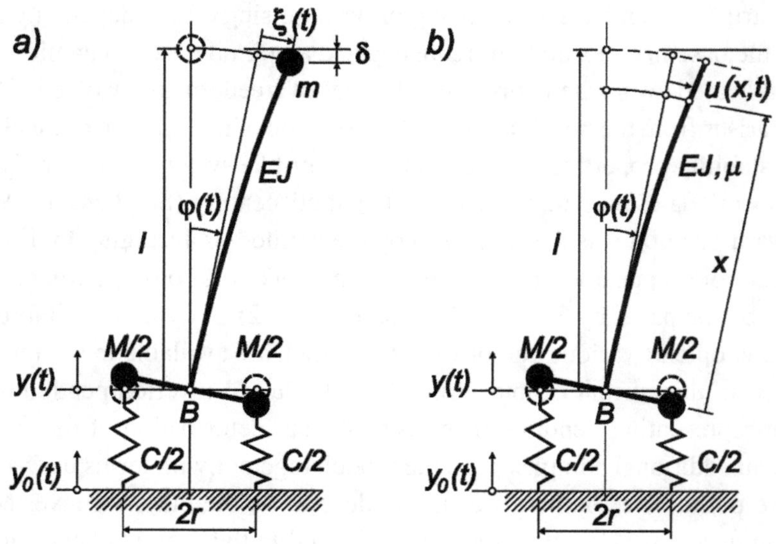

$$T = \frac{1}{2}M\left(\dot{y}^2 + r^2\dot{\varphi}^2\right) + \frac{1}{2}\mu\int_0^l \left[\left(\dot{\varphi}x + \dot{u}\left(x,t\right)\right)^2 + \dot{y}^2 - 2\dot{y}\left(\dot{\varphi}x + \dot{u}\left(x,t\right)\right)\sin\varphi\right]\mathrm{d}x,$$

$$U = Mg\cdot y + \frac{1}{2}C\left(\left(y-y_0\right)^2 + r^2\varphi^2\right) + \mu g\int_0^l \left[y - x\left(1-\cos\varphi\right) - u\left(x,t\right)\sin\varphi\right]\mathrm{d}x$$
$$+\frac{1}{2}EJ\int_0^l u''^2\left(x,t\right)\mathrm{d}x. \qquad (2)$$

The following notation has been introduced:

- $y = y(t)$: vertical displacement of the B point;
- $y_0 = y_0(t)$: kinematic excitation (seismic process);
- x :length coordinate along the console;
- $\varphi = \varphi(t)$: angular rotation of the system in the B point;
- $\xi = \xi(t), u(x,t)$: bending deformation of the vertical console;
- m, M :foundation and structure effective masses;
- C, EJ : subsoil stiffness, console bending stiffness;
- η_c, η_e : viscous components of the C, EJ stiffness (Kelvin);
- r, l: geometric parameters.

Non-dimensional response and excitation components are useful to be introduced:

$$\zeta_0\left(t\right)=\frac{y_0(t)}{l},\ \zeta(t)=\frac{y(t)}{l},\ \varphi(t),\ \xi=\frac{x}{l},\ \rho=r\,/\,l,\ m=\mu l \tag{3}$$

and for individual models:

Model A: $\psi\left(\xi,t\right)=\varphi(t)+\dfrac{\xi(t)}{l}$

Model B: $\psi\left(\xi,t\right)=\dfrac{u(x,t)}{l}$ (4)

The expression $\psi(\xi,t)$ for model A comprises non-dimensional deformation together with the rocking component $\varphi(t)$ because this substitution leads to advantageous formulation of the model.

When the Lagrange equations are deduced the approximation reflecting an early post-critical state can be adopted:

$$\sin\varphi\approx\varphi\,;\quad \cos\varphi\approx 1. \tag{5}$$

For the lumped mass 3DOF model A, the governing differential system can be formulated as follows:

$$\ddot{\zeta}-\kappa_0\frac{\mathrm{d}}{\mathrm{d}t}\left(\dot{\psi}\varphi\right)+\omega_0^2\left(\zeta-\zeta_0+\eta_c\left(\dot{\zeta}-\dot{\zeta}_0\right)\right)=0, \tag{6a}$$

$$\ddot{\varphi}-\kappa_{1A}\frac{\mathrm{d}}{\mathrm{d}t}\left(\dot{\zeta}\varphi\right)+\kappa_{1A}\ddot{\psi}+\kappa_1\dot{\zeta}\dot{\psi}-\kappa_{1A}\omega_2^2\psi+\omega_{1A}^2\left(\varphi+\eta_c\dot{\varphi}\right)=0, \tag{6b}$$

$$\ddot{\psi}-\frac{\mathrm{d}}{\mathrm{d}t}\left(\dot{\zeta}\varphi\right)+\omega_{3A}^2\left(\psi-\varphi+\eta_e\left(\dot{\psi}-\dot{\varphi}\right)\right)=0, \tag{6c}$$

Where both dot and $\dfrac{\mathrm{d}}{\mathrm{d}t}$ mean differentiation according to time and:

$$\kappa_0=\frac{m}{M+m},\quad \kappa_{1A}=\frac{m\cdot l^2}{M\cdot\rho^2},$$
$$\omega_0^2=\frac{C}{M+m},\quad \omega_{1A}^2=\frac{C}{M},\quad \omega_2^2=\frac{g}{l},\quad \omega_{3A}^2=\frac{6EJ}{m\cdot l^3}. \tag{7}$$

In model B, due to assumption of proportional damping of the console, the deformation can be expressed in the form of a convergent series:

$$u(x,t)=\sum_{i=1}^{n}\alpha_i(t)\cdot\psi_i(x)\text{ or dimensionless: }\psi(\xi,t)=\sum_{i=1}^{n}\alpha_i(t)\cdot\chi_i(\xi) \tag{8}$$

where the basis functions $\chi_i(\xi)$ are eigen-functions (mode shapes) of the differential equation:

$$\chi_i^{\mathrm{iv}}(\xi)+\lambda_i\chi_i(\xi)=0,\quad (\lambda_i/l)^4=\mu\omega_i^2/EJ \tag{9}$$

with boundary conditions valid for a console beam:

$$\chi_i(0)=0,\ \chi_i'(0)=0, \chi_i''(1)=0, \chi_i'''(1)=0\,.$$

Under assumption of proportional damping are the time coordinates $\alpha_i(t)$ independent and so the phase shift of each mode shape is constant over the whole definition interval if the damping is sub-critical.

Thus, the system of Lagrangian equations of model B reads:

$$\ddot{\zeta}-\frac{1}{4}\kappa_0\frac{\mathrm{d}^2}{\mathrm{d}t^2}(\varphi^2)-\kappa_0\sum_{i=1}^{n}\left[\frac{\mathrm{d}}{\mathrm{d}t}(\varphi\dot{\alpha}_i)\theta_{0,i}\right]+\omega_0^2\left[\zeta-\zeta_0+\eta_c\left(\dot{\zeta}-\dot{\zeta}_0\right)\right]=0 \tag{10a}$$

$$\ddot{\varphi}-\frac{1}{2}\kappa_{1B}\ddot{\zeta}\varphi+\kappa_{1B}\sum_{i=1}^{n}\left[\ddot{\alpha}_i\theta_{1,i}+\left(\dot{\zeta}\dot{\alpha}_i-\omega_2^2\alpha_i\right)\theta_{0,i}\right]+\omega_{1B}^2\left(\varphi+\eta_c\dot{\varphi}\right)=0 \tag{10b}$$

$$\ddot{\alpha}_i\theta_{2,i}+\ddot{\varphi}\theta_{1,0}-\left[\frac{\mathrm{d}}{\mathrm{d}t}(\dot{\zeta}\varphi)+\omega_2^2\varphi\right]\cdot\theta_{0,i}+\omega_{3B}^2\left(\alpha_i+\eta_e\dot{\alpha}\right)\theta_{3,i}=0 \tag{10c}$$

where in addition to (7) it has been denoted:

$$\kappa_{1B}=\frac{m}{M\rho^2+m/3},\qquad \omega_{1B}^2=\frac{C\rho^2}{M\rho^2+m/3},\qquad \omega_{3B}^2=\frac{EJ}{ml^3}, \tag{11}$$

$$\theta_{0,i}=\int_0^1\chi_i(\xi)\mathrm{d}\xi,\qquad \theta_{1,i}=\int_0^1\xi\chi_i(\xi)\mathrm{d}\xi,\qquad \theta_{2,i}=\int_0^1\chi_i^2(\xi)\mathrm{d}\xi,\quad \theta_{3,i}=\int_0^1\left(\chi_i''(\xi)\right)^2\mathrm{d}\xi.$$

Regarding parameters $\theta_{j,i}$, the eigen-functions of Equation (9) with respective boundary conditions have detailed form as follows:

$$\chi_i(\xi)=\left(C_1\cdot\cos\lambda_i\xi+C_2\cdot\sin\lambda_i\xi+C_3\cdot\mathrm{ch}\lambda_i\xi+C_4\cdot\mathrm{sh}\lambda_i\xi\right)$$

$$\begin{aligned} C_1 &= \sin\lambda_i \mathrm{sh}\lambda_i, & C_2 &= -\sin\lambda_i \mathrm{ch}\lambda_i - \cos\lambda_i \mathrm{sh}\lambda_i, \\ C_3 &= -\sin\lambda_i \mathrm{sh}\lambda_i, & C_4 &= \sin\lambda_i \mathrm{ch}\lambda_i + \cos\lambda_i \mathrm{sh}\lambda_i, \end{aligned} \tag{12}$$

where $\lambda_i = 1.8751,\ 4.6941,\ 7.8548,\ 10.9955$, etc., are solutions of the transcendent equation:

$$\mathrm{ch}\lambda_i \cdot \cos\lambda_i + 1 = 0\,.$$

In principal, the analytical form of parameters $\theta_{j,i}$ can be carried out. However, the results are very complicated and do not provide any information important from physical point of view. They will be replaced by numerical values in particular cases.

The system (10) represents a simultaneous differential system for $\zeta(t), \varphi(t)$, and $\alpha_i(t)$; its size is related to a number of mode shapes (9) taken into account. Although the console bending is considered linear, components $\alpha_i(t)$ are non-linearly related with $\zeta(t), \varphi(t)$. Nevertheless, a mutual link of $\alpha_i(t)$ components is not complicated. This fact follows from the linearity of the bending component, proportionality of its damping and of the orthogonality of relevant mode shapes χ_i as well as their second derivatives χ_i'' in the meaning of Equation (9) and respective boundary conditions. Concerning the excitation process $\zeta_0(t)$, it will be considered as harmonic in the first step in order to investigate limits of stable semi-trivial and post-critical regimes. Later the random non-stationary character of $\zeta_0(t)$ could be respected.

STABILITY OF SEMI-TRIVIAL SOLUTION

The dimensionless excitation process $\zeta_0(t)$ in both models A, B will be considered as harmonic:

$$\zeta_0 = a_0 \cdot \sin \omega t\,. \tag{13}$$

If the stationary semi-trivial solution exists, its general form can be written as follows:

$$\zeta_s(t) = a_c \cdot \cos \omega t + a_s \cdot \sin \omega t\,, \quad \varphi = 0\,, \quad \alpha_i = 0\,. \tag{14}$$

The semi-trivial solution reflects neither rotation nor bending component, thus this form and properties is common to both assumed models. Substituting Equation (14) into the system (6) or (10) equations (b) and (c) are satisfied identically while the equations (a) in both cases provide the coefficients a_c, a_s:

$$a_c = -\frac{a_0\omega_0^2}{\delta}\omega^3\eta_c\,, \tag{15a}$$

$$a_s = \frac{a_0\omega_0^2}{\delta}\left(\omega_0^2 - \omega^2 + \omega_0^2\omega^2\eta_c^2\right) \tag{15b}$$

where $\delta = (\omega^2 - \omega_0^2)^2 + \omega_0^4\omega^2\eta_c^2$. Expression (14) together with coefficients (15) represents an approximate simple linear stationary solution of the single degree of freedom (SDOF) system moving in vertical direction being excited kinematically in the point B. The resonance curve of the response amplitude has the form:

$$R_0^2 = a_c^2 + a_s^2 = \frac{a_0^2\omega_0^4}{\delta}\left(1 + \omega^2\eta_c^2\right) \tag{16}$$

which can be seen in Figure 2. However, the solution characterized by this curve can be unstable for excitation amplitude exceeding a certain value a_0 in some intervals of the excitation frequency ω. For this reason the stability analysis must be carried out.

In order to assess the stability limits of the semi-trivial solution (14), the linear perturbation approach will be adopted. Hence, it can be written approximately in the arbitrarily small neighbourhood of the semi-trivial solution:

$$\zeta(t) = \zeta_s(t) + q(t) \tag{17a}$$

$$\varphi(t) = 0 + p(t) \tag{17b}$$

where the perturbations take the harmonic form:

$$q(t) = q_c(t)\cos\omega t + q_s(t)\sin\omega t$$

$$p(t) = p_c(t)\cos\frac{1}{2}\omega t + p_s(t)\sin\frac{1}{2}\omega t\,. \tag{18}$$

The bending component has to be treated separately for both assumed models:

$$\psi(t) = 0 + s(t) = s_c(t)\cos\frac{1}{2}\omega t + s_s(t)\sin\frac{1}{2}\omega t \tag{19a}$$

$$\alpha_i(t) = 0 + s_i(t) = s_{c,i}(t)\cos\frac{1}{2}\omega t + s_{s,i}(t)\sin\frac{1}{2}\omega t. \tag{19b}$$

The absolute values of the perturbation amplitudes are assumed as small compared to semi-trivial solution. The argument (t) will be omitted in further text whenever possible for $\zeta, \varphi, \alpha_i, q, q_c, q_s, \ldots$, etc.

Introducing expression (17a) into Equation (6a) or (10a) and taking into account that ζ_s represents its semi-trivial solution (14), following relation for perturbation q can be extracted:

$$\ddot{q} + \omega_0^2\left(q + \eta_c \dot{q}\right) = 0 . \tag{20}$$

Equation (20) is linear and homogeneous. It is obvious that $\lim_{t\to\infty} q_c, q_s = 0$ if $\eta_c > 0$ and thus stationary solution vanishes. For this reason, the vertical response component ζ remains independent and stable in the neighbourhood of the semi-trivial solution ζ_s (on the level of the linear perturbation approach).

Regarding the expressions b, c of (6) or (10), the relevant perturbations $q, p, s,$ or s_i should be introduced into (6a) or (10a), respectively. Keeping only the linear terms of perturbations p, s and respecting that $q \equiv 0$, the following differential system can be obtained for model A:

$$\begin{aligned} \ddot{p} + \left(\kappa_{1A}\omega_{3A}^2\eta_e + \omega_{1A}^2\eta_c\right)\dot{p} - \kappa_{1A}\left(\omega_{3A}^2\eta_e - \dot{\zeta}_s\right)\dot{s} + \left(\kappa_{1A}\omega_{3A}^2 + \omega_{1A}^2\right)p - \kappa_{1A}\left(\omega_2^2 + \omega_{3A}^2\right)s = 0 \\ \ddot{s} + \ddot{\xi}_s p - \left(\omega_{3A}^2\eta_e + \dot{\zeta}_s\right)\dot{p} + \omega_{3A}^2\eta_e\dot{s} - \omega_{3A}^2 p + \omega_{3A}^2 s = 0 \end{aligned} \tag{21}$$

and for model B:

$$\begin{aligned} \ddot{p} - \frac{1}{2}\kappa_{1B}\ddot{\zeta}_s p + \kappa_{1B}\sum_{i=1}^{n}[\ddot{s}_i\theta_{1,i} + \dot{\zeta}_s\dot{s}_i\theta_{0,i} - \omega_2^2 s_i\theta_{0,i}] + \omega_{1B}^2\left(p + \eta_c\dot{p}\right) = 0 \\ \ddot{s}_i\theta_{2,i} + \ddot{p}\theta_{1,i} - \left(\ddot{\zeta}_s p + \dot{\zeta}_s\dot{p}\right)\theta_{0,i} - \omega_2^2 p\theta_{0,i} + \omega_{3B}^2\left(s_i + \eta_e\dot{s}_i\right)\theta_{3,i} = 0 \end{aligned} . \tag{22}$$

The systems (21) and (22) is linear similarly like Equation (20). However, three coefficients include harmonic components due to $\ddot{\zeta}_s, \dot{\zeta}_s$ terms being given by Equation (14). Hence the system (22) is of the Mathieu type (with parametric excitation) and its solution stability should be verified (Abarbanel, Brown, & Kadtke, 1990; Xu & Cheung, 1994).

The harmonic balance procedure assumes perturbations in the harmonic form, Equations (18) or (19). The number of unknown functions (i.e., $q_c, q_s, p_c, p_s, s_s, s_c,$ or $s_{c,i}, s_{s,i},$) is doubled by this step. It is necessary to formulate corresponding number of additional conditions. The common choice in the harmonic balance procedure is to prescribe first derivatives of p, s, s_i in the form:

$$\dot{X}\left(t\right) = \frac{1}{2}X_s\left(t\right)\omega\cos\frac{1}{2}\omega t - \frac{1}{2}X_c\left(t\right)\omega\sin\frac{1}{2}\omega t \quad \text{for } X \in \left\{p, s, s_i\right\} . \tag{23}$$

In order to check stability of the semi-trivial solution, the perturbations *q, p, s,* or s_i are assumed to be small and almost constant within one period $\frac{2\pi}{\omega}$, so that the non-linear terms involving mutual multiplication or higher powers of the small components can be neglected. The harmonic balance procedure consists in multiplying of the respective equations by harmonic terms $\sin\frac{1}{2}\omega t$, $\cos\frac{1}{2}\omega t$ and integration over one period. The harmonic terms cancel themselves and the procedure leads to an algebraic relation:

$$Hu = 0 \tag{24}$$

where **u** is the vector of unknown functions, $\mathbf{u} = (p_s, p_c, s_s, s_c)^T$ or $\mathbf{u} = (p_s, p_c, s_s, 1, s_{c,1}, \ldots s_{s,n}, s_{c,n})^T$ in case of model A or B, respectively, and **H** is the coefficient matrix of dimensions 4×4 or $2(n+1) \times 2(n+1)$ for models A or B, respectively. Nonzero solution **u** cannot exist unless the matrix **H** is singular (i.e., its determinant vanishes, $\det(\mathbf{H}) = 0$). The zero determinant of the system matrix will lay out the shape of the stability limit.

Detailed Relations for Model A

For the lumped mass model A is the particular form of the system matrix **H** given in the Appendix, Equation (41). Its determinant can be evolved with respect to amplitude R_0^2 — see Equation (16) — and written in the form of the scalar homogeneous equation:

$$\kappa_{1A}^2 \omega^8 R_0^4 - 2\omega^4 \kappa_{1A} A_2(\omega) R_0^2 + A_0(\omega) = 0 \tag{25}$$

where the polynomial coefficients $A_i(\omega)$ are listed in the Appendix, Equation (43).

Equation (25) represents a quadratic equation for vertical amplitude R_0^2. Only a real positive solution is useful. Negative or complex conjugate roots should be avoided. Conditions for parameters (7) and variable frequency ω could be carried out to fulfil these requests. Consequently, $R_0^2 > 0$ as a function of ω consists in a general case either of two or one branches or does not exist. Number of real roots of (25) is governed by behaviour of the discriminant, which has the following form:

$$\left(B_{14}\omega^6 + B_{12}\omega^4 + B_{10}\omega^2 + B_8\right)\omega^8 . \tag{26}$$

The coefficient B_{14} is always negative; B_8 always positive. Thus, for small ω there exist two real values of R_0^2. On the other hand, for large ω is the discriminant negative and the semi-trivial solution is stable. According to actual values of the other coefficients, the real values of R_0^2 can occur for one or two intervals of ω.

Two typical examples have been plotted in Figure 2. Resonance curves following Equation (16) for a few excitation amplitudes a_0 are plotted using solid lines; stability limits described by the Equation (25) are shown in dashed lines for a set of fixed parameters of the system (all parameters used are written in the caption). The shape of resonance curves depends on structural parameters: subsoil stiffness and total weight of foundation and structure, its scale is mainly affected by the amplitude of the excitation process. On the other hand, the stability limits do not depend on load parameters. For given structural parameters and increasing excitation amplitude one or two intervals of unstable frequencies emerge.

Figure 2. Model A: Relation of resonance curves (solid) and stability limits (dashed); plot a): $M = 3990$, $m = 15$, $\eta_c = 0.2$, $\eta_e = 0.2$, $l = 20$, $g = 9.81$, $C = 4000 > gml / r^2 = 2943$; $EJ = 10000$, $a_0 = 0.2 \Rightarrow s_1 = 0.4548, s_2 = 0.5718, s_3 = 0.8671$,*plotb):* $m = 10$; $C = 4000 > gml / r^2 = 1962$; , $a_0 = 0.1 \Rightarrow s_1 = 0.7652$, $s_4 = 1.0396$

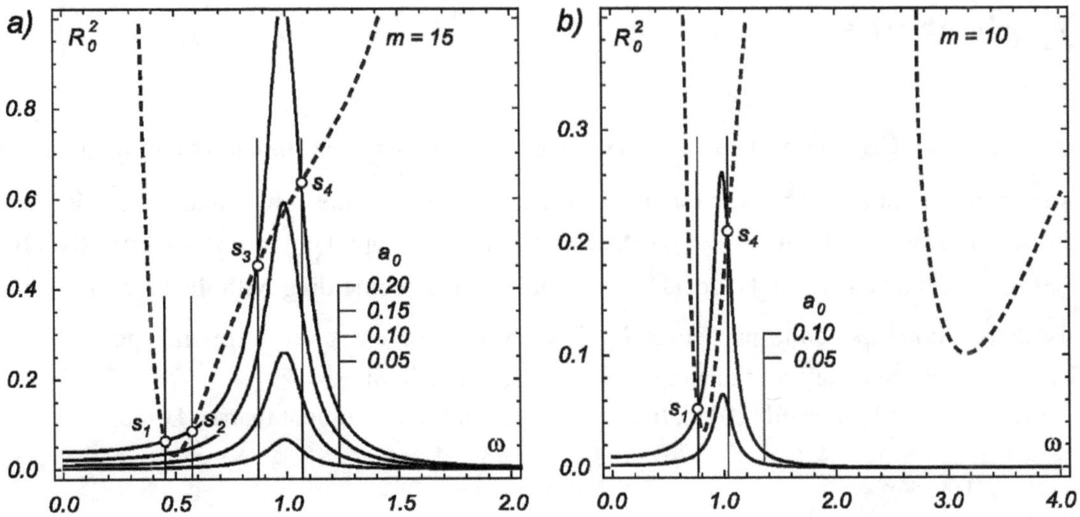

Points s_1, s_2 and s_3, s_4 in the left picture a) represent lower and upper limits of intervals where the semi-trivial solution becomes unstable and post-critical response should be investigated. It is obvious that the most sensitive interval where the stability loss can be expected is in the domain of the basic eigen-frequency ω_0 as can be seen in the left picture, interval $\omega \in (s_3, s_4)$, and also in the right picture b), interval $\omega \in (s_1, s_4)$. However, the stability limit indicates also another instability interval $\omega \in (s_1, s_2)$ in lower frequency range. This fact is related with a hypothetic "eigen-frequency" in the component φ. The right picture b) demonstrates two separate parts of the particular stability limit.

Detailed Relations for Model B

In case of model B, for given number of assumed mode shapes *n* the coefficient matrix $\mathbf{H}$ in (24) attains the following form:

$$\mathbf{H} = \frac{1}{2} \begin{pmatrix} \mathbf{P}, & \mathbf{S}_1, & \mathbf{S}_2, & \cdots & \mathbf{S}_n \\ \mathbf{S}_1, & \mathbf{D}_1, & 0, & \cdots & 0 \\ \mathbf{S}_2, & 0, & \mathbf{D}_2, & \ddots & \vdots \\ \vdots & \vdots & \ddots & \ddots & 0 \\ \mathbf{S}_n, & 0, & \cdots & 0, & \mathbf{D}_n \end{pmatrix} \tag{27}$$

where the sub-matrices $\mathbf{P}, \mathbf{S}_i, \mathbf{D}_i \in \mathbb{R}^{2\times2}$ are defined in (48) in Appendix. The special block form of the matrix (27) enables to eliminate the sub-vectors $\mathrm{s}_i = (s_{s,i}, s_{c,i})^T$ and the system in the compact version can be obtained:

$$\left(\mathbf{P} - \sum_{i=1}^{n} \mathbf{S}_i \cdot \mathbf{D}_i^{-1} \cdot \mathbf{S}_i\right) \cdot \mathbf{p} = 0 \tag{28}$$

where $\mathbf{p} = (p_s, p_c)^T$. Obtaining eigen-vectors $\mathbf{p}_{(j)}$ the sub-vectors $\mathrm{s}_{i(j)}$ can be subsequently easily derived by back substitution into the large version of the system (27). The compact formulation (28) is meaningful as the determinants of $\mathbf{D}_i$ are always positive whenever the damping η_e (console) is positive. If there is $\eta_e = 0$, one of the determinants $\det\left(\mathbf{D}_i\right)$ can vanish for ω coinciding with the eigen-frequency of the console as it corresponds to particular λ_i. This case, however, is very rare and should be treated specifically. It manifests itself as a turning point on a stability limit.

According to proof by Powell (2011), the determinant of the coefficient matrix $\mathbf{H}$ can be written as:

$$\det\left(\mathbf{H}\right) = \det\left(\mathbf{P} - \sum_{i=1}^{n} \mathbf{S}_i \cdot \mathbf{D}_i^{-1} \cdot \mathbf{S}_i\right) \prod_{i=1}^{n} \det\left(\mathbf{D}_i\right). \tag{29}$$

Because the individual $\det\left(\mathbf{D}_i\right)$ are positive, only the first term in the product (29) is significant. Its value governs possible existence of non-zero perturbations of the semi-trivial solution and thus its instability.

The sum in the compact form (28) can also serve as an indicator of the necessary number of mode shapes taken into account. Due to convergence properties of the mode shape expansion (8) the norm of individual matrix addends in the sum grow smaller with $n \to \infty$.

Influence of System Parameters: Model A

In order to get a certain overview regarding the influence of basic parameters m (mass of the structure) and C (stiffness of the subsoil) on the stability loss, dynamic response and other properties of the system for various sets of input parameters have been evaluated. Respective results are summarized in Figure *3* (increasing $m \in (10,160)$) and in Figure *4* (increasing $C \in (1000,15000)$). With increasing m, a certain drop of system sensitivity in the area of ω_0 against the stability loss is evident $m \in (10,30)$. The size of the secondary instability area is also decreasing in the same time. Starting at $m = 30$, the resonance domain ω_0 gains importance once again and for $m > 100$ both instability domains join together.

Although the cases for $m > 25$ demonstrated in Figure *3* are beyond the static stability and consequently their interpretation can be problematic, this series provides an obvious warning against an existence of an upper limit of m when other parameters remain fixed.

Figure 3. Model A - Resonance curves (solid) and stability limits (dashed) for an increasing value of the mass $m \in \left(10, 180\right)$ *(greyed item corresponds to the case indicated in Figure 2)*

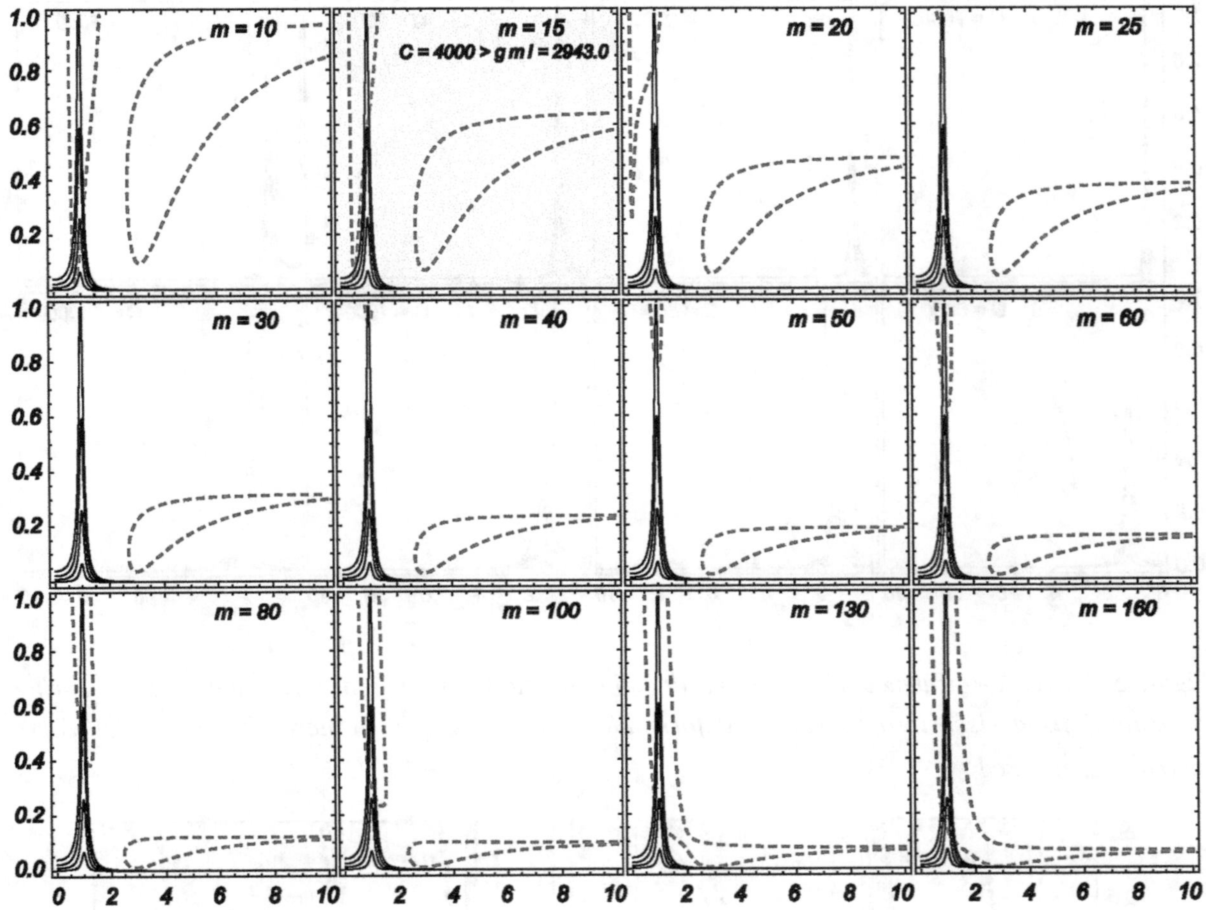

Just opposite tendency can be noted when increasing the stiffness C. The successive disappearance of the secondary instability area with increasing C is obvious. Taking into account the drop of the resonance curve in the resonance domain, one should conclude that increase of C stiffness contributes to the system stabilization.

Two types of damping have been taken into account in the basic mathematical model. The first is quantified by η_c and strongly influences the semi-trivial solution and the resonance curve shape (see Figure 5). The right part demonstrates the system sensitivity to η_c, especially in the ω_0 area. Concerning the left part of Figure 5, it can be seen that the influence of the viscosity of the vertical console is significant after the stability loss in the post-critical regime. Therefore, position and extent of the instability domain is weakly influenced by the parameter η_e. The shape of the secondary instability area changes regardless whether η_c or η_e is altered within the interval indicated.

Figure 4. Model A - Resonance curves (solid) and stability limits (dashed) for an increasing value of the spring stiffness $C \in (1\ 000, 15\ 000)$

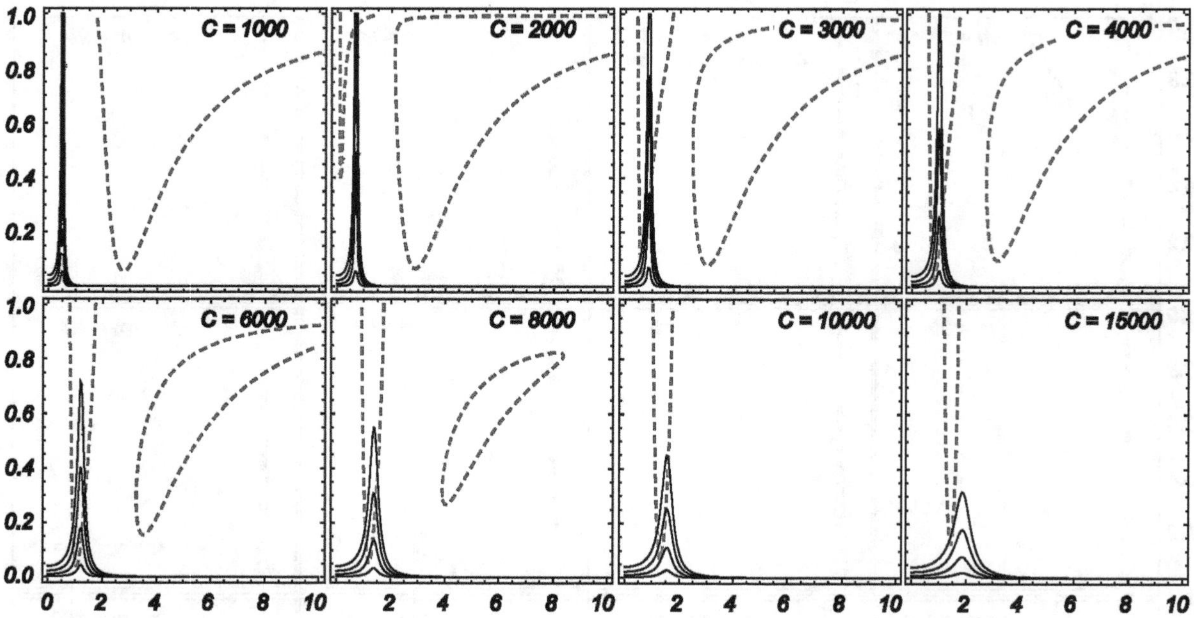

Figure 5. Model A - Influence of the viscous damping on the stability limits (dashed curves). a) spring viscosity η_c *fixed, viscosity of the vertical deformable console* $\eta_e \in (0, 0.4)$ *increasing; b):* $\eta_c \in (0, 0.4)$ *increasing,* η_e *fixed*

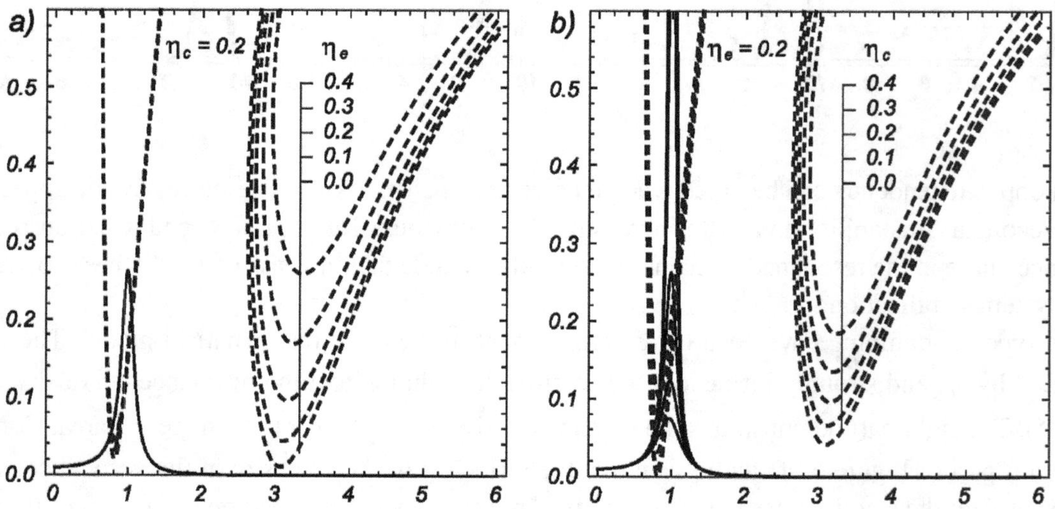

Influence of System Parameters: Model B

See Figure *6* for an overview about influence of system parameters onto the semi-trivial solution stability. The black dotted graphs represent resonance curves following Equation (16) for three excitation amplitudes a_0. They exhibit maximum in the neighbourhood of $\omega = 1.0$, the eigen-frequency of the semi-trivial solution. The solid curves stand in stability limits of the stiff structure when no bending is taken into account ($n = 0$). The dashed lines show the stability limits under circumstances that the console bending stiffness is employed by one or two mode shapes respectively. Pictures (a)-(f) are evaluated for six bending stiffness levels of the console. In principle, it is obvious that increasing number of mode shapes taken into consideration leads always to drop of the stability limit as the system is getting to be weaker. However, the dominant character of the first mode shape masks influence of the higher modes as the second and higher eigen-frequencies fall out of the frequency interval covered by the figure. In this frequency domain, the influence of the number of mode shapes taken into account is very weak.

Figure 6 (a) demonstrates that low bending stiffness leads to the stability loss being concentrated in the area around the first eigen-frequency of the console. Even for small excitation amplitudes a_0 one or two instability intervals have been ascertained for $\omega \in (s_1, s_2)$ and $\omega \in (s_3, s_4)$ or $\omega \in (s_1, s_4)$ depending

Figure 6. Model B - Stability limits of the semi-trivial solution including none, one or two console mode shapes ($n = 0,1,2$) for varying bending stiffness of the console; dampings $\eta_c = 0.05$, $\eta_e = 0.05$, ratio $\rho = 0.2$

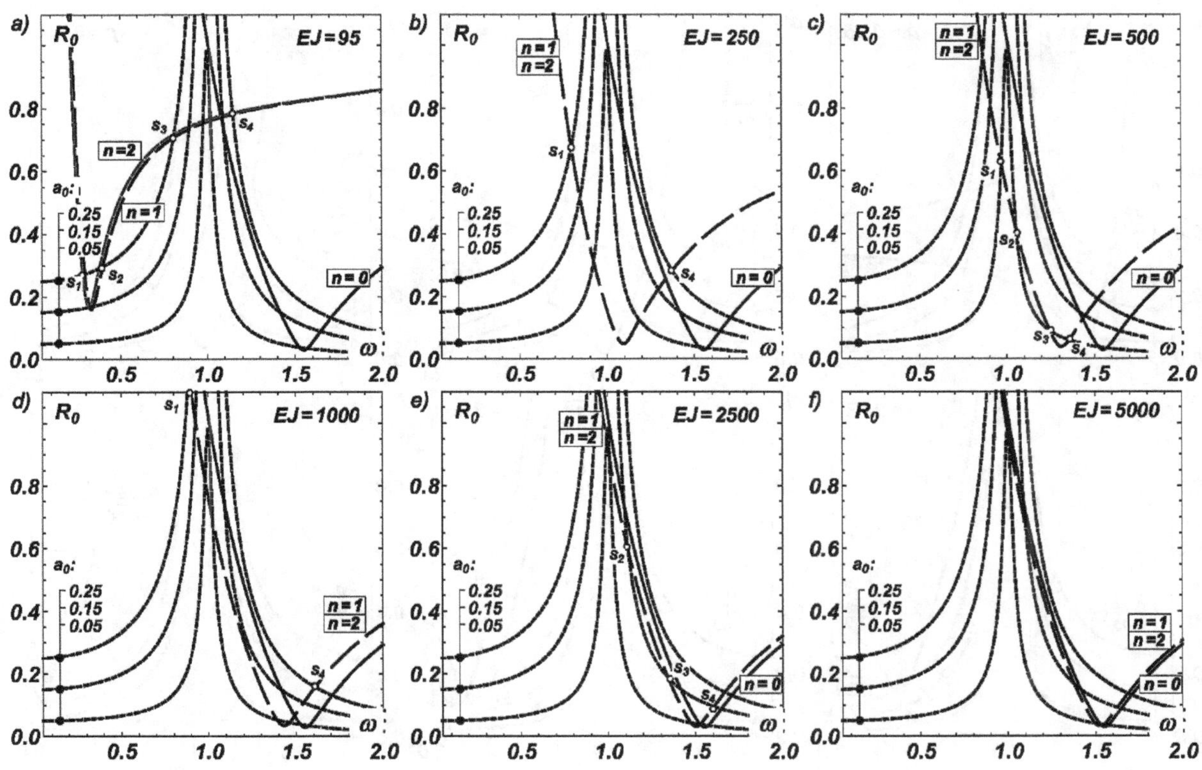

on the excitation amplitude a_0 and other system parameters. Picture (a) is the only case when the difference between stability limits for $n = 1$ and $n = 2$ is recognizable. In all other cases, the influence of the second mode shape is negligible and the curves for $n = 1$ and $n = 2$ coincide.

Instability intervals are concentrating mostly in proximity of frequencies $\omega_0, \omega_{1B}, \omega_{3B}$ (sub-soil and system basic properties) and $\omega_{4,5,6,...} = \omega_{3B} \cdot \lambda_{1,2,3,...}$ (console flexibility). Therefore, it is obvious that minimum of stability limits is moving to higher frequencies with increasing bending stiffness of the console. As a special case can be considered picture (b) where ω_0 and ω_5 nearly coincide and twofold eigen-frequency occurs. Thereafter for higher EJ the stability minimum exceeds $\omega = 1$ and approach minimum of the stiff case, see pictures (c)-(f). This knowledge can serve as an instruction for engineering practice.

Figure 7 is demonstrating an evolution of the stability limits when the ratio $\rho = r / l$ (i.e., ground width/console height is changing). The picture (b) represents an approximate boundary (exact value is $\rho_c = 0.086$ keeping other parameters) below which the static stability is violated. In other words, for $\rho < \rho_c$ the system is unstable even in a static state and collapses. Therefore, the dynamic problem is worthy to be investigated for $\rho > \rho_c$. Position of the static stability boundary in general is a function of all system parameters.

Figure 7. Model B - Stability limits of the semi-trivial solution ($n = 0,1$) for varying ratio $\rho = r / l$ (ground width/height); dampings $\eta_c = 0.05, \eta_e = 0.05$, stiffness $EJ = 500$

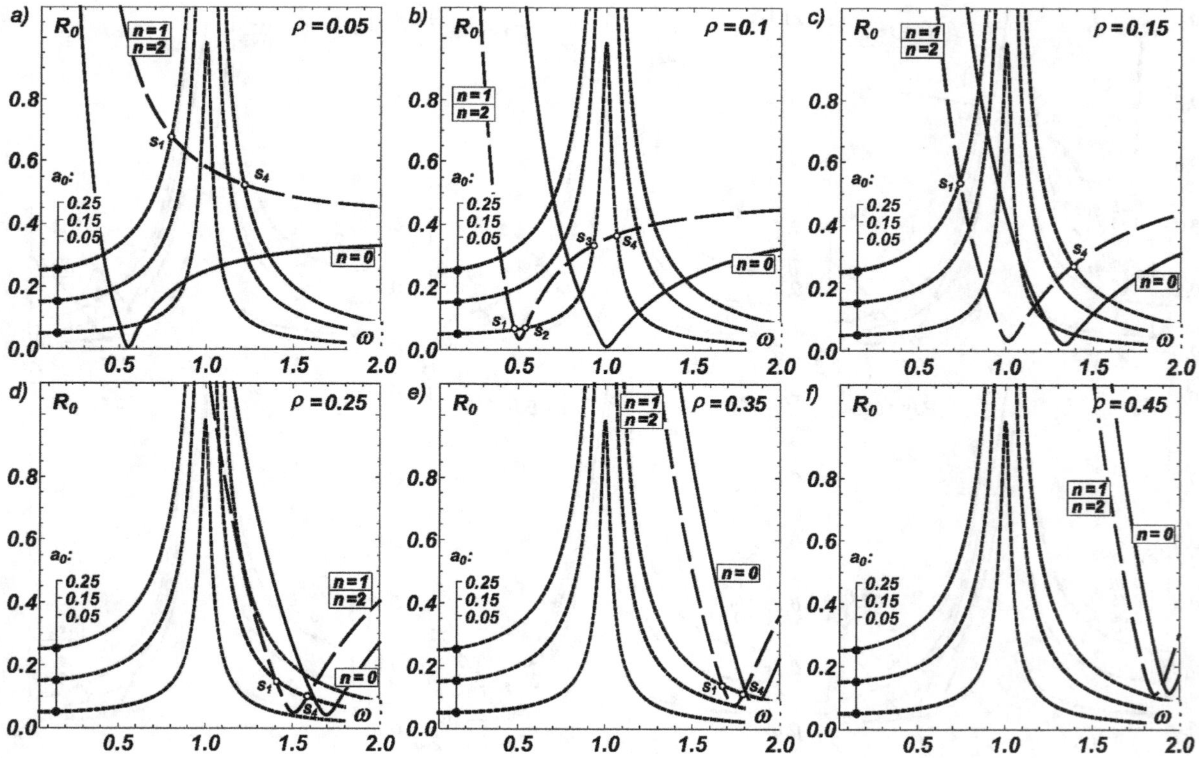

Position of minimum of the dynamic stability limits is well expressed on the frequency axis for each ratio ρ. The position is visibly rising with increasing ratio ρ abandoning resonance area of semi-trivial solution. Influence of the elasticity of the console decreases at the same time. As it follows from pictures (d)-(f), the stability loss is less and less probable even for higher amplitudes of excitation. Therefore, the broadband excitation is also less and less dangerous. This attribute should be taken into account in a practical engineering, despite its technical application is much more complex as adjusting of the console stiffness.

The third parameter significantly influencing the semi-trivial solution (or the system) stability is the sub-soil viscous damping. Although a lot of different models of the damping can be discussed, Voigt model is probably able to describe the principal properties of the system response with respect to the damping. It follows from Figure *8* that resonance curves of the semi-trivial system are rapidly dropping with increasing η_c parameter while the shape of stability limits does not change considerably. Instability area concentrates around frequency ω_0 and more or less keeps its position and extent. Therefore, for design practice it is recommended to try to increase the sub-soil viscosity as much as possible using some special additives for material treatment. Internal damping of the console η_e influences the stability limits as well (see Figure *9*). However, variation of this parameter did not lead to considerable changes in shape and character of respective stability limits provided that other system parameters are kept. Indeed, the interval $\eta_e \in (0.05; 0.30)$ where the system was tested is large enough to cover usual

Figure 8. Model B - Stability limits of the semi-trivial solution ($n = 0,1$) for various values of the sub-soil viscous damping; other parameters $\eta_e = 0.05$, $EJ = 500$, $\rho = 0.2$

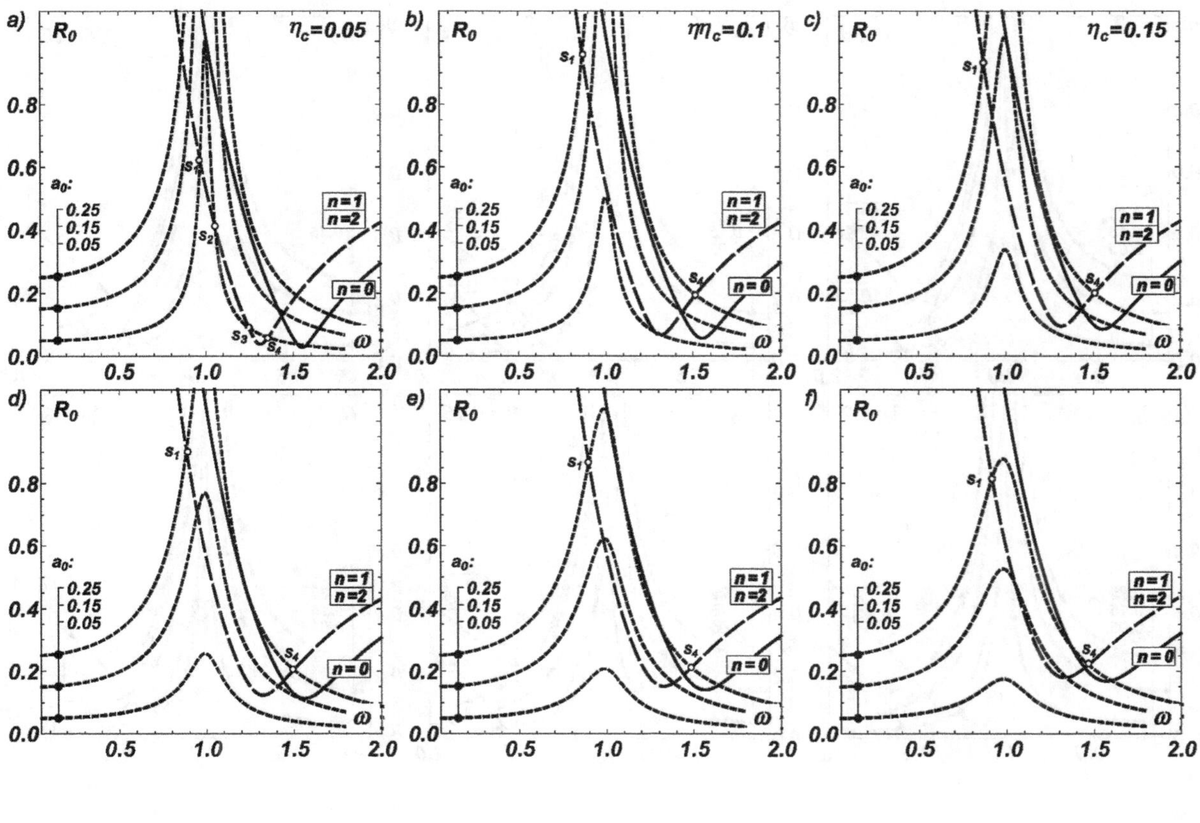

damping ratio values encountered in civil engineering regarding concrete or steel. In practice, $\eta_e \neq 0$ should be respected. However, this value is recommended by most standards. Slightly different is the case of the sub-soil damping ratio η_c, where the real values can approach the critical damping.

POSTCRITICAL RESPONSE

In order to get an overview concerning the system behaviour in frequency intervals of stable semi-trivial and post-critical solutions, several numerical analyses were performed for the model A (i.e., using the governing differential system (6)). The corresponding results for model B are similar. The input data correspond with those providing results plotted in Figure *2* (left picture) with the excitation amplitude $a_0 = 0.2$. Results of simulations are depicted in Figure *10*. In the left column, the vertical response component $\zeta(t)$ in dimensionless form — see Equation (3) — is presented. In the right column, both $\varphi(t)$ and $\psi(t)$ response components are plotted. Both curves nearly coincide in most cases. The initial conditions were zero, except for rocking component ψ, where a small initial value was prescribed for all excitation frequencies. Numerical integrations were performed using the Wolfram-Mathematica code. In particular, Gear - predictor-corrector procedure has been repeatedly applied. The stationary state (if any) has been tested after a certain time when the appropriate tests of stationarity passed. The phase plots for selected components of Figure *10* are shown in Figure *11*.

Figure 9. Model B - Stability limits of the semi-trivial solution ($n = 0,1$) for various values of the console material damping; other parameters $\eta_c = 0.05$, $EJ = 500$, $\rho = 0.2$

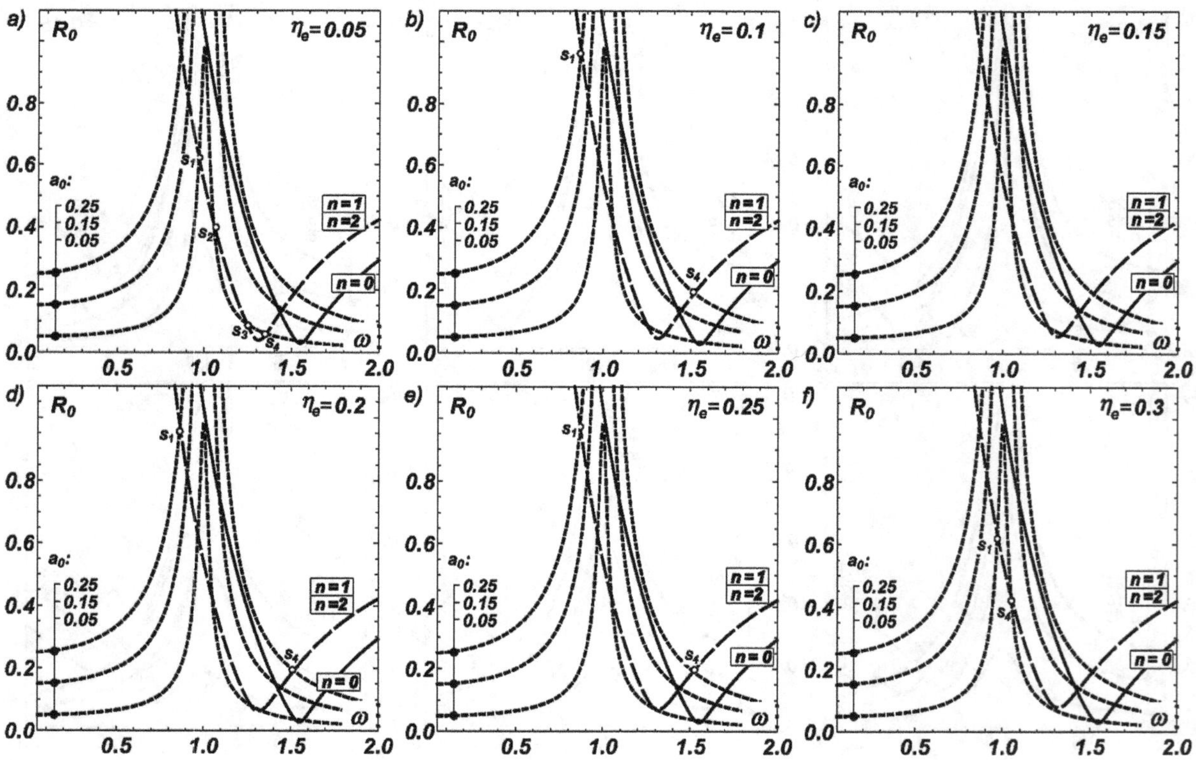

Figure 10. Time history of the system response; right column: $\varphi(t), \psi(t)$ (post-critical response components); left column: $\zeta(t)$ (vertical response component at point B); limits of semi-trivial solution stability/instability: $s_1 = 0.4548$, $s_2 = 0.5718$, $s_3 = 0.8671$, $s_4 = 1.0680$ (Model A)

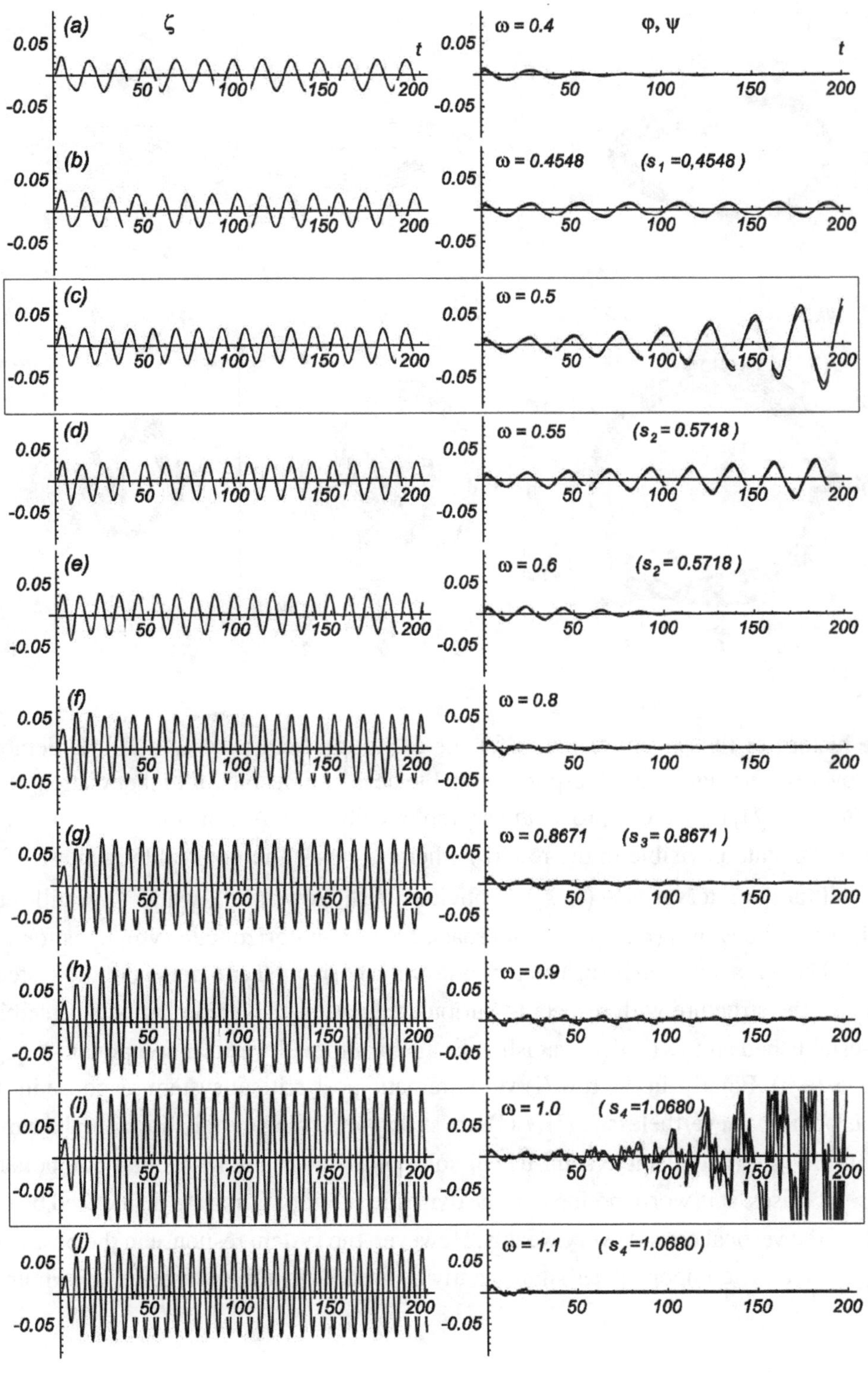

Figure 11. Phase plot of selected cases of the time history of the system response; left column: $\zeta(t)$, middle column: $\psi(t)$, right column: $\varphi(t)$. Time interval shown here is $t \in (800, 1000)$. Cf. corresponding rows (c) and (i) in Figure 10. (Model A).

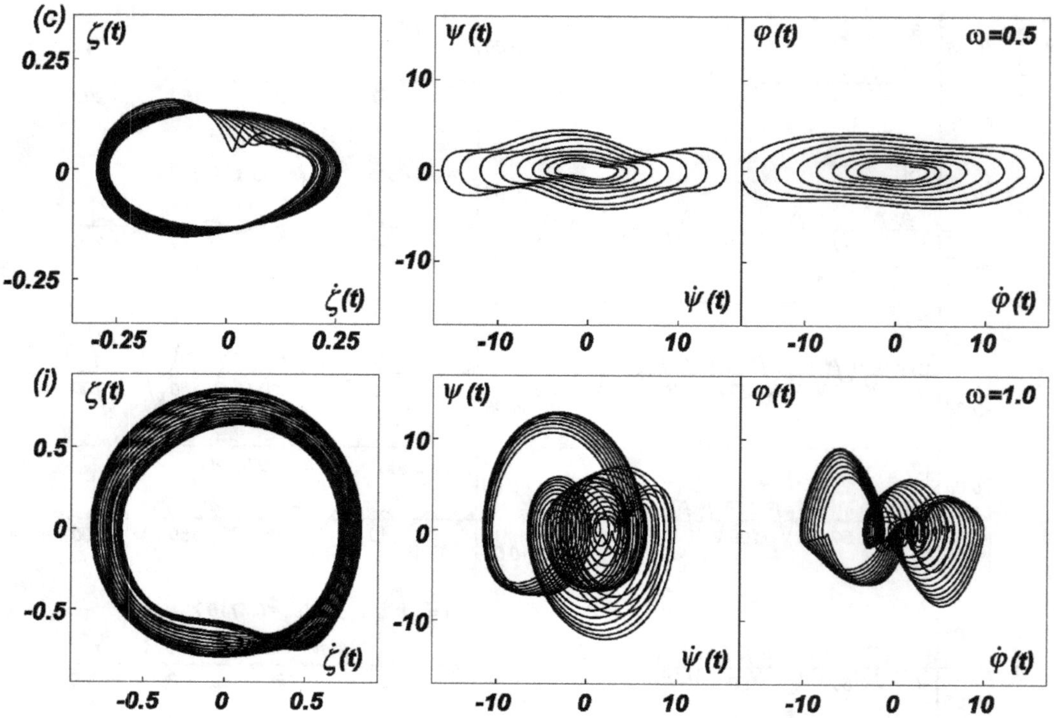

The time history of the system response for the excitation frequency $\omega = 0.4$ is demonstrated in Figure *10* – row (a). The semi-trivial response is stable despite of a small initial deviation of ψ component. Components $\varphi(t), \psi(t)$ are approaching asymptotically zero. A transitional state and a beginning of the post-critical state is visible in the row (b) where $\omega = s_1$. The noticeable increase of $\varphi(t), \psi(t)$ components within the interval $\omega \in (s_1, s_2)$ is obvious in the rows (c,d), while $\zeta(t)$ still remains on a similar level. Amplitudes of $\varphi(t), \psi(t)$ are approaching certain horizontal asymptotes for $t \to \infty$, cf. also Figure *11*. There is a question if such an increase of the horizontal component of the system response is admissible for the structure with respect to various regulations. Overcoming s_2, the semi-trivial response is re-established and $\varphi(t), \psi(t)$ vanish once again, although vertical component ζ significantly grows up, rows (e-h). Finally, in the row (i) very dramatic post-critical system response in the interval $\omega \in (s_3, s_4)$ is obvious, nevertheless, $\varphi(t), \psi(t)$ are still approaching certain horizontal asymptotes (see Figure *11*). In the row (j) the stable semi-trivial solution without any horizontal components for any arbitrary $\omega > s_4$ arises. It is worth noting that the dynamic stability limits are influenced by the bending stiffness EJ of the vertical console very weakly. However, the system response in the post-critical state, in particular $\varphi(t), \psi(t)$ components, is significantly influenced by this parameter (see Figure *10* – rows (c,i)).

The asymptotic behaviour of the response computed for the different excitation frequencies can be assessed also from the point of view of the Lyapunov characteristic exponents (LCE). Figure *12* shows the largest LCE for four excitation amplitudes a_0 depending on the excitation frequency. The used numerical procedure follows guidelines proposed by Sandri (1996). To relate the LCE with the semi-trivial solution, an obvious almost-zero initial condition has been supposed. Positive values of the LCE indicate the unstable part of the trajectory of the semi-trivial solution. However, even if the LCE are positive for $\omega \in (s_1, s_2) \cup (s_3, s_4)$ (see Figure *2*), their rather small absolute values permit the trajectories to remain within limits for the whole simulation time.

This preliminary numerical analysis shows that within resonance domains the semi-trivial response can remain in force although post-critical response type is rather typical. Three basic categories of post-critical response can be noticed:

1. Response amplitudes are either weakly variable or constant. In such a case, the system response can be studied by means of the harmonic balance method. It eliminates an oscillatory response character investigating (approximately) amplitudes only. This procedure assumes response in the (mono) harmonic form, similar to that introduced in perturbation analysis (18) and (19). However, dependence of amplitudes on time is respected in this approach. The resulting differential system can be symbolically written as:

$$\mathbf{M}\left(\mathbf{x}\right)\dot{\mathbf{x}} = \mathbf{K}\left(\mathbf{x}\right) \tag{30}$$

where $\mathbf{x}(t)$ represents unknown partial amplitudes. This system is meaningful only if the amplitudes are functions of a "slow time" (if their changes within one period $T = 2\pi / \omega$ are small or vanishing and individual steps of the harmonic balance operation are acceptable). In such a case, two response regimes corresponding to harmonic excitation can be encountered:

Figure 12. Values of the largest Lyapunov characteristic exponent λ depending on excitation frequency $\omega \in (0.4, 1.1)$, computed for excitation amplitudes $a_0 = 0.1, 0.15, 0.2, 0.25$ (Model A)

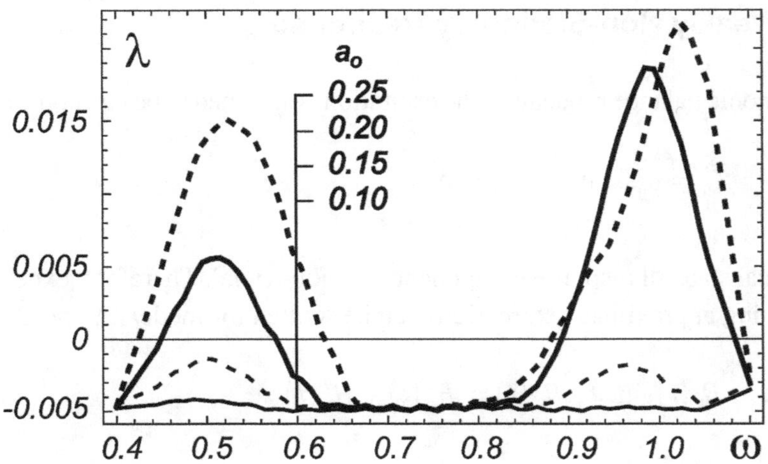

a. **Weakly Non-Stationary Response:** The obtained differential system cannot be solved analytically and a numerical procedure must be applied. As a feedback, the variability rate of the individual components is to be checked. If some periods are comparable (or even shorter) with T, results should be rejected. In general, shapes of amplitudes $\mathbf{x}(t)$ are deterministic and periodic with possibly small perturbations.

b. **Stationary Response (Special Case of the Previous One):** The time derivative of the amplitudes $\mathbf{x}(t)$ vanishes. Therefore, the remaining non-linear algebraic system should be solved:

$$\dot{\mathbf{x}} \equiv 0 \Rightarrow \mathbf{K}\left(\mathbf{x}\right) = 0 \tag{31}$$

Both cases (a), (b) can cover limit cycles typical for certain types of the post-critical behaviour. System (30) can provide stable as well as unstable limit cycles (Burton, 1982).

2. Response amplitudes are strongly variable. Provided that the response is of any other type than in category 1, numerical integration of the original system (6) or (10) is necessary. The following most important regimes of the response can be observed:
   a. **Transition Processes:** The system is asymptotically approaching the stationary state due to sudden energy supply/loss, etc. Response shape is rather deterministic.
   b. **Quasi-Periodic Regime:** Energy Periodic Transflux Between DOFS. Beating effects arose due to two or more DOFs interaction. A structure of one period can be complicated. However, the time history repeats and the length of individual periods fluctuates within very limited interval (Murphy, Bayly, Virgin, & Gottwald, 1994).
   c. **Chaotic Regime:** Its existence can be identified by means of Lyapunov exponent testing.
3. Response amplitudes blow up. This case can be treated as a special case of the previous category 2. Due to the unstable nature of the system under study, excitation that crosses certain limit leads to the collapse of the structure. However, it appears that the structure can withstand some time limited non-stationary response. Reasons supported by the numerical experiments will be presented in the last section.

Detailed comments regarding mentioned categories will be presented in the following subsections.

Stationary or Weakly Non-Stationary Response

To employ the harmonic balance procedure, the excitation is assumed to be harmonic as in Equation (13):

$$y_0 = A_0 \sin\omega t \quad \Rightarrow \quad \zeta_0 = a_0 \sin\omega t\,, \quad A_0 = a_0 \cdot l\,. \tag{32}$$

In the post-critical state all response components are non-trivial. Therefore, expecting a single mode response the following approximate expressions can be written for models A and B:

$$\zeta\left(t\right) = R_c\left(t\right)\cos\omega t + R_s\left(t\right)\sin\omega t\,,\; R^2\left(t\right) = R_c^2\left(t\right) + R_s^2\left(t\right) \tag{33a}$$

$$\varphi(t) = P_c(t)\cos\frac{1}{2}\omega t + P_s(t)\sin\frac{1}{2}\omega t\,,\; P^2(t) = P_c^2(t) + P_s^2(t) \tag{33b}$$

$$\psi(t) = S_c(t)\cos\frac{1}{2}\omega t + S_s(t)\sin\frac{1}{2}\omega t\,,\; S^2(t) = S_c^2(t) + S_s^2(t) \tag{33c}$$

$$\alpha_i(t) = S_{c,i}(t)\cos\frac{1}{2}\omega t + S_{s,i}(t)\sin\frac{1}{2}\omega t\,,\; S_i^2(t) = S_{c,i}^2(t) + S_{s,i}^2(t)\,. \tag{33c'}$$

To make a decision between single- and multi-harmonic approximation is a delicate question. For some interesting experiences, see Ren and Beards (1994). Another opinion can be found in the work by Lee and Hsu (1994), where the authors discuss conditions when the internal resonance or its proximity should be analysed.

Considering the model A, the approximations (33) can be put in the system (6) with the additional conditions (23). Going through the harmonic balance procedure, a differential system (30) for amplitudes:

$$\mathbf{x}(t) = \left(R_c(t), R_s(t), P_c(t), P_s(t), S_c(t), S_s(t)\right)^T \tag{34}$$

can be obtained. The detailed structure of the matrix $\mathbf{M}(\mathbf{x}) \in \mathbb{R}^{6\times6}$ and the vector $\mathbf{K}(\mathbf{x}) \in \mathbb{R}^{6}$ is shown in the Appendix, Equation (44). From that relation it can be seen that matrix $\mathbf{M}(\mathbf{x})$ depends on amplitude P only. Its determinant can be easily enumerated:

$$\det\left(\mathbf{M}(\mathbf{x})\right) = \frac{1}{256}\omega^6\left(\kappa_0 P^2(t) - 4\right)^2. \tag{35}$$

It is clear from Equation (35) that the determinant is non-negative for any ω and P, regularity of $\mathbf{M}$ (for $\omega > 0$) depends only on $P^2(t)$:

$$P^2(t) \neq 4\frac{M}{m}\left(1 + \frac{m}{M}\right).$$

Thus, with exception of the indicated value the inverse matrix $\mathbf{M}(\mathbf{x})^{-1}$ exists, its actual form is given in Appendix, Equation (45). If used for analysis, the actual value of amplitude $P(t)$ should be checked with respect to ratio m / M.

Knowing the exact form of $\mathbf{M}(\mathbf{x})^{-1}$, the normal form of the differential equation (30) can be easily established. However, as long as the matrix $\mathbf{M}(\mathbf{x})$ is regular and its determinant does not change sign the original right hand side $\mathbf{K}(\mathbf{x})$ can be studied equivalently.

If the stationary response of the system is considered, the derivatives $\dot{\mathbf{x}}$ vanish and the right hand side $\mathbf{K}(\mathbf{x})$ has to vanish too (see the response category 1b above). In this case, Equation (30) degenerates to a non-linear algebraic relation:

$$\mathbf{K}(\mathbf{x}) = 0. \tag{36}$$

Thus, to identify the stationary solutions, the zero solution points of $\mathbf{K}(\mathbf{x})$ should be traced. In the same time, signum the zero points of the Jacobi determinant of $\mathbf{K}(\mathbf{x})$ have to be checked as they indicate stability/instability of the stationary solution and existence of bifurcation points.

For the particular data, the dependence of $R^2(t), S^2(t)$ and $P^2(t)$ on excitation frequency ω is shown in the left hand part of Figure *13*. Three curves are present for each variable, they correspond to excitation amplitudes $a_0 = 0.2, 0, 4, 0.6$. Regarding the topmost part of Figure *13*a, it shows the resonance curve of the semi-trivial solution. For amplitudes $a_0 > 0.2$ are the curves enriched by the lover stable branches in the resonance region. Each of these non-linear branches begins in the intersection point with the ascending part of the semi-trivial resonance curve (denoting the corresponding frequency as ω_a for the curve for to $a_0 = 0.4$) and extends beyond the intersection point with the descending part of the linear resonance curve (ω_b) up to certain point ω_c. Another interesting point is the fact that position and

Figure 13. Left: a) stable and unstable branches of the resonance plots for generalised amplitudes R^2, P^2, S^2. *Right top: b) Position of the bifurcation points with respect to excitation frequency* ω *and amplitude* a_0. *Right bottom: c) Values of the Jacobi determinant* $\det(\mathbf{JK}(\mathbf{x}))$ *along the zero solutions* $\mathbf{K}(\mathbf{x}) = 0$. *Model data:* $M = 1990$, $m = 10$, $l = 20$, $C = 2000$, $\rho = 1$, $\eta_c = \eta_e = 0.32$, $EJ = 10^4$, $g = 9.81$. *(Model A)*

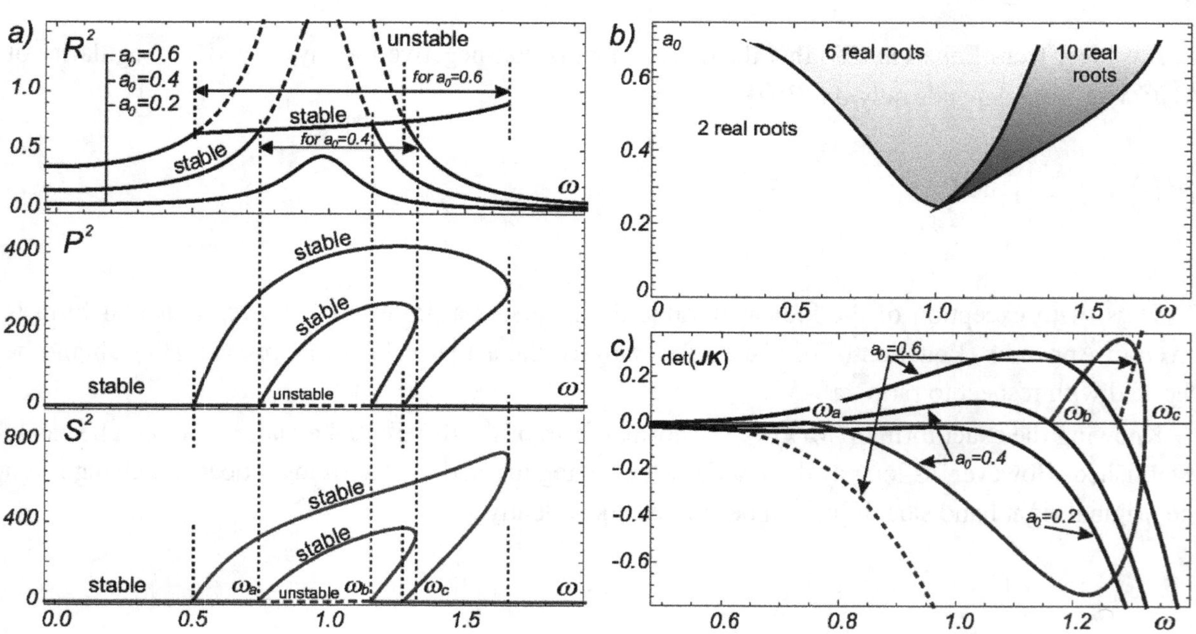

shape of the non-linear stable branch coincides for all different amplitudes a_0. The individual curves differ by their length; see the dotted lines connecting corresponding points for R^2, P^2, and S^2. Between both intersections ω_a and ω_b, the upper branch R^2 is unstable and the lower R^2 branch is stable. For P^2 and S^2 holds the opposite case. The cantilevered part of the non-linear branch R^2 (between ω_b and ω_c) is twofold, consisting of both stable and unstable parts.

Figure 13b shows the areas in the (ω, a_0) plane where the right-hand side $\mathbf{K}(\mathbf{x})$ has 2, 6, or 10 real roots. Boundaries between these areas are formed by bifurcation points of the equation $\mathbf{K}(\mathbf{x}) = 0$ or, equivalently, by singular points of the Jacobian. Course of the Jacobi determinant $\det\left(\mathbf{JK}(\mathbf{x})\right)$ for individual zero solutions from Figure *13*a is shown in Figure *13*c. It is worth to note that the negative sign of the Jacobi determinant corresponds to the stable branch of the solution. The black solid lines correspond to the semi-trivial solutions for excitation amplitudes $a_0 = 0.2, 0, 4, 0.6$. Their corresponding non-linear branches (dashed lines) begin and end in the zero points of the semi-trivial branches. The non-linear branches are twofold, which fact gives six or ten real zero solutions in the part between ω_a, ω_b, and ω_b, ω_c, respectively.

It is worth to pay attention to the case, when semi-trivial resonance curve loses its stability and the non-linear branch arises. The amplitude corresponding to this particular case forms the upper limit of the design value for any real structure. The critical amplitude would be characterised by the following three conditions:

$$\mathbf{K}\left(\mathbf{x}\right) = 0; \quad \det\left(\mathbf{JK}\left(\mathbf{x}\right)\right) = 0; \quad \frac{\mathrm{d}}{\mathrm{d}\omega}\det\left(\mathbf{JK}\left(\mathbf{x}\right)\right) = 0. \tag{37}$$

Solving (37) numerically for data from the example, the following values can be obtained: $\omega = 1.0242$, $a_0 = 0.261$. Such obtained configuration corresponds to the critical state, where the stability loss occurs.

Typical behaviour of the system can be demonstrated in Figure *14* as follows: Resuming for model A, the parameter set $C = 10000$, $EJ = 10000$, $g = 9.81$, $M = 3990$, $m = 15$, $\eta_c = 0.2$, $\eta_e = 0.2$, $l = 20$ being used in Figure 2, the basic features of the response can be outlined. The excitation with the amplitude $a_0 = 0.20$ can be considered as large providing the response amplitude plotted by a bold curve. The system is unstable within intervals $\omega \in (s_1, s_2)$ and $\omega \in (s_3, s_4)$, see hatched (squared) areas in the figure. In order to assess a real behaviour, three types of the analyses have been done.

The respective results are shown in Figure *14*. The left part concerns amplitudes R of the vertical component $\zeta(t)$, while right part amplitudes S of the dimensional-less horizontal (or angular) component $\psi(t)$. The curves plotted in dashed lines have been obtained using algebraic system (36) thick solid curves have been provided by means of the differential system (30) and finally thin solid curves represent results of direct simulations on the basis of the differential system (6) (Chossat & Lauterbach, 2000; Iooss & Adelmeyer, 1992).

As the simulations proved to have considerable chaotic component especially in the interval $\omega \in (s_3, s_4)$, variance of the amplitude has been evaluated and corresponding areas shown as hatched (squared) in Figure *14*. It is obvious that an applicability of results obtained by semi-analytical procedures, Equation

Figure 14. a) System response amplitudes evaluated by means of three different techniques; left: vertical component $\zeta(t) \rightarrow R$, b) horizontal component $\psi(t) \rightarrow S$ (Model A)

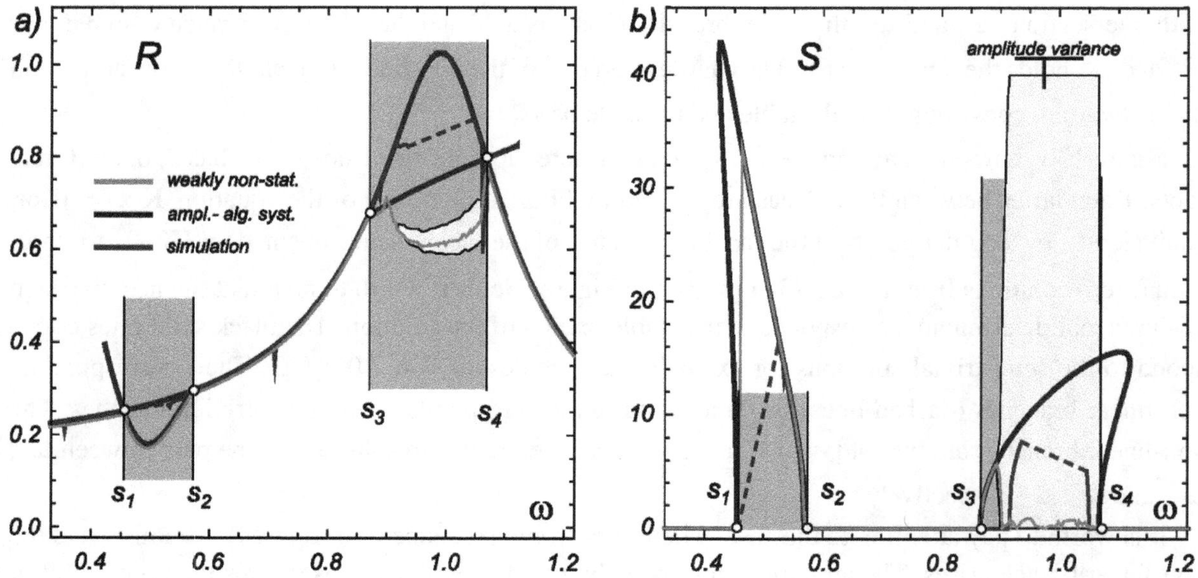

(30) and Equation (36) is very limited. They can be accepted in the lower interval $\omega \in (s_1, s_2)$, where solution of differential system (30) provides the most realistic results, see also time history in Figure *10*c-d. The algebraic system leads here to twofold solution and the unstable branch below s_1 should be eliminated. However, simulation gives similar results especially in the upper part of the interval.

In the upper interval $\omega \in (s_3, s_4)$ the high degree of instability is visible, cf. Figure *10*h-j. It applies mainly to the S component. Obviously, analytically obtained static amplitudes (dashed curves) are not realistic and only the simulation results providing chaotic results are representative. This is the case of response of the category 3 as discussed above, see also the corresponding dark parts of Figure *15*. The part of the dashed curve above s_4 is not accessible neither numerically nor experimentally being unstable (only SDOF nonlinear systems can deal with this branch having no possibility to escape into the stable state due some other degree of freedom).

In case of model B, the form of matrix $\mathbf{M}(\mathbf{x})$ and the vector $\mathbf{K}(\mathbf{x})$ in the general differential relation (30) will be generally similar to that of model A. Due to linearity of the mode shape expansion, the derivation can be performed for a single mode shape and simply extended into the full case. Dimensions of the complete problem will be $(4+2n)$, where n is number of mode shapes used. The block structure of matrix $\mathbf{K}(\mathbf{x})$ will be close to that in Equation (27) enriched by a pair of rows/columns, which correspond to response component $\zeta(t)$.

Thus, the explicit formulation of the particular form will be omitted here.

Adoption of an assumption of stationary vertical response ζ_s, which was used in the semi-trivial solution (14) and (15) leads to a simplified version of the general relation (30). Using (33b) and (33c') for $\varphi(t)$ and $\alpha_i(t)$, respectively, and denoting $\mathbf{x}(t) = \left(P_c(t), P_s(t), S_{c,1}(t), S_{s,2}(t), \ldots, S_{c,n}(t), S_{s,n}(t)\right)^T$ (for model B) the harmonic balance procedure leads to a linear differential system, where general non-linear

Figure 15. Sensitivity of the system (model A) on the frequency and amplitude of the excitation. Light grey area corresponds to the non-zero stationary response in φ and ψ components. Dark grey area exhibits areas of non-stationary (growing) numerical response of the rocking components.

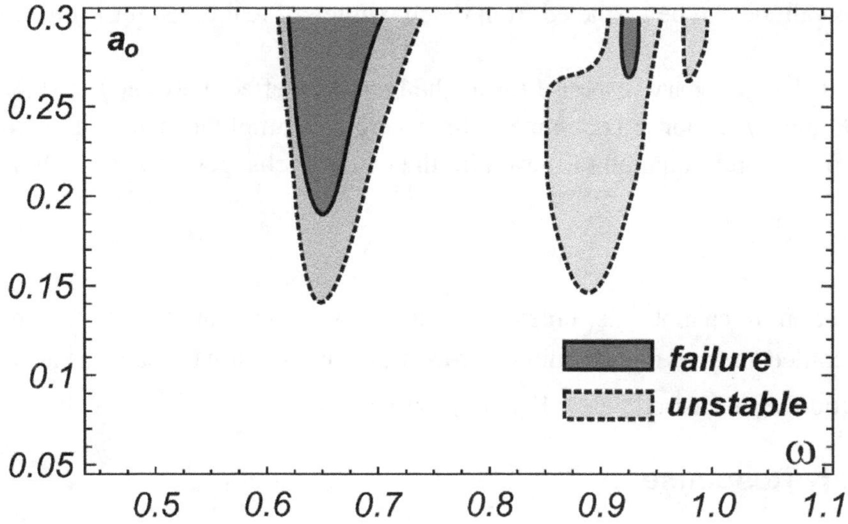

function $\mathbf{K}(\mathbf{x})$ reduces itself to matrix-vector multiplication $\mathbf{H}\cdot\mathbf{x}$ and both coefficient matrices $\mathbf{M}$ and $\mathbf{H}$ depend only on amplitudes of the stationary vertical response ζ_s:

$$\mathbf{M}\dot{\mathbf{x}} = \mathbf{H}\mathbf{x}\,; \qquad \mathbf{M},\mathbf{H} \in \mathbb{R}^{(2n+2)\times(2n+2)}. \tag{38}$$

The matrix $\mathbf{H}$ is the same as that which has been derived for assessment of stability of the semi-trivial solution and is given in (27). Detailed form of $\mathbf{M}$ is given in the Appendix, Equation (46). As in the previous case, its determinant can be enumerated and is given in Equation (49) in the Appendix. For fixed n and $\omega > 0$ is the determinant $\det(\mathbf{M})$ nonzero and does not change its sign so that the matrix $\mathbf{M}$ is regular and invertible. Its inversion in a form of a block matrix is presented in the Appendix, Equation (50). Thus, the differential system (38) can be rewritten in a normal form:

$$\dot{\mathbf{u}} = \mathbf{M}^{-1}\mathbf{H}\mathbf{u} \tag{39}$$

and its stability can be checked using eigen-value analysis. Let symbols Λ_i denote the eigen-values of the matrix $\mathbf{M}^{-1}\mathbf{H}$. Three particular cases can be pointed out:

1. The stability region where all the eigen-values Λ_i have negative real part will be delimited by zero values of the determinant $\det(\mathbf{H}) = 0$;
2. if the real part of an eigen-value Λ_i is positive but small, the system response remains within certain limits while the excitation is active and system can return to standstill when excitation stops;

3. Real part of an eigen-value Λ_i is positive and large, the system collapses before the excitation stops. The actual value of the limit between "small" and "large" eigen-value depends on the structure properties and duration and intensity of the excitation. However, the rate of increase of the response amplitude can be predicted from the maximum of real parts of the eigen-values.

Appearance of the stationary response forms the boundary effect between the stable and unstable response. If a stationary response is assumed, the generalized amplitudes $\mathbf{u}$ can be considered to be constant. Thus, $\dot{\mathbf{u}} = 0$ and Equation (38) even for the model B changes into an algebraic equation:

$$\mathbf{Hu} = 0. \tag{40}$$

Non-zero solution $\mathbf{u}$ cannot exist unless the matrix $\mathbf{H}$ is singular, i.e., its determinant vanishes $\det(\mathbf{H}) = 0$. Detailed analysis of determinant properties is in this case the same as it was regarding the semi-trivial solution, as discussed below Equation (29).

Non-Stationary Response

The system under study is in fact an inverted pendulum; as such, its stability is limited. Excitation that crosses certain limit inevitably leads to the collapse of the structure. It seems that the dangerous rocking movement needs some time interval to fully develop. This fact works for the sake of the structure: if the excitation stops within a certain time, then the structure can return to a stable state.

Figure 15 can serve as a rough illustration. It is based on numerical integration of the Lagrange equations of the model A, Equation (6). Three different levels of amplitude of the response φ (rocking component) are shown for excitation frequencies $\omega \in (0.4, 1.1)$ and amplitudes $a_0 \in (0.05, 0.3)$: white area corresponds to negligible rocking component φ, grey area shows nonzero but limited amplitude of φ. Return to the stable state is still possible if relevant conditions are regained. The dark grey sections of the figure correspond to the cases where a numerical overflow occurs, i.e., $\varphi \to \infty$. The final collapse is inevitable. Coincidence of the infected ranges of ω with instability intervals (s_1, s_2) and (s_3, s_4) is evident (see Figure 2).

A more detailed illustration of the process of a stability loss and of a transition from the semi-trivial to post-critical response can be demonstrated using following example. The parameters C and η_c of the system (model A) examined in previous sections have been slightly increased (subsoil stiffness and internal damping). The actualized parameter set reads now: $C = 16000$, $EJ = 10000$, $g = 9.81$, $M = 3990$, $m = 15$, $\eta_c = 0.6$, $\eta_e = 0.2$, $l = 20$. Outline of instability domains for various amplitudes of excitation and corresponding bifurcation diagrams can be found in Figure *16*. Figure *16*a represents resonance curves (solid) of the semi-trivial solution for amplitudes $a_0 \in (0.3, 0.6)$. Stability limit (dashed) represents the set of bifurcation points delimiting intervals ω where two solutions appear: the unstable semi-trivial solution (dotted) and stable post-critical solution (solid line).

Evolution of the response amplitudes $\mathrm{R, P, S}$ for frequency $\omega = 1.1700$ (roughly the middle of the "instability interval") with increasing excitation amplitude a_0 is demonstrated in Figures *16*b-c. With respect to testing techniques mentioned above (Lyapunov exponent, structure of the response frequency, etc.) it can be shown that Equation (30) or Equation (31) are suitable to evaluate response amplitudes

Figure 16. Process of the stability loss and of the post-critical response; a) instability domains for various amplitudes of excitation, b) bifurcation diagram of the vertical response component amplitude R, c), d) bifurcation diagrams of horizontal (rocking) components P and S. (Model A)

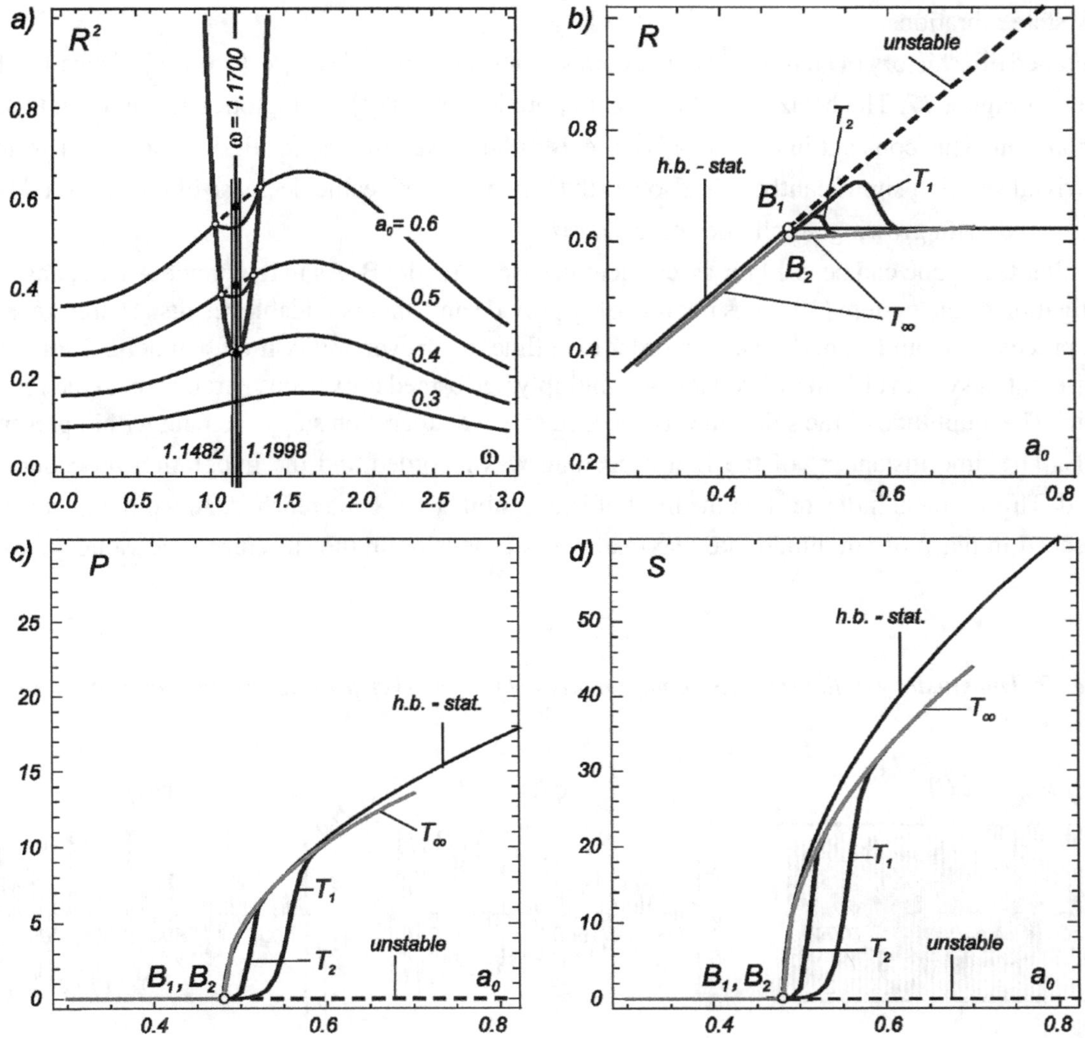

for $\omega = 1.1700$. In Figures *16*b-c, starting from the bifurcation point B_1, two branches evaluated by means of Equation (31) can be found. Dotted curves represent unstable part corresponding to semi-trivial solution, while long-dashed curves (nearly horizontal for R and rising curvilinear for P, S) demonstrate stable post-critical solution.

Evaluating the system response for the same parameters and homogeneous initial conditions by means of numerical simulation, nearly the same results are provided, when a long time elapsed $T \to \infty$ (the curve T_∞) and a steady state is concerned. If the period of excitation is limited, an effect of transition from the semi-trivial until post-critical stable state proceeds, see curves T_1 and T_2 in Figures *16*b-d. For short excitation period (curve T_1) the excitation amplitude a_0 can significantly exceed the bifurcation point B_2 and the response has still the semi-trivial character, although increasing a_0 even more, the response is approaching the steady state (long-dashed curves, T_∞). The medium time of excitation leads to some intermediate response type (T_2 curves). It can be ascertained that for time limited excitations, e.g., earthquake shock, higher excitation amplitudes can be admissible than those obtained for the steady state vibrations.

Detailed time history of all three response components for amplitudes $a_0 \in (0.4, 0.6)$ discussed above is given in Figure *17*. The horizontal response amplitudes $\varphi(t), \psi(t)$ asymptotically approach the non-zero constant. This constant in case of $a_0 = 0.4$ is rather small and the result does not differ from the semi-trivial solution significantly. It is also worth to mention noticeable decrease of the amplitude $\zeta(t)$ for $a_0 = 0.6$ with growing amplitudes of $\varphi(t), \psi(t)$.

Similar technique can be used for assessment in case of model B. As in the previous case, analytical investigation of the system (10) does not probably provide any understandable results. Therefore, simulation processes should be undertaken in order to outline the irreversibility limit. Numerical solution of the differential system (10) in full version was multiply performed for various parameters in the unstable domain. The amplitude of the solution was checked in each integration step to remain within prescribed bounds. The time instant τ_c of the first exceeding was recorded and the integration procedure was stopped. This moment indicates that the limit of irreversibility has been reached. Any other quantity can be checked in place of amplitude like maximal stress or tension in the structure. The value τ_c can be

Figure 17. Time history of the response components $\zeta(t), \varphi(t), \psi(t)$ for increasing excitation amplitude (Model A)

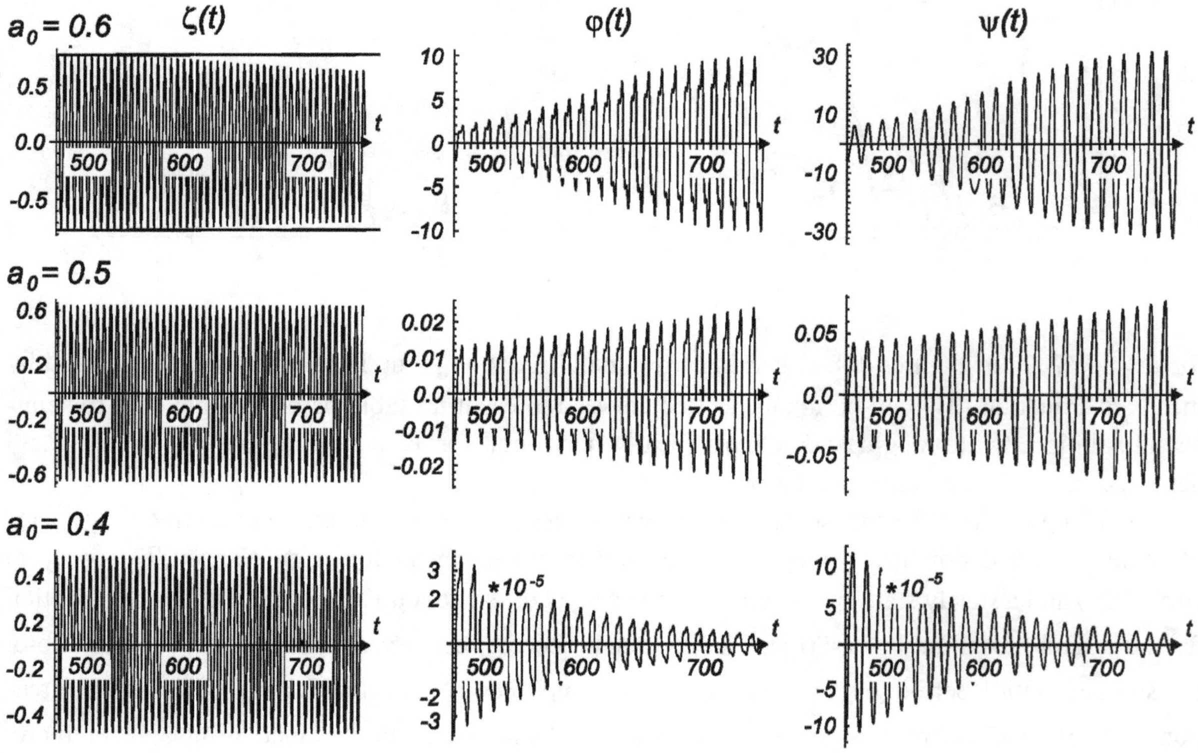

Figure 18. Outer stability limit (limit of irreversibility) of the system for the increasing console bending stiffness. Dashed green line – stability limit of the semi-trivial solution ($n = 3$), solid curves – limit of irreversibility for $\tau_c = 800, 300, 100$; other parameters $\eta_e = 0.05$, $\rho = 0.2$ (Model B)

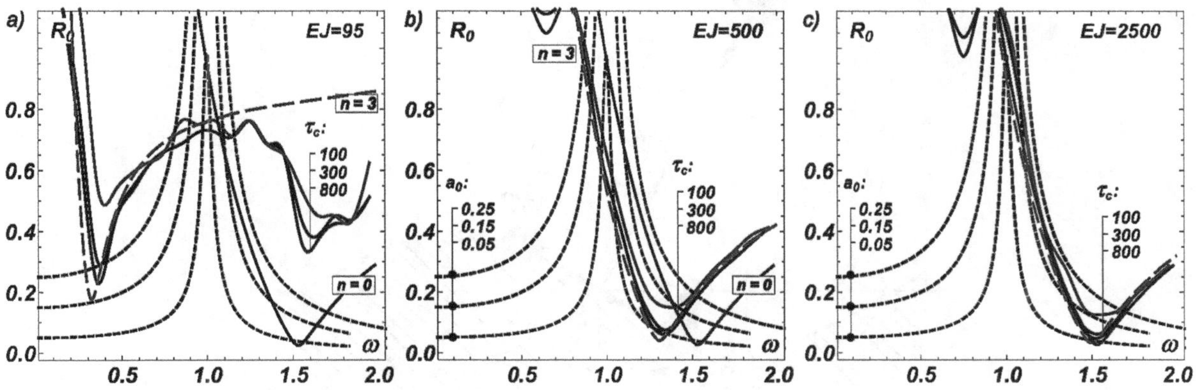

interpreted as a limit of duration of a critical excitation. The actual limit value is structure dependent and should be specified as a part of the design process.

Some results are plotted in Figure *18* for three console bending stiffnesses. The stability limit of the semi-trivial solution has been plotted for $n = 3$ (three mode shapes considered) together with the stiff case ($n = 0$, solid red line). Solid colour curves represent limits of irreversibility, dashed green line stands in the corresponding stability limit of the semi-trivial solution. Three limits are plotted in every picture (a)-(c). Each one demonstrates interconnection of points with the same value $\tau_c = 100, 300, 800$ s (time, when the failure criterion has been reached). It is obvious that increasing τ_c the results approach the stability limits of the semi-trivial solution, especially for higher values of the bending stiffness of the console. The only exception is the most elastic case (a), where the both stability limits, $n = 0$ and $n = 3$, should be taken into account.

Figure 19. Maximal real part of the system matrix (39) eigen-values for the increasing console bending stiffness. Solid black curves show the individual levels of $max\left(\Re(\Lambda_i)\right) = 0.05, 0.1, 0.15, 0.2$, dashed green curve stands in semi-trivial stability limit for $n = 3$ (and $\Lambda = 0$); other parameters $\eta_e = 0.05, \rho = 0.2$ (Model B)

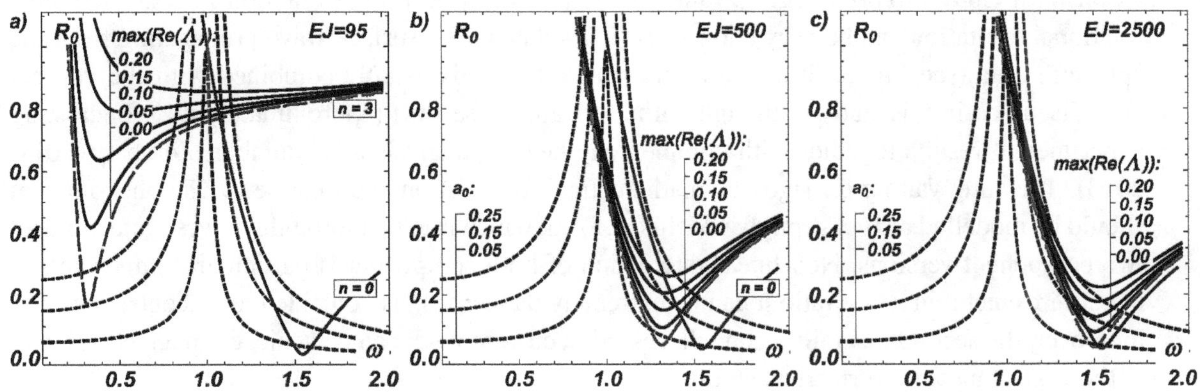

Figure 20. Rotating excitation due to surface wave propagation

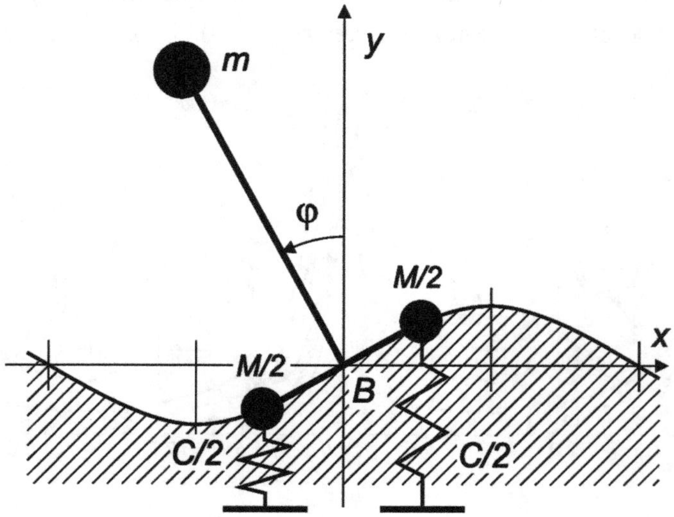

The numerically obtained values can be compared to an approximate criterion based on maximal real parts of eigen-values Λ_i of the normalized differential system given by Equation (39). Figure *19* presents the distribution of the eigen-values Λ_i in the (ω, R_0) plane for increasing bending stiffness of the console. The dashed and solid lines have the same meaning as in the previous pictures. The dashed line moreover coincides with the curve for $\max\left(\Re(\Lambda_i)\right) = 0$ as it represents zero of the determinant of the system matrix (39). It is worth to mention, that the eigen-value contour lines resemble the numerical limits of irreversibility for different τ_c in Figure *18*. The maximal real part of the eigen-values can serve as a measure of a critical duration of the excitation, which oversteps the stability limit.

FUTURE RESEARCH DIRECTIONS

The topics treated in the chapter include a large potential of further research. There are a couple of directions worthy to be followed. Some of these are pointed out below:

- **Rotational Components of Excitation:** In addition to the vertical excitation in a seismic area, rotational excitation can be very dangerous. It is related with surface wave propagating from the epicentre area (see Figure *20*). The rotational excitation is usually combined with the vertical one (discussed in this study) although with a certain phase shift (approximately π/2). Interesting experimental results together with complex engineering analysis are available (Zembaty, 2011, 2013). They are waiting for rigorous mathematical formulation and treatment. The phenomenon should be modelled also in strongly non-linear form, which is to be thoroughly investigated in one/two component versions. Non-linear interaction of both component is particularly important, as in certain synchronization ratio it can be extremely dangerous also outside the epicentre area and moreover, the second instability zone discussed in current work can be achieved much easier than in the case of a solely vertical excitation.

- **Time Variability of Subsoil Properties During a Seismic Shock:** Results of experimental measurements show that the stiffness, internal damping, and other features of the subsoil can change substantially during a seismic shock. This evolution is amplified by structure-subsoil interaction. This process can change spectral properties of the system and, consequently, its sensitivity to acting excitation. These changes do not proceed too fast. Therefore, evolution can be regarded as a function of "slow" time, similarly like partial amplitudes of the response. They can be analysed using procedures of multiple scales.
- **Sub-/Supra-Harmonic Resonance:** The model with continuous mass distribution at the vertical console should be subdued to thorough analysis concerning sub-/supra-harmonic resonances effects. These phenomena produce dangerous multi-harmonic resonance which cannot be seen at linear systems. However, in non-linear state they can emerge and strongly amplify effects of the basic resonance. They are particularly risky at structures with weak bending stiffness of the vertical console.
- **Random Excitation:** Response due to random or other type of broadband excitation in one and two components (vertical and rotational) corresponding to seismic processes is worth to be include into model. Significant non-stationary character of a seismic process should be respected. Although non-stationarities are of several type, in principal besides the fully general type two main categories are to be discussed: amplitude modulation and frequency modulation. Advanced semi-analytical and numerical methods based on Markov processes and Fokker-Planck equation are recommended for this purpose as the most effective (Náprstek & Král, 2014).
- **Vibration Damping:** A large variety of passive and active vibration damping devices exist and are widely used in practice. Besides a vast number of paper, many monographs are available (Liang, Lee, Dargush, & Song, 2012). The most frequent absorbers are based on various pendulums (Den Hartog principle), auto-parametric principle (Tondl principle), and hydraulic systems (Tondl, 1991; Tondl et al., 2000). Nevertheless, they are studied either independently from real structures or in a linear system as a whole. Therefore, the non-linear structure, including a non-linear damper, should be investigated as an integral system exhibiting all features indicated in the present chapter.
- **Spatial Approach:** While the formulations in the chapter are considered in 2D, real structures have space character. This should be respected when a particular structure is developed and intended for realization.

It is worth noting that phenomena reflecting the nature of the inverted pendulum can be encountered in more places than structures exposed to a seismic shock in civil engineering. Many engineering and physical systems suffer from these types of effects born by fluid structure excitation, neighbourhood of traffic artery, physical fields influence (plasma processes, nano-tubes, particle interaction, etc.), and others related with the stochastic resonance and other general phenomena (Náprstek, 2015b).

CONCLUSION

In this chapter, non-linear analysis of a simplified slender structure has been presented. It has been shown that the auto-parametric resonance is a potentially dangerous effect, which cannot be captured using the traditional linear approach. In such cases, interaction of individual degrees of freedom takes place due to non-linear terms in post-critical regime only. Non-linear stability analysis was shown using two simplified models of an idealized slender vertical structure on elastic subsoil exposed to strong vertical excitation. Hamiltonian functionals were formulated. Subsequently, the respective Lagrangian governing systems were carried out for both models. Differential systems showed that horizontal and vertical response components are independent on the linear level. Existence of certain intervals of the excitation frequency was proven where semi-trivial solution loses its dynamic stability and strong horizontal response components become decisive from the point of view of various standards. In post-critical state, the bending effects in the structure foundation and deformability of the vertical console are dominant. Various forms and internal structure of stability limits with respect to principal system parameters were found. Post-critical states were thoroughly analysed with respect to excitation frequency and excitation amplitude. All such states enable steady state, quasi-periodic, or irregular time history. Two generally different types of the post-critical regimes were shown:

1. Although in the post-critical state, a certain area in the neighbourhood of the stable state exists from where the structure response can return back to the stable state when the stability conditions are regained; sensitivity of the system parameters concerning auto-parametric stability loss was analysed;
2. Beyond the primary area of the instability the rocking response component rises rather exponentially leading inevitably to the failure of the structure. Consequently, presence of the horizontal component in the system response does not automatically mean inevitable collapse of the structure. Such a response can keep stationary (or at least bounded) character and can disappear, if the excitation is removed. However, if the limit of the irreversibility is overstepped, horizontal response components rise beyond any limits and the structure collapses.

ACKNOWLEDGMENT

The kind support of the Czech Science Foundation project No. 15-01035S and the institutional support No. RVO 68378297 are gratefully acknowledged.

REFERENCES

Abarbanel, H. D. I., Brown, R., & Kadtke, J. B. (1990). Prediction in chaotic nonlinear systems: Methods for time series with broadband Fourier spectra. *Physical Review A.*, *41*, 1742–1807. PMID:9903289

Ambraseys, N., & Douglas, J. (2000). *Reappraisal of the effect of vertical ground motions on response.* ESEE Report No. 00-4. Imperial College, Civil and Environmental Engineering Dept., London.

Arnold, V. I. (1978). *Mathematical methods of classical mechanics*. New York: Springer.

Bolotin, V. V. (2013). *Random vibrations of elastic systems*. Springer.

Bozorgnia, Y., & Campbell, K. W. (2004). The vertical-to-horizontal response spectral ratio and tentative procedures for developing simplified V/H and vertical design spectra. *Journal of Earthquake Engineering*, *8*(2), 175–207.

Bozorgnia, Y., Niazi, M., & Campbell, K. W. (1995). Characteristics of free-field vertical ground motion during the Northridge earthquake. *Earthquake Spectra*, *11*, 515–525.

Broderick, B. M., & Elnashai, A. S. (1995). Analysis of the failure of Interstate 10 freeway ramp during the Northridge earthquake of 17 January 1994. *Earthquake Engineering & Structural Dynamics*, *24*, 189–208.

Burton, T. D. (1982). Non-linear oscillator limit cycle analysis using a time transformation approach. *International Journal of Non-linear Mechanics*, *17*(1), 7–19.

Carydis, P., Castiglioni, C., Lekkas, E., Kostaki, I., Lebesis, N., & Drei, A. (2012). The Emilia Romagna, May 2012 earthquake sequence. The influence of the vertical earthquake component and related geo-scientific and engineering aspects. *Ingegneria Sismica*, *29*(2-3), 31–58.

Chossat, P., & Lauterbach, R. (2000). *Methods in equivalent bifurcations and dynamical systems*. Singapore: World Scientific Publishing.

Elnashai, A. S., & Papazoglou, A. J. (1996). Analytical and field evidence of the damaging effect of vertical earthquake ground motion. *Earthquake Engineering & Structural Dynamics*, *25*, 1109–1137.

Guckenheimer, J., & Holmes, P. (1983). *Nonlinear oscillations, dynamical systems and bifurcations of vector fields*. New York: Springer.

Hatwal, H., Mallik, A. K., & Ghosh, A. (1983). Forced nonlinear oscillations of an autoparametric system. *ASME Journal of Applied Mechanics*, *50*(3), 657–662.

Haxton, R. S., & Barr, A. D. S. (1974). The autoparametric vibration absorber. *ASME Journal of Applied Mechanics*, *94*, 119–125.

Iooss, G., & Adelmeyer, M. (1992). *Topics in bifurcation theory and applications*. London: World Scientific Publishing.

Kalkan, E., Adalier, K., & Pamuk, A. (2004). Near source effects and engineering implications of recent earthquakes in Turkey.*Fifth International Conference on Case Histories in Geotechnical Engineering*, 13-17.

Kapitsa, P. L. (1951). Dynamic stability of a pendulum with a vibrating point of suspension. Journal of Experimental and Theoretical Physics, 21, 588–598. (In Russian)

Lee, W. K., & Hsu, C. S. (1994). A global analysis of an harmonically excited spring-pendulum system with internal resonance. *Journal of Sound and Vibration*, *171*(3), 335–359.

Liang, Z., Lee, G. C., Dargush, G. F., & Song, J. (2012). *Structural damping – Applications in seismic response modification*. Boca Raton, FL: CRC Press – Taylor & Francis.

Lin, Y. K., & Cai, G. Q. (1995). *Probabilistic structural dynamics - Advanced theory and applications*. New York: McGraw-Hill.

Murphy, K. D., Bayly, P. V., Virgin, L. N., & Gottwald, J. A. (1994). Measuring the stability of periodic attractors using perturbation-induced transients: Applications to two non-linear oscillators. *Journal of Sound and Vibration*, *172*(1), 85–102.

Nabergoj, R., & Tondl, A. (1994). A simulation of parametric ship rolling: Effects of hull bending and torsional elasticity. *Nonlinear Dynamics*, *6*, 265–284.

Nabergoj, R., Tondl, A., & Virag, Z. (1994). Autoparametric resonance in an externally excited system. *Chaos, Solitons, and Fractals*, *4*, 263–273.

Náprstek, J. (1996). Stochastic exponential and asymptotic stability of simple non-linear systems. *International Journal of Non-linear Mechanics*, *31*(5), 693–705.

Náprstek, J. (1998). Non-linear self-excited random vibrations and stability of an SDOF system with parametric noises. *Meccanica*, *33*, 267–277.

Náprstek, J. (2015b). Stochastic resonance - Challenges to engineering dynamics. In B. H. V. Topping & J. Kruis (Eds.), *Computational technology reviews* (pp. 53–101). Civil-Comp.

Náprstek, J., & Fischer, C. (2008). Non-linear auto-parametric stability loss of a slender structure due to random non-stationary seismic excitation. In *Proceedings of 14th Word Conference on Earthquake Engineering*. Chinese Assoc. Earthq. Eng.

Náprstek, J., & Fischer, C. (2009). Auto-parametric semi-trivial and post-critical response of a spherical pendulum damper. *Computers & Structures*, *87*, 1204–1215.

Náprstek, J., & Fischer, C. (2011). Auto-parametric stability loss and post-critical behavior of a three degrees of freedom system. In M. Papadrakakis et al. (Eds.), *Computational methods in stochastic dynamics* (pp. 267–289). Springer.

Náprstek, J., & Fischer, C. (2012). Dynamic stability of a non-linear continuous system subjected to vertical seismic excitation. In B. H. V. Topping (Ed.), *Proceedings of the Eleventh International Conference on Computational Structures Technology* (paper 179). Civil-Comp Press.

Náprstek, J., & Král, R. (2014). Finite element method analysis of Fokker-Plank equation in stationary and evolutionary versions. *Advances in Engineering Software*, *72*, 28–38.

Náprstek, J., & Pirner, M. (2002). Non-linear behaviour and dynamic stability of a vibration spherical absorber. In A. Smyth et al. (Eds.), *Proceedings of 15th ASCE Engineering Mechanics Division Conference*. New York: Columbia University.

Nayfeh, A. H., & Mook, D. T. (2008). *Nonlinear Oscillations*. New York: John Wiley & Sons.

Nayfeh, A. H., Mook, D. T., & Marshall, L. R. (1973). Nonlinear coupling of pitch and roll modes in ship motion. *Journal of Hydronautics*, *7*, 145–152.

Newmark, N. M., Blume, J. A., & Kapur, K. K. (1973). Seismic design spectra for nuclear power plants. *Journal of the Power Division*, *99*, 287–303.

Newmark, N. M., & Hall, W. J. (1978). *Development of criteria for seismic review of selected nuclear power plants. NUREG/CR-0098*. Nuclear Regulatory Commission.

Powell, P. D. (2011). *Calculating determinants of block matrices*. arXiv:1112.4379 [math.RA]

Pugachev, V. S., & Sinitsyn, I. N. (1987). *Stochastic differential systems - Analysis and filtering*. Chichester, UK: J. Wiley.

Ren, Y., & Beards, C. F. (1994). A new receptance-based perturbative multi-harmonic balance method for the calculation of the steady state response of non-linear systems. *Journal of Sound and Vibration*, *172*(5), 593–604.

Salazar, A. R., & Haldar, A. (2000). Structural responses considering the vertical component of earthquakes. *Computers & Structures*, 131–145.

Sandri, M. (1996). Numerical calculation of Lyapunov exponents. *Mathematica Journal*, *6*, 78–84.

Shih, T.-Y., & Lin, Y. K. (1982). Vertical seismic load effect on hysteretic columns. *ASCE Journal of Engineering Mechanics Div.*, *108*, 242–254.

Silva, W. J. (1997). Characteristics of vertical strong ground motions for applications to engineering design. In I. M. Friedland et al. (Eds.) *Proceedings of the FHWA/NCEER Workshop on the National Representation of Seismic Ground Motion for New and Existing Highway Facilities* (pp. 205-252). Technical Report NCEER-97-0010.

Smith, H. J. T., & Blackburn, J. A. (1992). Experimental Study of an Inverted Pendulum. *American Journal of Physics*, *60*, 909–911.

Stephenson, A. (1908). On a new type of dynamical stability. *Memoirs of the Manchester Literary and Philosophical Society*, *52*, 1–10.

Thompson, J. M. T., & Stewart, H. B. (2002). *Nonlinear dynamics and chaos* (2nd ed.). Chichester, UK: Wiley.

Tondl, A. (1991). *Quenching of self-excited vibrations*. Prague: Academia.

Tondl, A. (1997). To the analysis of autoparametric systems. *Zeitschrift für Angewandte Mathematik und Physik*, *77*(6), 407–418.

Tondl, A., & Nabergoj, R. (1994). Non-periodic and chaotic vibrations in a flow induced systems. *Chaos, Solitons, and Fractals*, *4*, 2193–2202.

Tondl, A., Ruijgrok, T., Verhulst, F., & Nabergoj, R. (2000). *Autoparametric resonance in mechanical systems*. Cambridge, UK: Cambridge University Press.

van der Burgh, A. H. P. (1968). On the asymptotic solutions of the differential equations of the elastic pendulum. *Journal de Mécanique*, *7*(4), 507–520.

Xu, Z., & Cheung, Y. K. (1994). Averaging method using generalized harmonic functions for strongly non-linear oscillators. *Journal of Sound and Vibration*, *174*(4), 563–576.

Zembaty, Z. (2011). How to model rockburst seismic loads for civil engineering purposes? *Bulletin of Earthquake Engineering*, *9*(5), 1403–1416.

Zembaty, Z. (2013). Numerical analyses of seismic ground rotations from the wave passage effects. *Geotechnical. Geological and Earthquake Engineering*, *24*, 15–28.

ADDITIONAL READING

Bajaj, A. K., Chang, S. I., & Johnson, J. M. (1994). Amplitude modulated dynamics of a resonantly excited autoparametric two degree of freedom system. *Nonlinear Dynamics*, *5*, 433–457.

Baker, G. L. (1995). Control of the chaotic driven pendulum. *American Journal of Physics*, *63*, 832–838.

Bartuccelli, M. V., Gentile, G., & Georgiou, K. V. (2001). On the dynamics of a vertically-driven damped planar pendulum. *Proceedings of the Royal Society of London. Series A*, *457*, 1–16.

Bartuccelli, M. V., Gentile, G., & Georgiou, K. V. (2002). On the stability of the upside-down pendulum with damping. *Proceedings of the Royal Society of London. Series A*, *458*, 255–269.

Bartuccelli, M. V., Gentile, G., & Georgiou, K. V. (2003). KAM Theorem and Stability of the Upside-Down Pendulum. *Discrete and Continuous Dynamical Systems*, *9*, 413–426.

Benettin, G., Galgani, L., Giorgilli, A., & Strelcyn, M. (1980). Lyapunov characteristic exponents for smooth dynamical systems and Hamiltonian systems: A method for computing all of them. Part 2, Numerical application. *Meccanica*, *15*, 21–30.

Cameron, T. M., & Griffin, J. H. (1989). An alternating frequency/time domain method for calculating the steady state response on nonlinear dynamic systems. *ASME Journal of Applied Mechanics*, *56*, 149–154.

Chetayev, N. G. (1961). *The stability of motion*. Oxford: Pergamon Press.

El Rifai, K., Haller, G., & Bajaj, A. K. (2007). Global dynamics of an autoparametric spring-mass-pendulum system. *Nonlinear Dynamics*, *49*, 105–116.

Hammel, S. M. (1990). A noise reduction method for chaotic systems. *Physics Letters. [Part A]*, *148*, 421–428.

Lichtenberg, A. J., & Lieberman, M. J. (1983). *Regular and stochastic motion*. New York: Springer.

Manuca, R., & Savit, R. (1996). Stationarity and nonstationarity in time series analysis. *Physica D. Nonlinear Phenomena*, *99*(2-3), 134–161.

Mazzino, A., Musacchio, S., & Vulpiani, A. (2004). Multiple-scale analysis and renormalization for preasymptotic scalar transport. *Physical Review E: Statistical, Nonlinear, and Soft Matter Physics*, *71*, 011113. PMID:15697586

Moser, J. (1973). *Stable and random motions in dynamical systems*. Princeton University Press.

Náprstek, J. (2015a). Combined analytical and numerical approaches in dynamic stability analyses of engineering system. *Journal of Sound and Vibration*, *338*, 2–41.

Ott, E. (2002). *Chaos in dynamical systems* (2nd ed.). Cambridge: Cambridge University Press.

Rosenstein, M. T., Colins, J. J., & De Luca, C. J. (1993). A practical method for calculating largest Lyapunov exponents from small data sets. *Physica D. Nonlinear Phenomena*, *65*, 117–134.

Schuster, H. G., & Just, W. (2005). *Deterministic chaos: An introduction* (4th ed.). Weinheim: Wiley.

Skokos, C. (2009). *The Lyapunov characteristic exponents and their computation.* Cornell University arXiv, http://arxiv.org/abs/0811.0882v2

KEY TERMS AND DEFINITIONS

Auto-Parametric Resonance: The state of an auto-parametric system, when the vibrations of the primary system act as parametric excitation of the secondary systems. The secondary systems then do not remain at rest any longer.

Auto-Parametric Systems: Consist of two or more sub-systems. Only primary part is subjected to external excitation. The excitation structure is not arbitrary and must be in accordance with the structure of the system itself and must respect its auto-parametric character. The primary as well as the secondary sub-systems can be either linear or nonlinear. Nevertheless, the links between them are always non-linear. The dynamic behaviour of the individual sub-systems is independent until the system response is in the sub-critical regime. Under such circumstances, the sub-systems or parts do not influence each other. Overstepping a certain limit, the response becomes non-trivial for the whole system. This state activates nonlinear links of the primary and secondary sub-systems. The system response enters a regime of auto-parametric resonance.

Harmonic Excitation: The function describing one-dimensional simple harmonic motion, solution of undamped vibration equation $m\ddot{x} = -kx$. In particular, $a \sin \omega t$ is used in the present text.

Internal Resonance: A phenomenon which can emerge at non-linear systems when ratio of (at least two) eigen-frequencies is a rational number (optimal ratio is 1:2). In such a case the sub-/super-harmonic synchronization (internal resonance) can take place. The system response exhibits two or more harmonic components despite the excitation is mono-harmonic.

Quasi-Periodic Response: Can emerge in a non-linear system, provided the frequency of the harmonic excitation differs slightly from the eigen-frequency of the system. It exhibits beating effects due to interaction of forced and self-excited vibration. Frequency of beatings starts from zero and rises until infinitum in a small interval around the eigen-frequency. Simultaneously, the amplitude of beatings decays until zero.

Semi-Trivial Solution of an Auto-Parametric System: The solution of primary part is non-trivial, though that of the secondary parts remain trivial (zero), until the stability limit is reached.

Stability Limit: Stability limit for given auto-parametric system is a boundary in the space of system parameters (damping, excitation frequency or amplitude, etc.) between regions leading to semi-trivial response or auto-parametric resonance.

Stationary Response: Stationary response of an auto-parametric system is its solution whose components can be fully described using its Fourier expansion with constant Fourier coefficients (amplitudes).

APPENDIX

The coefficient matrix $\mathbf{H}$ present in Equation (24) for model A are given as:

$$\mathbf{H} = \begin{pmatrix} h_1 & h_2 & h_3 & \kappa_{1A}\omega^2 a_s - h_4 \\ -h_2 & h_1 & \kappa_{1A}\omega^2 a_s + h_4 & -h_3 \\ -4\omega_{3A}^2 + \omega^2 a_c & -2\omega_{3A}^2\eta_e + \omega^2 a_s & 4\omega_{3A}^2 - \omega^2 & 2\omega\omega_{3A}^2\eta_e \\ 2\omega_{3A}^2\eta_e + \omega^2 a_s & -4\omega_{3A}^2 - \omega^2 a_c & -2\omega\omega_{3A}^2\eta_e & 4\omega_{3A}^2 - \omega^2 \end{pmatrix} \tag{41}$$

where:

$$h_1 = 4\left(\kappa_{1A}\omega_{3A}^2 + \omega_{1A}^2\right) - \omega^2,\ \mathrm{h}_2 = 2\left(\kappa_{1A}\omega_{3A}^2\eta_e + \omega_{1A}^2\eta_c\right),$$

$$h_3 = \kappa_{1A}\left[\omega^2 a_c - 4\left(\omega_2^2 + \omega_{3A}^2\right)\right] \quad ,\ \mathrm{h}_4 = 2\kappa_{1A}\omega_{3A}^2\eta_e. \tag{42}$$

Polynomial coefficients of determinant of the system matrix $\mathbf{H}$ given by (41) read:

$$\begin{aligned} A_0(\omega) = \omega^8 + 4\omega^6 &\left[\left(\omega_1^2\eta_c + \kappa_1\omega_3^2\eta_e\right)^2 - \left(2\kappa_1 + 1\right)\left(1 - \omega_3^2\eta_e^2\right)\omega_3^2 - 2\omega_1^2 - \omega_3^2\right] + \\ &+ 16\omega^4\left[\omega_1^2\omega_3^2\eta_c\left(\omega_1^2\eta_c\left(\omega_3^2\eta_e^2 - 2\right) + 2\kappa_1\omega_2^2\eta_e\right) + \right. \\ &\left. +2\gamma\omega_3^2\left(1 - \left(\kappa_1 + 1\right)\omega_3^2\eta_e^2\right) + \left(\left(\kappa_1 + 1\right)\omega_3^2 + \omega_1^2\right)^2\right] + \\ &+ 64\omega^2\left[\omega_1^4\omega_3^4\eta_c^2 + \gamma^2\omega_3^4\eta_e^2 - 2\gamma\omega_3^2\left(\left(\kappa_1 + 1\right)\omega_3^2 + \omega_1^2\right)\right] + 256\gamma^2\omega_3^4 \end{aligned}$$

$$\begin{aligned} A_2(\omega) = \omega^4 - 4\omega^2 &\left[\omega_{1A}^2\left(1 - \omega_{3A}^2\eta_c\eta_e\right) + \omega_{3A}^2\left(\kappa_{A1}\left(1 - 2\omega_{3A}^2\eta_e^2\right) + 1\right)\right] + \\ &+ 8\left[\kappa_1\left(3\omega_{3A}^4 + \left(\omega_2^2 + \omega_{3A}^2\right)^2\right) + 2\omega_1^2\omega_{3A}^2\right] \end{aligned} \tag{43}$$

with $\gamma = \omega_{1A}^2 - \kappa_{1A}\omega_2^2$.

Detailed structure of the matrix $\mathbf{M}(\mathbf{x}) \in \mathbb{R}^{6\times6}$ and the vector $\mathbf{K}(\mathbf{x}) \in \mathbb{R}^6$ according to general relation (30) for model A:

$$\mathbf{M}(\mathbf{x}) = \frac{1}{2}\omega\begin{pmatrix} 0 & 2 & 0 & 0 & -\frac{1}{2}\kappa_0 P_s & -\frac{1}{2}\kappa_0 P_c \\ -2 & 0 & 0 & 0 & \frac{1}{2}\kappa_0 P_c & -\frac{1}{2}\kappa_0 P_s \\ \kappa_1 P_s & -\kappa_1 P_c & 0 & 1 & 0 & \kappa_1 \\ \kappa_1 P_c & \kappa_1 P_s & -1 & 0 & -\kappa_1 & 0 \\ P_s & -P_c & 0 & 0 & 0 & 1 \\ P_c & P_s & 0 & 0 & -1 & 0 \end{pmatrix}$$

$$\mathbf{K}(\mathbf{X}) = \frac{1}{4}\begin{pmatrix} 4\left[a_0\eta_c\omega\omega_0^2 + \left(\omega^2 - \omega_0^2\right)R_c - \eta_c\omega\omega_0^2 R_s\right] + \kappa_0\omega^2\left(P_s S_s - P_c S_c\right) \\ 4\left[a_0\omega_0^2 + \left(\omega^2 - \omega_0^2\right)R_s - \eta_c\omega\omega_0^2 R_c\right] - \kappa_0\omega^2\left(P_s S_c + P_c S_s\right) \\ \left[P_c\left(\omega^2\left(1-\kappa_1 R_c\right) - 4\omega_1^2\right) - 2\omega\eta_c P_s\omega_1^2\right] - \kappa_1\left[\left(\left(R_c - 1\right)\omega^2 - 4\omega_2^2\right)S_c + \omega^2 R_s\left(P_s + S_s\right)\right] \\ \left[P_s\left(\omega^2\left(1+\kappa_1 R_c\right) - 4\omega_1^2\right) + 2\omega\eta_c P_c\omega_1^2\right] + \kappa_1\left[\left(\left(R_c + 1\right)\omega^2 + 4\omega_2^2\right)S_s - \omega^2 R_s\left(P_c + S_c\right)\right] \\ \left(S_c - P_c R_c - P_s R_s\right)\omega^2 + 2\omega_3^2\left(2(P_c - S_c) + \omega\eta_e\left(P_s - S_s\right)\right) \\ \left(S_s + P_s R_c - P_c R_s\right)\omega^2 + 2\omega_3^2\left(2(P_s - S_s) - \omega\eta_e\left(P_c - S_c\right)\right) \end{pmatrix} \quad (44)$$

Inverse matrix $\mathbf{M}(\mathbf{x})^{-1}$ can be written as:

$$\mathbf{M}(\mathbf{x})^{-1} = 2\begin{pmatrix} 0 & 0 & 0 & -\omega^{-1} & 0 & \kappa_1\omega^{-1} \\ 0 & 0 & \omega^{-1} & 0 & -\kappa_1\omega^{-1} & 0 \\ 0 & 2\delta & 0 & 0 & \delta\kappa_0 p_s(t) & \delta\kappa_0 p_c(t) \\ -2\delta & 0 & 0 & 0 & -\delta\kappa_0 p_c(t) & \delta\kappa_0 p_s(t) \\ -2\delta p_s(t) & 2\delta p_c(t) & 0 & 0 & 0 & 4\delta \\ -2\delta p_c(t) & -2\delta p_s(t) & 0 & 0 & -4\delta & 0 \end{pmatrix} \quad (45)$$

where $\delta = \omega^{-1}\left(\kappa_0\left(p_c(t)^2 + p_s(t)^2\right) - 4\right)^{-1}$.

The coefficient matrices present in Equation (38) for model B are given as:

$$\mathbf{M} = \omega \begin{pmatrix} \frac{1}{\kappa_{1B}}\mathbf{J}, & \theta_{1,1}\mathbf{J}, & \theta_{1,2}\mathbf{J}, & \cdots & \theta_{1,n}\mathbf{J} \\ \theta_{1,1}\mathbf{J}, & \theta_{2,1}\mathbf{J}, & 0, & \cdots & 0 \\ \theta_{1,2}\mathbf{J}, & 0, & \theta_{2,2}\mathbf{J}, & 0, & \vdots \\ \vdots & \vdots & & \ddots & 0 \\ \theta_{1,n}\mathbf{J}, & 0, & \cdots & 0, & \theta_{2,n}\mathbf{J} \end{pmatrix} \tag{46}$$

$$\mathbf{H} = \frac{1}{2} \begin{pmatrix} \mathbf{P}, & \mathbf{S}_1, & \mathbf{S}_2, & \cdots & \mathbf{S}_n \\ \mathbf{S}_1, & \mathbf{D}_1, & 0, & \cdots & 0 \\ \mathbf{S}_2, & 0, & \mathbf{D}_2, & \ddots & \vdots \\ \vdots & \vdots & \ddots & \ddots & 0 \\ \mathbf{S}_n, & 0, & \cdots & 0, & \mathbf{D}_n \end{pmatrix} \tag{47}$$

where sub-matrices $\mathbf{P}, \mathbf{S}_i, \mathbf{D}_i, \mathbf{J} \in \mathbb{R}^{2\times2}$ in (27-29) and (46-47) have a form as follows:

$$\mathbf{P} = \begin{pmatrix} -\omega\left(\omega a_s + 2\omega_{1B}^2\eta_c\kappa_{1B}^{-1}\right), & \left(\omega^2 - 4\omega_{1B}^2\right)\kappa_{1B}^{-1} - \omega^2 a_c \\ \left(\omega^2 - 4\omega_{1B}^2\right)\kappa_{1B}^{-1} + \omega^2 a_c, & -\omega\left(\omega a_s - 2\omega_{1B}^2\eta_c\kappa_{1B}^{-1}\right) \end{pmatrix},$$

$$\mathbf{S}_i = \begin{pmatrix} -\omega^2 a_s\theta_{0,1}, & \theta_{1,1}\omega^2 + \left(4\omega_2^2 - \omega^2 a_c\right)\theta_{0,1} \\ \theta_{1,1}\omega^2 + \left(4\omega_2^2 + a_c\omega^2\right)\theta_{0,1}, & -\omega^2 a_s\theta_{0,1} \end{pmatrix},$$

$$\mathbf{D}_i = \begin{pmatrix} -2\omega\eta_e\omega_{3B}^2\theta_{3,i}, & \omega^2\theta_{2,i} - 4\omega_{3B}^2\theta_{3,i} \\ \omega^2\theta_{2,i} - 4\omega_{3B}^2\theta_{3,i}, & 2\omega\eta_e\omega_{3B}^2\theta_{3,i} \end{pmatrix}, \quad \mathbf{J} = \begin{pmatrix} 1 & 0 \\ 0 & -1 \end{pmatrix}. \tag{48}$$

Determinant of matrix $\mathbf{M}$ (46) can be written as:

$$\det\left(\mathbf{M}\right) = (-1)^{n+1}\frac{\omega^{2n+2}}{\kappa_{1B}^2}\left(\kappa_{1B}\sum_{l=1}^{n}\theta_{1,l}^2\left(\prod_{k=1,k\neq l}^{n}\theta_{2,k}\right) - \prod_{k=1}^{n}\theta_{2,k}\right)^2. \tag{49}$$

Inversion of matrix $\mathbf{M}$ (46) can be presented in a form of a block matrix:

$$\mathbf{M}^{-1} = A_n^{-1}\left[\mathbf{N}_{i,j}\right], \qquad i,j = 0,\ldots,n \tag{50}$$

where the block elements $\mathbf{N}_{i,j} \in \mathbb{R}^{2\times 2}$ and constants A_i are given as:

$$\mathbf{N}_{0,0} = -\kappa_1 \left(\prod_{k=1}^{n} \theta_{2,k} \right) \mathbf{J},$$

$$\mathbf{N}_{0,i} = \mathbf{N}_{i,0}^T = \kappa_1 \theta_{2,i} \left(\prod_{k \neq i, k=1}^{n} \theta_{2,k} \right) \mathbf{J},\ i = 1,\ldots,n,$$

$$\mathbf{N}_{i,j} = -\kappa_1 \theta_{1,i} \theta_{1,j} \left(\prod_{k \neq i, k \neq 1}^{n} \theta_{2,k} \right) \mathbf{J},\ i \neq j; i,j = 1,\ldots,n,$$

$$\mathbf{N}_{i,i} = \left(\kappa_1 \sum_{l \neq i; l=1}^{n} \theta_{1,l}^2 \prod_{k \neq i, l; k=1}^{n} \theta_{2,k} - \prod_{k \neq i, k=1}^{n} \theta_{2,k} \right) \mathbf{J},\ i,j = 1,\ldots,n,$$

$$A_0 = \omega\left(\kappa_1 \theta_{1,1}^2 - \theta_{2,1}\right),\ A_i = A_{i-1}\theta_{2,i} + \kappa_1 \omega \theta_{1,i}^2 \left(\prod_{j=1}^{i-1} \theta_{2,j} \right) \quad (51)$$

Chapter 7
A Review of the Accuracy of Force- and Deformation-Based Methods in Determining the Seismic Capacity of Rehabilitated RC School Buildings

Orkun Gorgulu
Istanbul Technical University, Turkey

Beyza Taskin
Istanbul Technical University, Turkey

ABSTRACT

This chapter focuses on the comparison of the conventional linear force-based method with the advanced nonlinear deformation-based method that are commonly preferred to investigate the seismic performances of the existing RC school buildings. School buildings which have different structural characteristics and RC infill wall index are generated from an existing school's layout plan. During the nonlinear dynamic analysis, seven recorded earthquake motions which are scaled in accordance with the Turkish Earthquake Code are employed. Seismic performances of the school buildings against the two different earthquake hazard level are evaluated considering not only various RC infill wall indexes but also different material strengths and number of stories in terms of limit states specified in the code. In order to determine the most appropriate method related to material strength, floor level and RC infill wall index for the seismic strengthening of the existing RC school buildings, the obtained linear forced and nonlinear deformation based analyses results are compared to each other.

DOI: 10.4018/978-1-5225-2089-4.ch007

INTRODUCTION

Earthquakes are one of the most powerful natural disasters tragically affecting human lives and environment. In many seismically vulnerable regions on earth, hundreds of thousands of people remain homeless, buildings are being destroyed and worst, lots of people are losing their lives after a destructive shake. Therefore, improving the seismic capacity of existing structures became a vital field of investigation in these countries. It is a known fact that the lack of sufficient amount of attention paid during the design, construction and control phases accompanied by poor structural materials; result many of the reinforced concrete (RC) buildings to have inadequate structural resistance against the expected earthquake demand.

The two main issues during the rehabilitation of RC buildings are the decision of the retrofitting technique to be employed and the analysis method to be utilized. Although many retrofitting techniques such a jacketing of columns; steel bracing; addition of RC infill walls or FRP wrapping are proposed as alternatives for seismic improvement, introducing RC infill walls into the existing structural system is a widely accepted and preferred retrofitting technique due to its effectiveness, relative ease, and lower overall project cost compared to the other methods. In this technique, sufficient amounts of the existing RC frames are converted into RC infill walls, so that an adequate improvement in the seismic capacity is achieved by increasing the existing structural lateral stiffness and strength rapidly and easily.

During the recent years, many analytical researches have been conducted to investigate the seismic behavior of existing RC buildings. Lopez et al. (2004) obtained structural seismic capacity of RC buildings by means of a nonlinear analysis as a part of a project to reduce seismic risks in schools in Venezuela. Postelnicu et al. (2005) studied on the existing RC buildings in Romania and determine the seismic performances of them by using the linear and nonlinear analyses methods. Kuran et al. (2007) preferred linear and nonlinear analyses procedures to determine the seismic performance of existing 1502 type disaster buildings in Bingol (Turkey), which were built after 1971 Bingol Earthquake. Taskin and Tugsal (2014) investigated the seismic performances of the existing RC school buildings subjected to 2011 Van earthquakes by using the linear analyses methods. Ferraioli (2015) performed a case study of seismic performance assessment of RC buildings in Italy considering the nonlinear analyses method approaches. As it is seen, independent from the strengthening technique, linear force-based and nonlinear performance-based analysis procedures have become prominent methods for the determination of the seismic capacity and performance of the retrofitted RC buildings against the seismic demand.

In the conventional force-based seismic performance evaluation methods, it is generally assumed that the structure responds governed by its fundamental (generally first mode) mode during an earthquake. Therefore, seismic loads are applied to the analytical model by considering the first modal shape and the elements' structural responses are controlled in terms of the residual capacities. In this sense, force-based method gives a rather simpler solution providing time with cost efficiency when compared to the nonlinear performance-based methods. On the other hand, procedures based nonlinear dynamic analyses are more realistic compared to the conventional force-based methods, since the incremental dynamic behavior of the structure is simultaneously being considered during the analysis. Many procedures are proposed by researchers for nonlinear analyses (Chopra & Goel, 2002; Gupta & Kunnath, 2000; Elnashai, 2000; Antoniou & Pinho, 2004; Jan et al., 2004). Nevertheless, time history analyses by using the real recorded ground motion is more realistic due to including of site and ground motion characteristic such as frequency content of the input motion or site amplification. In this method, dynamic response of a structure subjected to a specific ground motion ensemble is obtained at each time increment. Thus, capability of producing results with relatively low uncertainty is increased with requisition of large quantities of data and also time. Therefore, the effectiveness of the simple and fast but approximate force-based methods with respect to the complicated and time consuming but realistic performance-based methods should be comparatively investigated in terms of reflecting the structural behavior adequately.

This chapter involves in the inspection of the accuracy and the precision of linear force and nonlinear deformation based evaluation methods for the seismic rehabilitation of existing RC buildings. For this purpose, due to their importance amongst the rest of the existing buildings which also serve as post-disaster shelters, RC school buildings retrofitted by the inclusion of RC infill walls are taken into consideration. School buildings, which have different structural materials, number of floors and RC infill wall indexes, are analytically generated from an existing RC school building. The generated school buildings include the common negative features encountered in many existing sub-standard schools or other RC buildings, such as:

- Low concrete strength,
- Inadequate stiffness,
- Insufficient amount of confinement and
- Improper detailing.

Thus, the most realistic reflection of the existing building stock is taken into account during the analyses.

Later, seismic performance evaluation of the generated RC school buildings is conducted by using the force-based and deformation-based methods. Force based analyses are performed by utilizing frame and shell element models in 3D, while the nonlinear dynamic analyses are conducted with the mathematical model constituted by using the fiber element modeling approach in 2D. Time history analyses are achieved by employing seven recorded earthquake motions that are scaled to satisfy the demand spectrum provided is the seismic code, so that the comparison became meaningful. Furthermore, in order to reflect the cyclic behavior more accurately, fiber element based mathematical models of the RC school buildings are represented by using a specific hysteretic model calibrated by two selected well-known large-scale RC infill wall experiments conducted by Strepelias et al.(2012) and Canbay et al.(2003). According to the consecutive analyses, seismic performance levels computed by both analysis methods are utilized for the entire school buildings in terms of the limit states such as:

- Interstory drifts,
- Shear strength demand/capacity ratios,
- Concrete and steel strains.

Thus, accuracy and reliability of force and deformation-based methods are investigated more precisely in seismic rehabilitation assessment of RC school buildings strengthened with various RC infill wall indexes.

DESCRIPTIONS OF THE INVESTIGATED RC SCHOOL BUILDINGS

The structural system of most RC public school buildings in Turkey is moment-resisting frames with a few shear walls mainly around the staircase locations. In order to provide an easily accessible system for education, these typical school buildings have limited number of stories. Unfortunately, most of the

Figure 1. Typical small- (left) and large-scale (right) RC school buildings in Turkey
Pictures taken by B.Taskin

existing sub-standard RC school buildings include common negative features: low concrete strength; insufficient stiffness, strength and ductility due to the lack of confinement zones and even unsymmetrical plans causing torsional irregularities in the structural system since they were mainly constructed in the 1960s and early 1970s. Figure 1 shows the photos of small and large-scale typical school buildings that can be commonly met anywhere in Turkey.

During the numerical investigations, an existing small-scale school building is selected as the primary structural model. Typical structural layout plan including the reinforcement details for the cross-sections of columns and beams are exhibited in Figure 2. As illustrated in the figure, the selected school building has a frame type structural system including six bays in the longitudinal and three bays in the transversal directions, respectively. The building has a rectangular floor plan by the dimensions of 36.45 m × 18.45 m with a standard story height of 3.40 m in each floor. The slab thicknesses are 200 mm which are the same at each floor level. Selected school building does not have any major irregularities through the elevation or in planar sections.

According to the site investigations and also as-built projects, transverse and longitudinal reinforcement are determined to be round plain bars denoted by S220 grade with a characteristic yield strength of

Figure 2. Typical layout plan (left) and reinforcement schemes for columns and beams

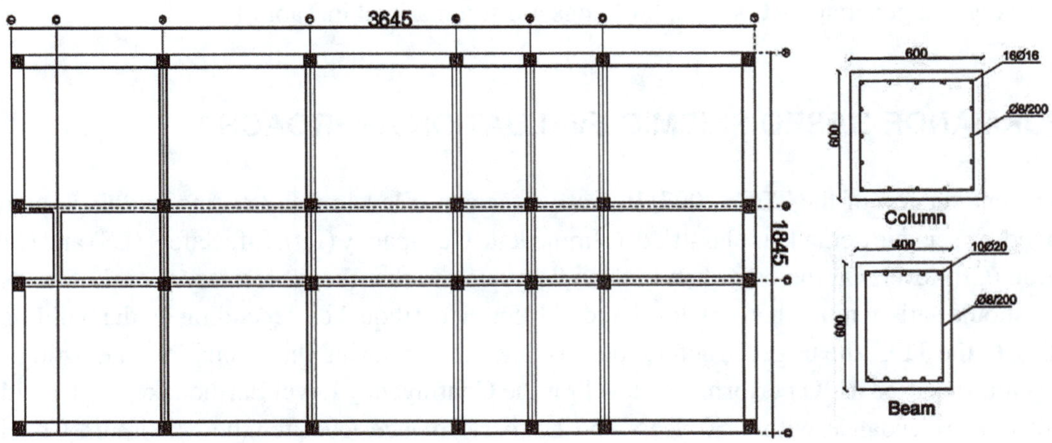

f_{yk}=220 MPa. On the other hand, in order to obtain existing concrete strength, one directional compression tests carried out on 10 core specimens taken from the structural elements. As a result of these tests, the average concrete compressive strength f_{ck} is found out to be 10 MPa. All columns in the structure have typical square cross-sections with the side length of 600mm, while the beams have flanged cross-sections with dimensions 400 ^{mm}/ 600 ^{mm}. Longitudinal reinforcement for the columns are arranged as 16 pieces of diameter 16mmrebars (16ϕ16) which resulted in a longitudinal reinforcement ratio of ρ_l=0.9%. Top and bottom reinforcement of the beams are Equal and 10ϕ20 plain bars making ρ_l=1.13%, totally. The transverse reinforcement in the beams and the columns are averagely spaced at 200mm by using the ϕ8 bars. Upon the minimum conditions according to the Turkish Earthquake Code (TEC) regulations, the existing longitudinal reinforcement ratio in the columns is slightly less than the enforced minimum of 1.0%, while it is significantly greater than the minimum in the beams. On the other hand, the stirrups used both in the columns and beams are less than the minimum ratio as well as their spacing being twice as much defined for the confinement zones in the TEC.

The existing RC school building settles on the highest seismic zone defined in the TEC, therefore an effective ground acceleration of A_0=0.40g is taken into account for the conducted seismic analyses. According to the geotechnical investigations report, local site class is defined to be *Z*2 with spectrum characteristic periods T_A=0.15s and T_B=0.40s.

Employing the existing RC school building's structural characteristics as the basis, several RC school buildings are generated with different characteristics including the number of stories, material strength and RC infill wall ratios. Thus, obtained results for the investigation of the RC school buildings are enhanced to represent more recent constructions. In this regard, number of story is increased to five. Additionally, the existing reinforcing steel grade of S220 is altered to be S420 having f_{yk}=420 MPa. On the other hand, most of the school buildings in Turkey have low concrete strength; henceforth the existing concrete compressive strength is taken as is (10 MPa) for the entire generated school buildings, which is the approximate average strength according to Tasdemir and Ozkul (1999). The last parameter in the numerical calculations is the RC infill wall ratio (ρ) that is to be used for the strengthening of the school buildings. For this purpose, some of the frames are converted into the RC infill walls and different RC infill wall ratios ranging from 0.35% to 2.40% are obtained, as provided in Figure 3. The concrete compressive strength is taken as 30 MPa and bars denoted by S420 are used as the transverse and the longitudinal reinforcement in each RC infill wall. The clear height and thickness of the additional RC infill walls in each story are selected to be 2.80 m and 300 mm, respectively. The structural characteristics of the investigated generated RC school buildings are summarized in Table 1.

PERFORMANCE-BASED SEISMIC EVALUATION APPROACH

In the current version of the TEC (2007), target performance level of an existing building against the corresponding seismic demand is classified as Immediate Occupancy (IO), Life Safety (LS) and Collapse Prevention (CP) level. On the other hand, it can be requested that the performance level of an existing structure should satisfy more than one level under different earthquakes depending on the building type. According to the TEC, the target structural performance levels during the evaluation of existing school buildings are specified as IO performance level for the Contingency Level Earthquake (CLE) with 10% probability of exceedance within 50 years and LS performance level for the Maximum Considered Earthquake (MCE) with 2% of probability of exceedance. Meanwhile, MCE level can be calculated

Table 1. Structural characteristic of the generated school buildings

RC Infill Wall Thickness (mm)	Beams (mm/mm)	Columns (mm×mm)	Concrete Class	Steel Grade	# of Stories	RC Infill Wall Ratio (ρ)	H_w / L_w
300	400/600	600×600	C10	S220 S420	3 5	0.35%	1.60 2.70
						0.70%	
						1.05%	
						1.40%	
						1.70%	
						2.10%	
						2.40%	

Figure 3. Structural layouts for the generated school buildings with the locations of RC infill walls having various ratios (ρ) in the transversal direction

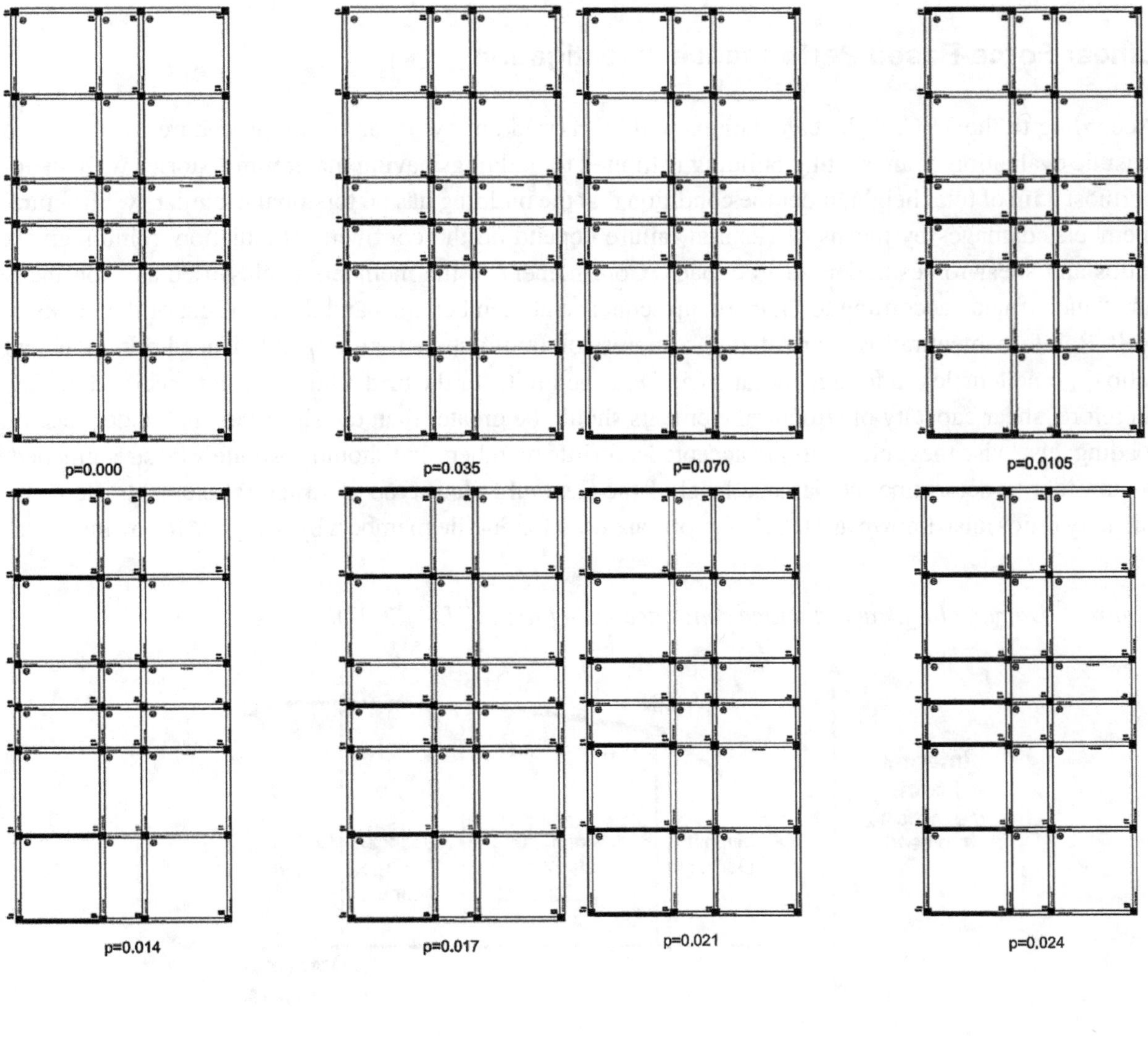

to be 1.5 times of the CLE level in terms of the design acceleration spectrum. Member damage levels to determine the structural capacity are introduced as Minimum Damage (MN), Safety Level (SF) and Collapse (CL) Level as illustrated in Figure 4.

According to the performance definitions for the IO level, maximum 10% of beams are let to perform in the significant damage zone, while the rest of the structural elements should remain within the minimum damage zone in any story. In order to achieve the LS performance level, maximum 30% of beams and columns may be in the advanced damage zone with the condition that the shear force resisted by these columns are limited to 20% of the entire story shear force. Moreover, the rest of the structural components should remain within the minimum or the significant damage zones. Earthquake demands are obtained by utilizing the seismic load reduction factor and importance factor as 1.00. All capacities are calculated by employing the existing material strengths. Regarding to the obtained information from the existing structure, the knowledge level coefficient varying from 0.75 to 1.00 is determined. During the strength capacity calculations, knowledge level is assumed to satisfy the comprehensive level requirements. Therefore knowledge factor is taken as 1.0 and characteristic strength of the material is substituted into the expressions instead of the design strength. In addition, cracked moment of inertia for the sections should be used during the seismic analyses.

Linear Force-Based Performance Investigation

According to the TEC, if the calculations will be carried out by linear elastic procedures, force-based seismic evaluation of an existing building is limited to buildings having maximum 8 stories with an uppermost 25m of total height under the condition that the building has no torsional irregularity. Structural members' damages by means of flexural failure depend on the confinement situation, reinforcement ratios and shear forces under seismic loads. Confinement of the members is classified as "confined" and "unconfined" according to their stirrup sequence at member ends and the detailing of the hooks as well. Reinforcement ratios are related to compression reinforcement ratio (ρ'); balanced reinforcement ratio (ρ_b) and tensile reinforcement ratio (ρ). Damage limits for the brittle failure is not specified in TEC therefore, shear capacity of structural elements should be greater than the shear demand under seismic loading; else wise these elements are accepted as brittle members and should absolutely be strengthened.

In order to determine the damage level of the flexural failure type of structural elements, demand/capacity ratios, also known as "DCR" factors, are used for ductile members by using the following Equa-

Figure 4. Damage limits and damage zones according to the TEC (2007)

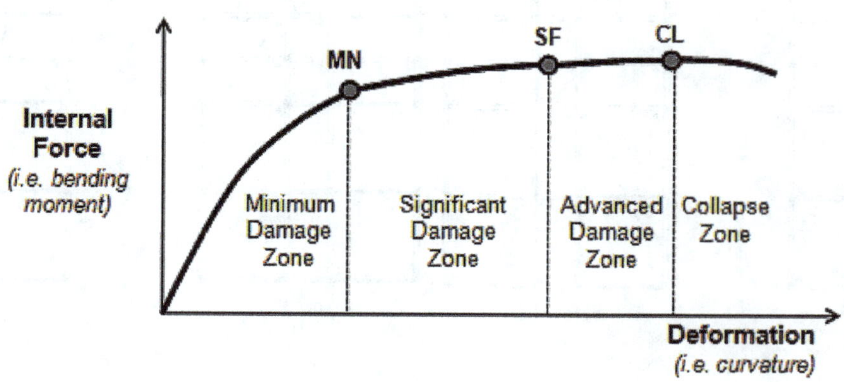

tion (1). In this expression, DCR depends on the earthquake demand (M_E); residual moment capacity (M_A); total flexural capacity (M_K) and moments under vertical loads (M_{G+Q}). Obtained DCR factors are then compared with the limits given in Table 2, Table 3, and Table 4 to determine the member damage region defined in accordance with the TEC. Moreover, relative story drift values and corresponding damage zones are also limited in terms of the damage zones as given in Table 5.

$$r = \frac{M_E}{M_A} = \frac{M_E}{M_K - M_D} = \frac{M_E}{M_K - M_{G+Q}} \leq r_s \tag{1}$$

Force-based seismic performances of each school building are evaluated by using the Equivalent force distribution pattern corresponding to the design spectrum. According to the TEC, the spectrum coefficient ($S_{(T)}$) shall be determined by Equation (2), depending on the local site conditions and the building's natural vibration period(T), as illustrated in Figure 5. Spectrum characteristic periods, T_A and T_B are set to 0.15s and 0.40s respectively, according to the geotechnical investigation report. The spectral acceleration coefficient ($A_{(T)}$) to be considered for determining the seismic loads is given by Equation (3). In this Equation, importance factor (I) is taken as 1.5 corresponding to the MCE level. In order to

Table 2. Shearwall damage limits specified for linear force-based method

Ductile Shearwalls	Performance Level		
Confinement	IO	LS	CP
Exists	3	6	8
N/A	2	4	6

Source: (TEC, 2007)

Table 3. Column damage limits specified for linear force-based method

Ductile Columns			Performance Level		
$\frac{N}{A_c f_c}$	Confinement	$\frac{V}{b_w d f_{ct}}$	IO	LS	CP
≤0.1	Exists	≤0.65	3	6	8
≤0.1	Exists	≥1.30	2.5	5	6
≥0.4 and ≤0.70	Exists	≤0.65	2	4	6
≥0.4 and ≤0.70	Exists	≥1.30	2	3	5
≤0.1	N/A	≤0.65	2	3.5	5
≤0.1	N/A	≥1.30	1.5	2.5	3.5
≥0.4 and ≤0.70	N/A	≤0.65	1.5	2	3
≥0.4 and ≤0.70	N/A	≥1.30	1	1.5	2
≥0.7	-	-	1	1	1

Source: (TEC, 2007)

Table 4. Beam damage limits specified for linear force-based method

Ductile Beams			Performance Level		
$\frac{\rho-\rho'}{\rho_b}$	Confinement	$\frac{V}{bwdfct}$	IO	LS	CP
≤0.0	Exists	≤0.65	3	7	10
≤0.0	Exists	≥1.30	2.5	5	8
≥0.5	Exists	≤0.65	3	5	7
≥0.5	Exists	≥1.30	2.5	4	5
≤0.0	N/A	≤0.65	2.5	4	6
≤0.0	N/A	≥1.30	2	3	5
≥0.5	N/A	≤0.65	2	3	5
≥0.5	N/A	≥1.30	1.5	2.5	4

Source: (TEC, 2007)

Table 5. Relative story limits specified for linear force-based method

Relative Story Limits	Performance Level	
	IO	LS
δ_{ji}/h_{ji}	0.01	0.03

Source: (TEC, 2007)

Figure 5. Design Spectrum in the TEC (2007)

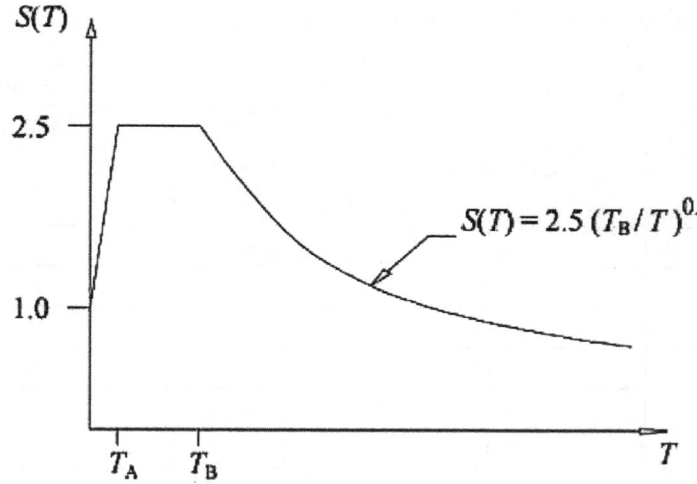

obtain CLC level, importance factor should be taken as 1.0. The elastic spectral acceleration ($S_{ae(T)}$), which is defined as the ordinate of 5% damped elastic design acceleration spectrum, is equal to spectral acceleration coefficient times the acceleration of gravity, (g), as given in Equation (4). The graphical representation of the elastic spectral acceleration depending on the local site condition and the vibration period is exhibited in Figure 6 for MCE level.

$$S_{(T)} = 1.+1.5\frac{T}{T_A} \qquad (0 \le T \le T_A)$$

$$S_{(T)} = 2.5 \qquad (T_A < T \le T_B) \tag{2}$$

$$S_{(T)} = 2.5\left(\frac{T_B}{T}\right)^{0.8} \qquad (T_B < T)$$

$$A_{(T)} = A_0 I S_{(T)} = 0.40x1.5xS_{(T)} = 0.60S_{(T)} \tag{3}$$

$$S_{ae(T)} = A_{(T)g} \rightarrow \max S_{ae(T)} = 0.60x2.5 = 1.50g \tag{4}$$

Figure 6. Elastic spectral acceleration corresponding to MCE level

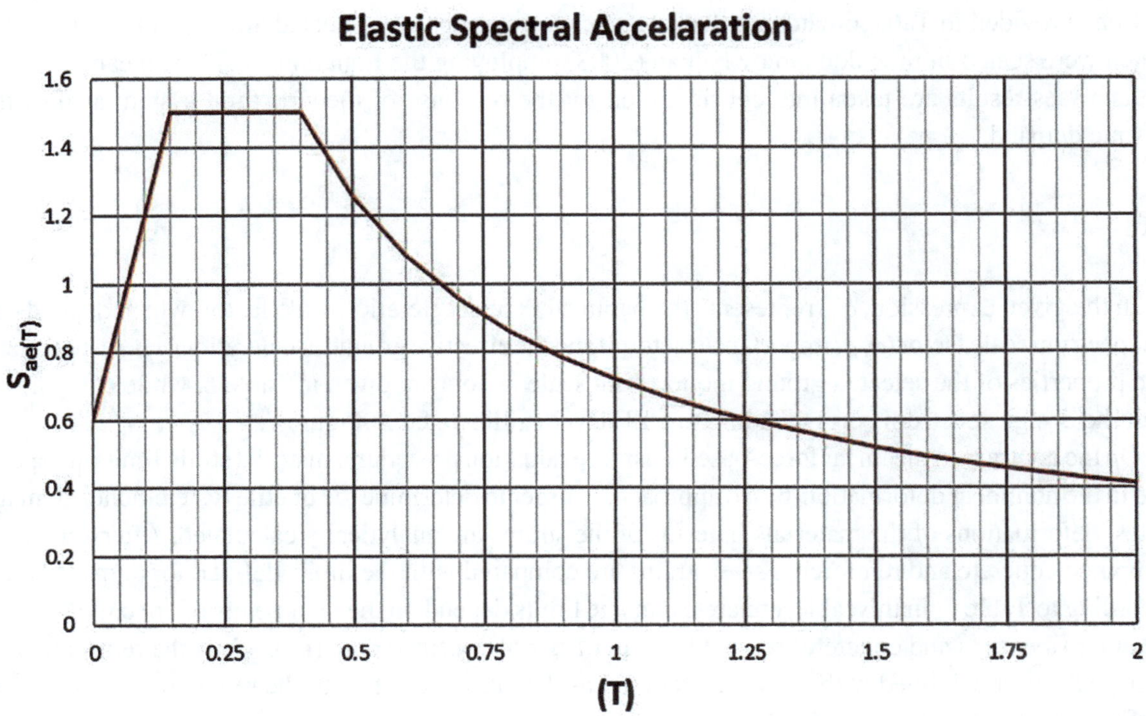

Nonlinear Deformation-Based Performance Investigation

TEC permits to use nonlinear static and time history analyses during the deformation-based seismic investigation of existing structures. By using the nonlinear static analysis; namely pushover analysis, direct evaluation of the structural performance can be utilized at each limit states very efficiently (Tehranizadeh & Moshref, 2011). On the other hand, considering only the fundamental vibration mode and assuming the structural response remains unchanged after the yielding of the structure are the main limitations for the static pushover analysis (Chopra & Goel, 2002). Though multi-mode pushover analysis are also encouraged in the TEC for such systems, for more accurate and realistic results, nonlinear time history analyses are utilized under the effect of a ground motion ensemble during the deformation-based performance evaluation of existing school buildings.

According to the TEC, maximum of the structural responses should be considered if the ground motion ensemble consists of three ground motions and at least seven ground motions should form an ensemble if the mean values of the responses are to be considered. Ground motions can be obtained from three main sources:

- Artificially generated,
- Simulated or
- Real accelerograms.

Among these sources, usage of real recorded ground motions for dynamic analysis is more realistic due to the site and ground motion characteristics such as frequency content or site amplification (Ergun et al., 2014). Therefore, recorded accelerograms are preferred during the nonlinear dynamic analyses of the RC school building. According to this method, selected real ground motions should be scaled in time or frequency domain to match the design spectrum characteristics. For this purpose, seven earthquake motions provided in Table 6matching the Z2 site characteristics, are selected amongst the real records which were scaled in time domain by Fahjan (2008) employing the Equation (5).Thus, mean values of the analyses results are taken into consideration for the response of the structural system against the seismic demand.

$$\alpha_{AT} = \alpha_{ST}.A_0.I \tag{5}$$

In the given expression, α_{AT} represents the parameters of acceleration scale factor, which depends on the spectrum scale factor (α_{ST}); importance factor(I) and the effective ground acceleration(A_0), respectively. The properties of the selected ground motion with scale factors are given in Table 6, while comparison with the design spectrum corresponding to CLE level is illustrated in Figure 7.

On the contrary to the linear force-based seismic evaluation procedure, material strain limits are specified in the nonlinear deformation-based approach. In order to determine the ductile RC elements' damage states, deformations of the materials in terms of the strains are analytically calculated. Afterwards, the computed concrete and reinforcing steel strains are compared with the limits defined for damage zones according to Table 7. In this table, concrete damage limits depend on the parameters of concrete strain at the outer fiber (ε_{cu}) and concrete strain at the outer fiber of confined core (ε_{cg}), while the reinforcement damage limits are defined with the steel strain (ε_s) as the ratio of existing to the required amounts of the confinement reinforcement (ρ_s / ρ_{sm}).

Table 6. Selected ground motions and scale factors

PEER Record No		Earthquake	Date	Station	Record Name	Duration	PGA	Scale F.	Spec.Acc. Scale F.	
New	Old					(sec)	(g)	(α_{ST})	CLE	MCE
NGA0175	P0017	Imperial Valley	15.10.1979	931 El Centro Array #2	H-E12230	39.00	0.116	8.31	3.324	4.986
NGA0175	P0012	Imperial Valley	15.10.1979	931 El Centro Array #12	H-E12240	39.00	0.143	7.01	2.804	4.206
NGA0728	P0730	Superstation Hill (B)	24.11.1987	11369 Westmorland Fire St	B-WSM090	40.00	0.172	5.10	2.040	3.060
NGA0970	P0898	Northridge	17.01.1994	90066 El Monte-Fairview Av	FA1095	34.99	0.122	8.99	3.596	5.394
NGA0832	P0856	Landers	28.06.1992	21081 Amboy	ABY090	50.00	0.146	5.76	2.304	3.456
NGA1000	P0967	Northridge	17.01.1994	24612 LA - Pico & Sentous	PIC090	40.00	0.103	9.34	3.736	5.604
NGA0161	P0020	Imperial Valley	15.10.1979	5060 Brawley Airport	H-BRA315	37.80	0.220	4.58	1.832	2.748

Figure 7. Comparison of the acceleration spectra of the scaled ground motions and the TEC design spectrum for the CLE level

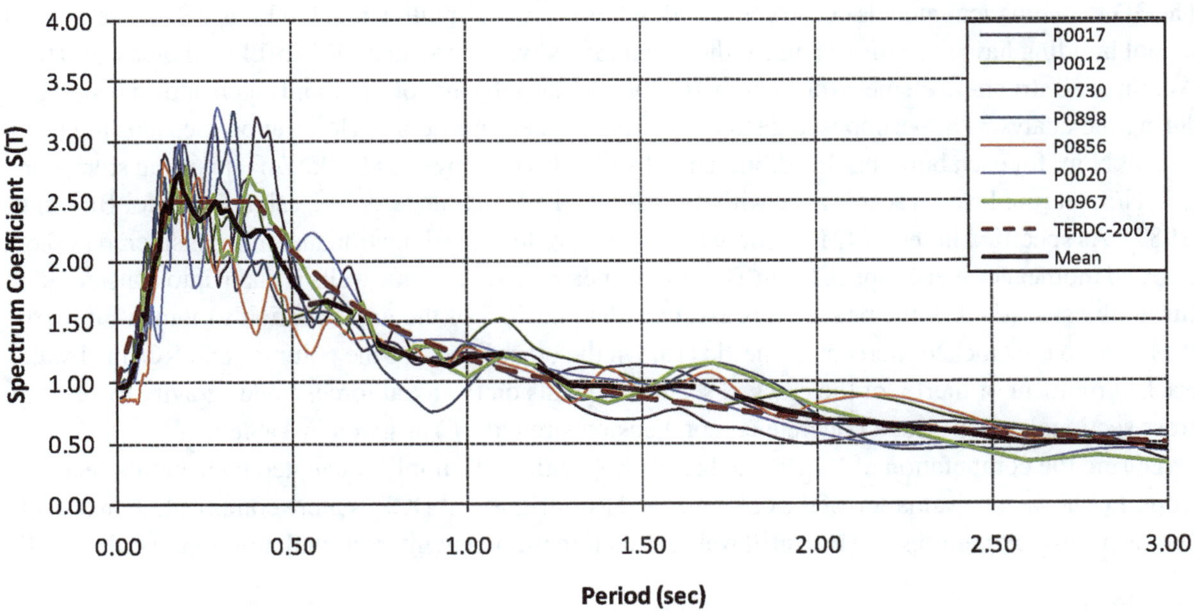

Performance and damage zone identification in terms of relative story drift ratios are limited to linear analysis procedures in the TEC. Therefore during the examination of the nonlinear analyses results, drift ratios are compared with the limitations given in ASCE/SEI-41 (2006) which are summarized in Table 8. For all other limit states, seismic performances of each school building are evaluated by employing the regulations given in the TEC.

Table 7. Material strain limits for nonlinear deformation-based analyses

Damage Limit	Concrete Strain	Steel Strain
MN	ε_{cu}=0.0035	ε_s=0.010
SF	ε_{cg}= min[0.0035 + 0.010(ρ_s/ρ_{sm}) ;0.0135]	ε_s = 0.040
CL	ε_{cg}= min[0.0040 + 0.014(ρ_s/ρ_{sm}) ;0.0180]	ε_s = 0.060

Source: (TEC, 2007)

Consequently, after locating the RC members' damage zones by inspecting the concrete and steel strains; then by checking the corresponding upper and the lower bounds defined for drifts and finally by evaluating the existence of brittle failure or not, performance level against the target demand is achieved for each school building.

MATHEMATICAL MODEL DETAILS FOR THE SEISMIC INVESTIGATION OF THE EXISTING RC SCHOOL BUILDINGS

Mathematical Model Assumptions for Linear Analyses

Force-based linear seismic performance investigation of the existing school buildings are carried out by utilizing widely-used commercial software (ideCAD Static), which is also compatible with the TEC. The 3D mathematical models constructed within this software are illustrated in Figure 8 for the as-built school building having 3 stories and for the strengthened version with an RC infill wall index of 0.014. Weight of all structural elements are automatically calculated by the software and taken into consideration during the analyses. Superimposed dead loads from plaster, mortar and flooring on the slabs are taken as 2.0 kN/m^2 for each building. In addition, the live load is employed as 5.0 kN/m^2 at normal stories and 1.5 kN/m^2 at roof level in accordance with the TS-498: Turkish Standard for Design Loads for Buildings (1987). As specified in the TEC for school buildings, live load participation factor is considered as 0.60. Cracked moment of inertia specified in TEC for the beams and columns/ walls is taken into consideration during the analysis. For the beams, it is assumed that the 40% of the gross-section's moment of inertia (I_g) Equal to the cracked moment of inertia (I_c). On the other hand, for the columns and RC infill walls, cracked moment of inertia of the cracked section depends on the axial forces under gravity loads (N_D), gross sectional area (A_c) and concrete comprehension strength (f_c) as given in Table9.

During the computation of DCRs, the beams' capacities are simply calculated by using its sectional properties however, evaluation of the columns' and RC infill walls' DCR is more complicated. In order to find the ratios for columns and RC infill walls, axial force-moment interaction diagrams are individually

Table 8. Relative story limits specified for nonlinear deformation based method

Relative Story Limits	Performance Level	
	IO	LS
δ_{ji}/h_{ji}	0.0040	0.0075

Source: (ASCE/SEI-41, 2006)

Figure 8. 3D computer models for the 3-story as-built school building (left) and the strengthened version for p=0.014 (right)

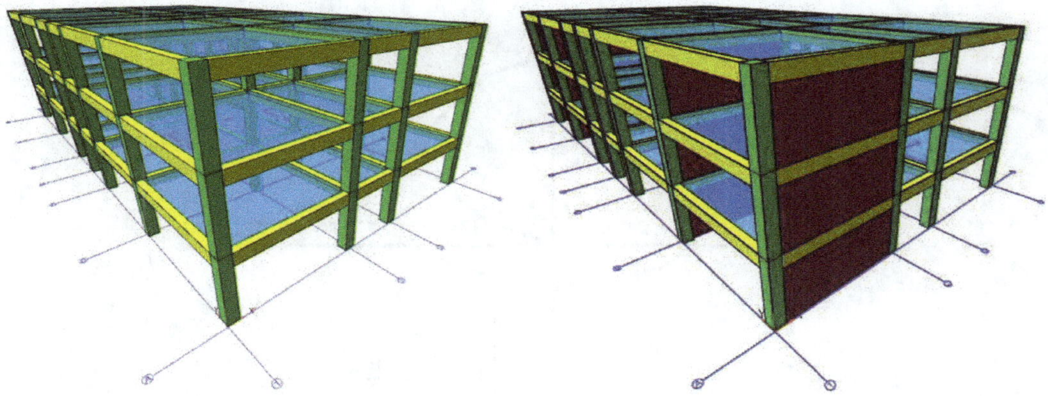

Table 9. Moment of inertia for cracked sections of beams, columns and RC infill walls

Structural Elements	Cracked Moment of Inertia
Beams	$0.40I_g$
Columns & RC Infill Walls	$N_D/(A_c f_c) \leq 0.10 \rightarrow 0.40I_g$ $N_D/(A_c f_c) \geq 0.40 \rightarrow 0.80I_g$

evaluated by using the section geometry and rebar orientation. Afterwards, internal moment and axial forces from gravity (M_D-N_D) and lateral earthquake loads (M_E-N_E) are calculated in accordance with the gravity loads on system and seismic demands. Assuming a linear pattern between these two characteristic points on the interaction diagram can be applied, the residual moment capacity and the corresponding axial force (M_K-N_K) is obtained numerically, as illustrated in Figure 9. Finally, demand/capacity ratios are evaluated and compared to the limits in order to find the seismic performance level of each section.

Mathematical Model Assumptions for Nonlinear Analysis

School buildings are modeled utilizing Perform-3D computer program, which is capable of handling the nonlinear dynamic analyses not only in less computer elapsed time but also negligible numerical round-off errors. Within the model, in order to consider the complex non-linear stress-strain behavior by defining the material properties of the reinforcement as well as the confined and unconfined concrete for the cross section, RC infill walls and columns are represented by fiber elements as shown in Figure 10. On the other hand, due to its predictable behavior under dynamic analyses, beams are modeled by frame elements. The interaction between the fiber based column and RC infill wall model with the frame elements are provided by using the fictitious imbedded beams which have high stiffness in bending and negligible stiffness in axial loading as illustrated in Figure 11.

In order to reflect the demand of the structural system more appropriate, a fiber-based material model calibrated by Görgülü and Taskin (2015), is used for the nonlinear dynamic analyses of existing school buildings. By using the calibrated model, the non-buckling inelastic steel material and the inelastic 1D

Figure 9. Graphical representation for the evaluation of the demand and capacity for the columns and RC infill walls according to the TEC (2007)

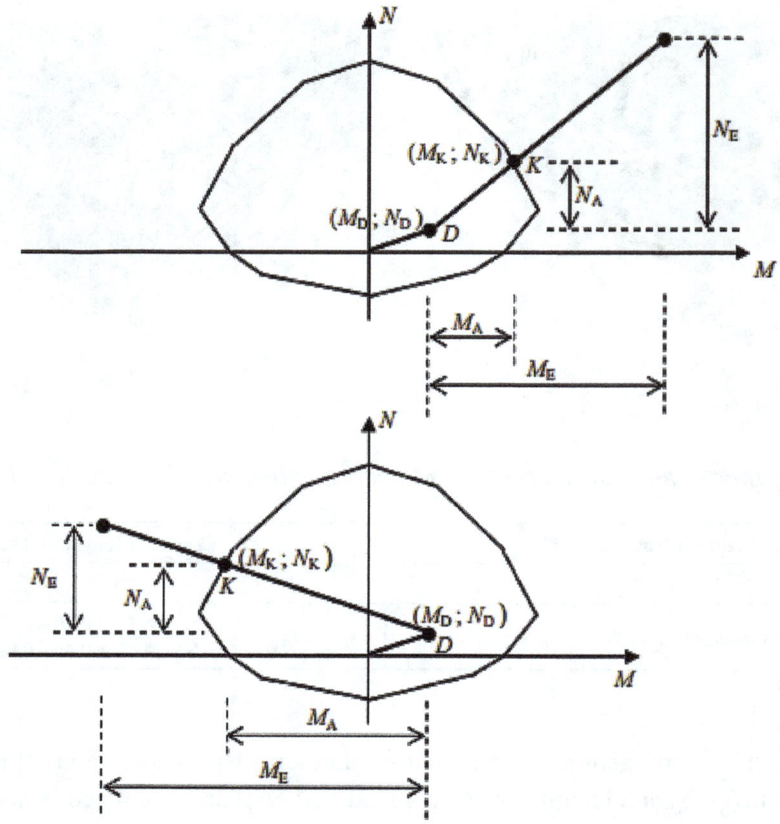

Figure 10. The configuration of the fiber layers in RC infill wall and columns

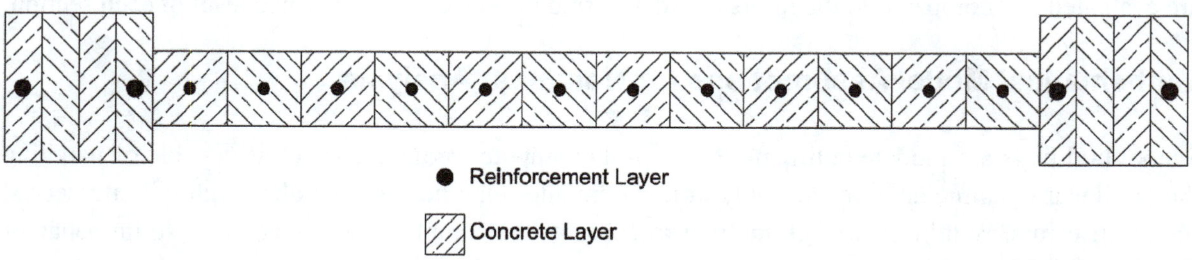

concrete material component are implemented into the program for the definition of reinforcement and concrete material, respectively. The skeleton for the stress-strain curve of the reinforcement is represented by using 3 points including strain hardening, while the concrete material model is defined throughout the 4 points as given in Figure 12, which considers the strength loss as well. For the fiber element representation, longitudinal bars are located in their exact locations while the stirrups are defined to the fiber sections by means of an area ratio that is described as the rebar area divided by the concrete area. In order to achieve the strain of the material over the plastic hinge length, axial strain gages are

Figure 11. Mathematical model representation of the structural system for the nonlinear analyses

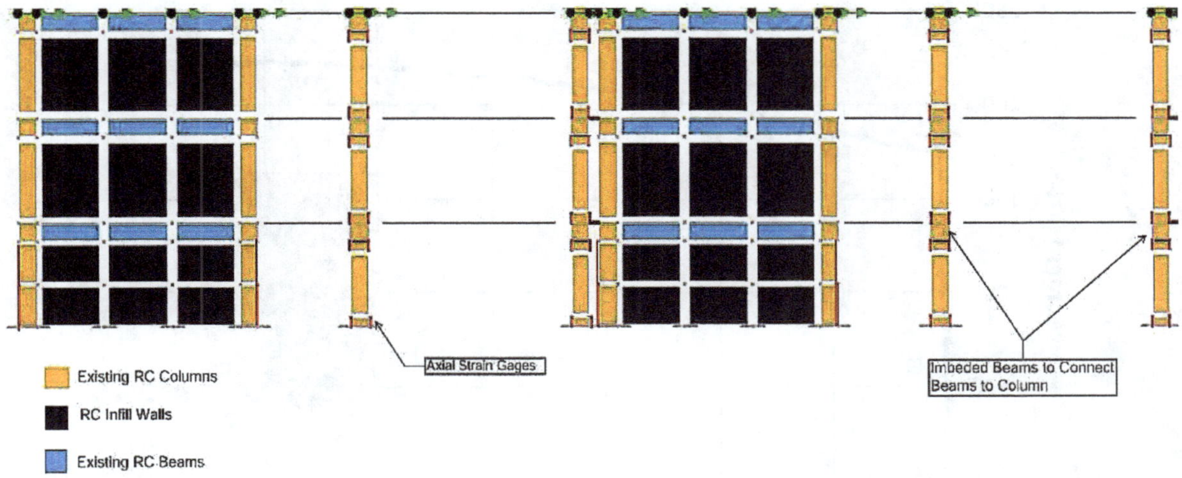

implemented into the program. For the calculation of material strains in beams, moment-rotation bi-linear curve is defined according to lumped plasticity theory as illustrated in Figure 13. In accordance with the TEC, the shear force strength of the section is introduced into the program and the hysteretic behavior of the shear stress-strain relation is represented by using the elastic-perfectly-plastic (e-p-p) model in the numerical analyses.

Weight of the slabs and the perpendicular beams are defined as additional nodal dead loads since the analyses are carried out in 2D. On the other hand, self-weight of the elements in the direction of the loading are automatically calculated and taken into account by the software. Superimposed dead and live loads including the live load participation factor are considered in the same manner with the linear analyses model assumptions. Due to all layers in the fiber element model are interacted with each other and the neutral axis shift is simulated, calculation of the cracked moment of inertia (I_c) for RC infill walls and columns is not required. Cracked moment of inertia is taken as the 40% of the gross-section's moment of inertia (I_g) for the beams.

Figure 12. Implemented constitutive material models: concrete (left) and steel (right)

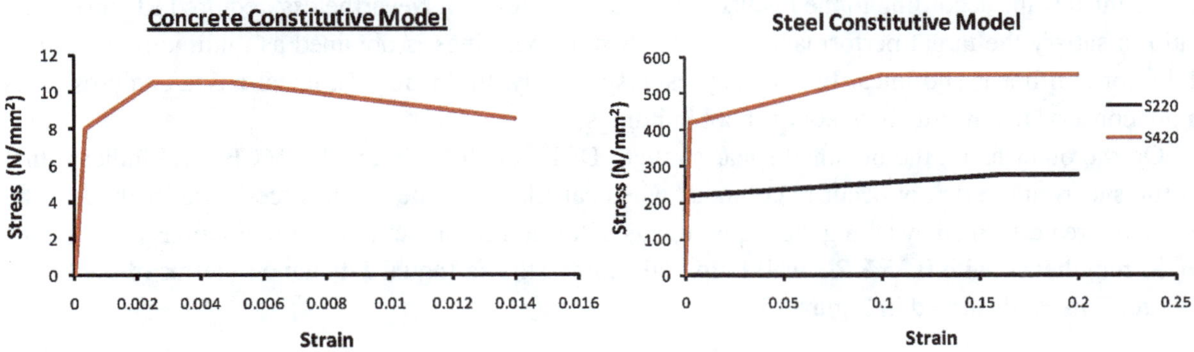

Figure 13. The typical moment-curvature relationship in RC beams

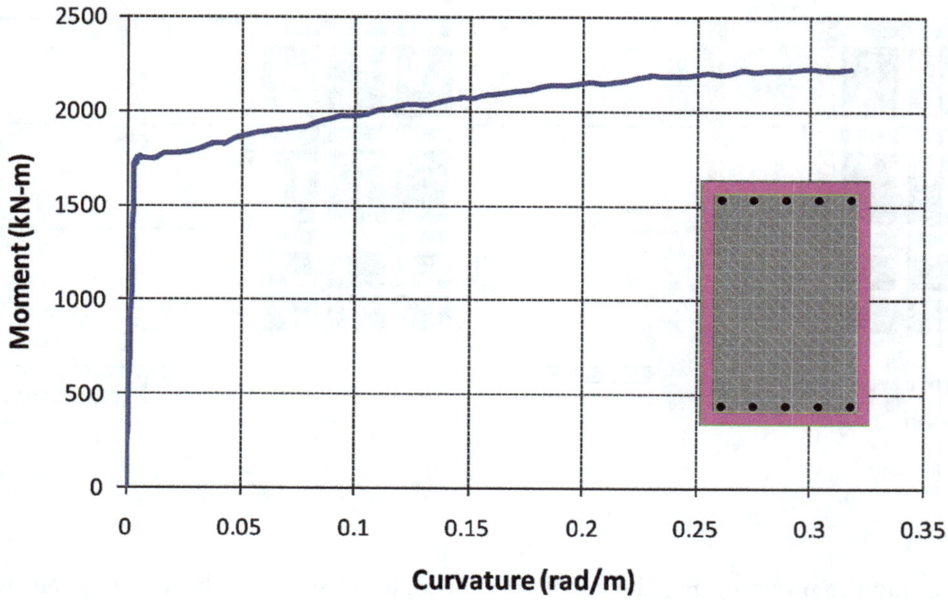

COMPARISONS OF THE LINEAR AND NONLINEAR ANALYSES RESULTS

Linear force- and nonlinear deformation-based analyses of the typical RC school building are conducted for CLE and MCE level earthquakes. According to the analyses results for each direction, no significant difference is occurred between the results of the two orthogonal directions therefore, only the results of the transversal direction are presented herein. In order to simplify the examination procedure, analyzed school buildings are classified with codes in the given tables and figures. The first two characters assigned in the codes show the total number of the stories, while the other parts describe the reinforcement material used in mentioned school building. For instance, 3KS220 demonstrate the 3 story school building detailed with S220 grade reinforcement, while 5KS420 designated as 5 story school building constructed with S420 grade reinforcement.

Methods based on linear and nonlinear analyses are initially compared in terms of the shear strength of column and RC infill wall elements. Since the demand of the MCE level is 1.5 times the CLE level, shear strength demand/capacity ratios is achieved and compared only for MCE level. In 3 story RC school buildings, minimum 1.0% of RC infill wall ratio is adequate to prevent the brittle shear failure of RC infill walls according to the results illustrated in Figure 14. Nevertheless, required RC infill wall ratio to satisfy the target performance level for 5 story buildings is obtained as minimum 1.40% and 2.4% for non-linear and linear based analyses, respectively. In this regard, linear based analyses gives more conservative results in 5 story school buildings.

On the other hand, the obtained shear strength DCRs of RC columns for MCE level indicate that brittle shear failure do not occur in columns, when calculated by the linear force-based method. If the analyses are performed by taking the requirements of the nonlinear methods, a few columns in RC building, strengthened with 0.35% RC wall ratio, fell short of the earthquake demand in terms of the brittle failure. This is illustrated in Figure 15.

Figure 14. Shear strength demand/capacity ratio of RC infill wall for MCE level

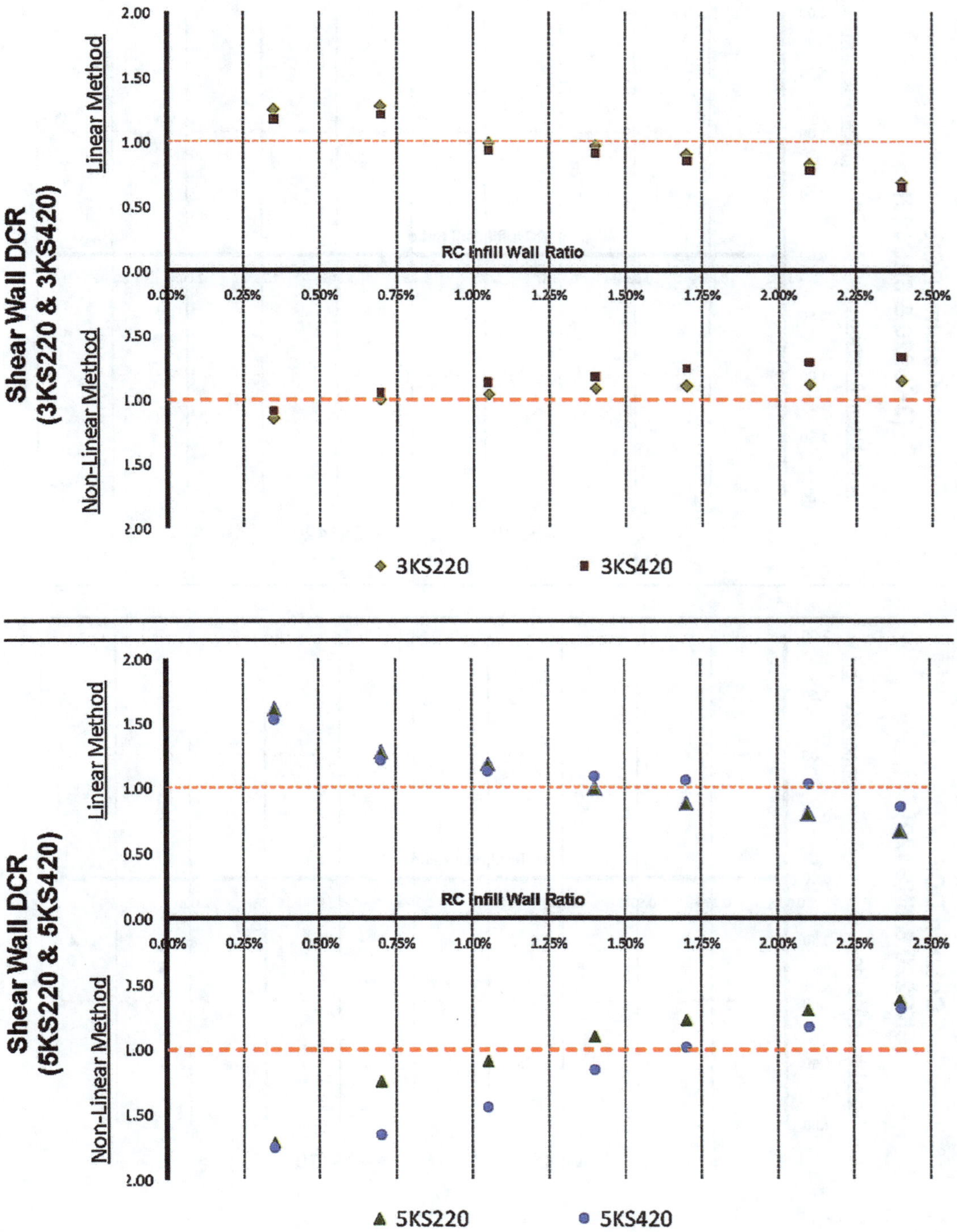

Figure 15. Shear strength demand/capacity ratio of RC column for MCE level

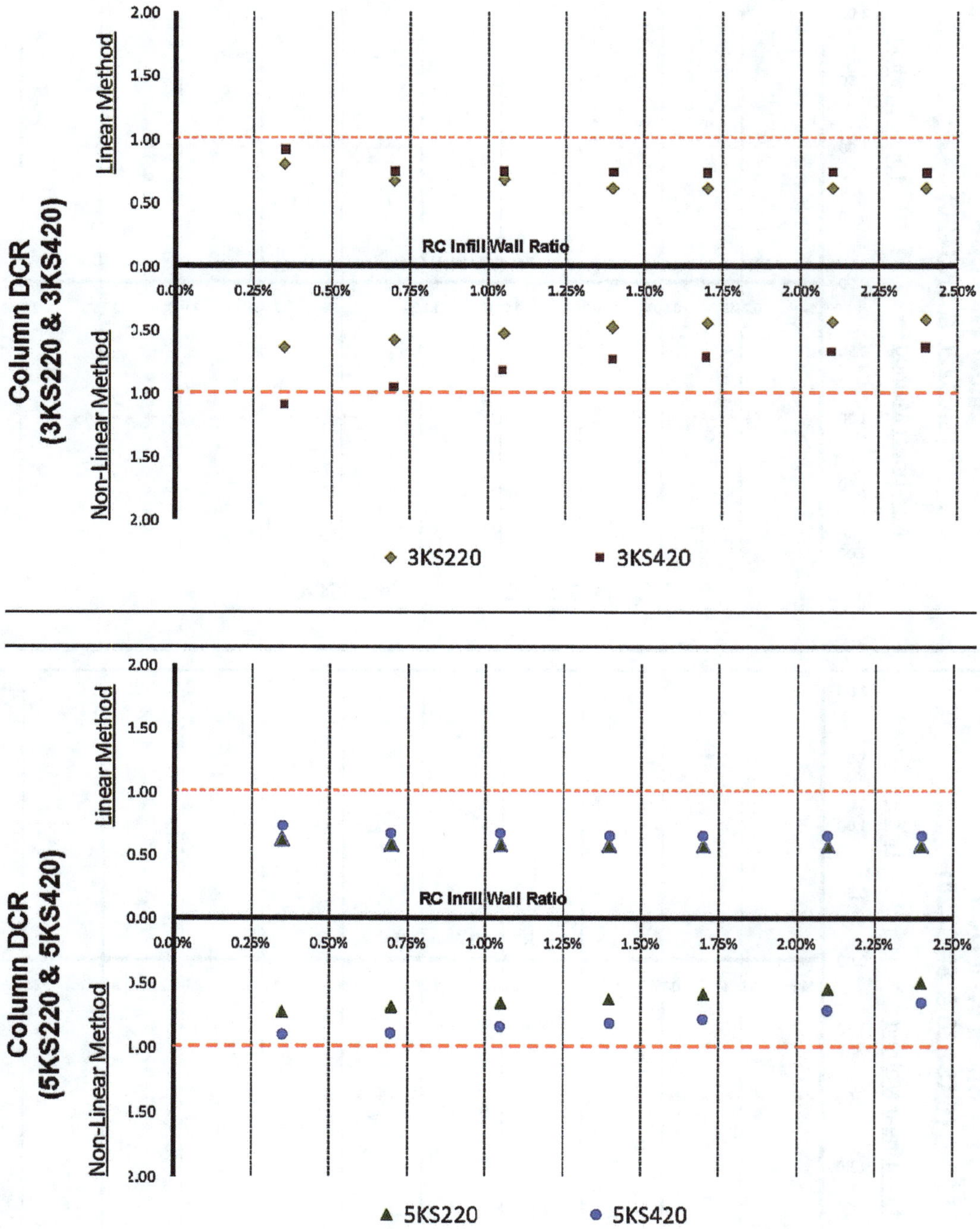

Analytically obtained damage states as well as the percentages with respect to the total number of elements and the corresponding seismic performance distribution of the RC beams are shown in Table 10 for both analyses methods. The boldface marks in the table indicate the target seismic performance level of the RC school building with respect to the earthquake level of concern. According to the table, percentage of RC beams in the damaged state and the performance level against the relevant earthquake level are close to each other for both analyses methods. However, computed damage states of the RC beams in some buildings are higher in nonlinear analyses when compared to the linear as given in the same table. This is because the beams are in plastic region and continue to absorb the seismic demand by balancing the kinetic energy of the earthquake with the strain energy contrary to the linear force-based analysis. Another reason for this phenomenon is the strictness of IO performance level requirements in TEC for nonlinear analyses. Therefore, instead of linear force-based method, if the nonlinear deformation-based analysis method is preferred, the encountered performance level will be lower in terms of the RC beam section capacity.

Damage states and corresponding seismic performance distribution of the RC columns for each RC school buildings is presented in Table 11. In this table, the row identified by bold characters shows the required performance level for the RC school building corresponding to the relevant earthquake demand. As stated in the table, adequate RC infill wall ratio to satisfy the target performance level is higher in linear analyses, when compared to the results obtained from the nonlinear analyses. This situation occurs due to the assumption that beam-column sections remain elastic during linear analyses. In other words, released kinetic energy due to the earthquake demand is absorbed proportional to the structural element's constant stiffness without redistribution of the internal forces. For this reason, expected seismic performance level of the RC columns will be lower in linear analyses in terms of the DCRs.

Variation of the relative story drift ratios compared with the limits for linear and nonlinear analyses are illustrated in Figure 16 for each investigated RC school building. According to the obtained results, all RC school buildings provide the required performance levels by means of limiting values defined in the TEC for linear analyses. On the other hand, most of the school buildings, especially the 5 story ones, exhibited inadequate performance levels to satisfy the regulations for story limits given in the ASCE/SEI-41. In this regard, if the nonlinear analyses methods are used for seismic investigation, a minimum of 1.05% of RC infill wall ratio is found out to be satisfactory for both performance levels in 5 story RC school buildings. The reason for the increasing of the necessary RC infill wall ratio in 5 story RC school buildings is the flexural shear wall-frame behavior instead of squat walls in 3 story buildings. Furthermore, linear analyses methods are not sensitive for the plastic deformation of the structural elements due to its force-based approach during a seismic investigation; therefore the computed story drift ratios are limited in allowable drift ranges, when the results are compared with the nonlinear deformation-based method.

CONCLUSION

In order to determine the most appropriate method for the seismic capacity of the rehabilitated RC school buildings, force-based linear and deformation-based nonlinear analyses are conducted and results obtained are compared to each other. During the nonlinear mathematical modeling of RC school buildings, a well calibrated fiber model is used to represent the cyclic seismic behavior of the structural sections more precisely.

The governing analysis method corresponding to the performance limits for entire school buildings is obtained and given in Table 12. As a result of the consecutive analyses, although some RC beams yield better seismic performance if the force-based seismic investigation analyses is conducted, structural performances of most of the beams for the two different earthquake levels are found out to be close to each other for the two different analyses methods. On the other hand, performance levels computed for the columns during the linear analyses are the most unfavorable due to the redistribution of internal forces and modification of the energy absorption points in nonlinear analyses.

Table 10. Damage states, percentages and seismic performance distribution of the RC beams according to the TEC requirements

Code	RC Infill Wall Ratio (%)	CLE Level										MCE Level									
		Damage States								Performance Level		Damage States								Performance Level	
		Minimum (%)		Significant (%)		Extreme (%)		Collapse (%)				Minimum (%)		Significant (%)		Extremen (%)		Collapse (%)			
		LA	NLA	LA	NLA	LA	NLA	LA	NLA	LA	NLA	LA	NLA	LA	NLA	LA	NLA	LA	NLA	LA	NLA
3KS220	0.35	81	57	5	24	0	19	14	0	C	LS	57	0	14	57	0	5	29	38	C	C
	0.70	100	75	0	25	0	0	0	0	**IO**	LS	86	40	14	40	0	10	0	10	**LS**	CP
	1.05	100	100	0	0	0	0	0	0	IO	**IO**	100	42	0	58	0	0	0	0	IO	**LS**
	1.40	100	100	0	0	0	0	0	0	IO	IO	100	83	0	17	0	0	0	0	IO	LS
	1.70	100	100	0	0	0	0	0	0	IO	IO	100	100	0	0	0	0	0	0	IO	IO
	2.10	100	100	0	0	0	0	0	0	IO	IO	100	100	0	0	0	0	0	0	IO	IO
	2.40	100	100	0	0	0	0	0	0	IO	IO	100	100	0	0	0	0	0	0	IO	IO
3KS420	0.35	97	95	3	5	0	0	0	0	**IO**	**IO**	10	48	0	52	0	0	0	0	**IO**	**LS**
	0.70	100	100	0	0	0	0	0	0	IO	IO	100	90	0	10	0	0	0	0	IO	IO
	1.05	100	100	0	0	0	0	0	0	IO	IO	100	95	0	5	0	0	0	0	IO	IO
	1.40	100	100	0	0	0	0	0	0	IO	IO	100	100	0	0	0	0	0	0	IO	IO
	1.70	100	100	0	0	0	0	0	0	IO	IO	100	100	0	0	0	0	0	0	IO	IO
	2.10	100	100	0	0	0	0	0	0	IO	IO	100	100	0	0	0	0	0	0	IO	IO
	2.40	100	100	0	0	0	0	0	0	IO	IO	100	100	0	0	0	0	0	0	IO	IO
5KS220	0.35	5	0	52	62	0	19	43	19	C	CP	5	0	9	57	43	5	43	38	C	C
	0.70	43	0	19	65	0	20	38	15	C	CP	10	0	38	60	14	0	38	40	C	C
	1.05	67	42	14	58	0	0	19	0	C	LS	43	0	14	58	5	0	38	42	C	C
	1.40	90	83	0	17	5	0	5	0	C	LS	52	6	29	72	0	11	19	11	C	CP
	1.70	90	94	10	6	0	0	0	0	**IO**	**IO**	67	18	24	71	9	11	0	0	**LS**	**LS**
	2.10	90	100	10	0	0	0	0	0	IO	IO	90	38	0	56	10	6	0	0	LS	LS
	2.40	100	100	0	0	0	0	0	0	IO	IO	100	67	0	33	0	0	0	0	LS	LS
5KS420	0.35	76	76	5	24	0	0	19	5	C	CP	48	0	19	76	5	19	28	5	C	CP
	0.70	100	80	0	20	0	0	0	0	**IO**	LS	71	20	10	80	0	0	19	0	C	**LS**
	1.05	100	84	0	16	0	0	0	0	IO	LS	100	63	0	37	0	0	0	0	**IO**	LS
	1.40	100	94	0	6	0	0	0	0	IO	**IO**	100	78	0	22	0	0	0	0	IO	LS
	1.70	100	100	0	0	0	0	0	0	IO	IO	100	82	0	18	0	0	0	0	IO	LS
	2.10	100	100	0	0	0	0	0	0	IO	IO	100	94	0	6	0	0	0	0	IO	IO
	2.40	100	100	0	0	0	0	0	0	IO	IO	100	100	0	0	0	0	0	0	IO	IO

LA: Linear Analyses ; NLA: Non-Linear Analyses

Table 11. Seismic performance distribution of the RC columns according to the TEC requirements

Code	RC Infill Wall Ratio	CLE Level				MCE Level			
		Linear Analyses		Nonlinear Analyses		Linear Analyses		Nonlinear Analyses	
		Damage States	Performance Level	Damage States	Performance Level	Damage States	Performance Level	Damage States	Performance Level
3KS220	0.35%	Extreme	CP	Significant	LS	Extreme	CP	**Significant**	**LS**
	0.70%	Significant	LS	**Minimum**	**IO**	Extreme	CP	Minimum	IO
	1.05%	Minimum	IO	Minimum	IO	**Significant**	**LS**	Minimum	IO
	1.40%	**Minimum**	**IO**	Minimum	IO	Minimum	IO	Minimum	IO
	1.70%	Minimum	IO	Minimum	IO	Minimum	IO	Minimum	IO
	2.10%	Minimum	IO	Minimum	IO	Minimum	IO	Minimum	IO
	2.40%	Minimum	IO	Minimum	IO	Minimum	IO	Minimum	IO
3KS420	0.35%	Extreme	CP	**Minimum**	**IO**	Extreme	CP	**Significant**	**LS**
	0.70%	**Minimum**	**IO**	Minimum	IO	**Significant**	**LS**	Minimum	IO
	1.05%	Minimum	IO	Minimum	IO	Minimum	IO	Minimum	IO
	1.40%	Minimum	IO	Minimum	IO	Minimum	IO	Minimum	IO
	1.70%	Minimum	IO	Minimum	IO	Minimum	IO	Minimum	IO
	2.10%	Minimum	IO	Minimum	IO	Minimum	IO	Minimum	IO
	2.40%	Minimum	IO	Minimum	IO	Minimum	IO	Minimum	IO
5KS220	0.35%	Extreme	CP	Extreme	CP	Collapse	C	Collapse	C
	0.70%	Extreme	CP	**Minimum**	**IO**	Collapse	C	Collapse	C
	1.05%	Significant	LS	Minimum	IO	Collapse	C	**Significant**	**LS**
	1.40%	**Minimum**	**IO**	Minimum	IO	Extreme	CP	Minimum	IO
	1.70%	Minimum	IO	Minimum	IO	**Significant**	**LS**	Minimum	IO
	2.10%	Minimum	IO	Minimum	IO	Minimum	IO	Minimum	IO
	2.40%	Minimum	IO	Minimum	IO	Minimum	IO	Minimum	IO
5KS420	0.35%	Extreme	CP	Significant	LS	Collapse	C	Collapse	C
	0.70%	Significant	LS	**Minimum**	**IO**	Collapse	C	Extreme	CP
	1.05%	**Minimum**	**IO**	Minimum	IO	Extreme	CP	**Significant**	**LS**
	1.40%	Minimum	IO	Minimum	IO	Extreme	CP	Minimum	IO
	1.70%	Minimum	IO	Minimum	IO	**Significant**	**LS**	Minimum	IO
	2.10%	Minimum	IO	Minimum	IO	Minimum	IO	Minimum	IO
	2.40%	Minimum	IO	Minimum	IO	Minimum	IO	Minimum	IO
IO: Immediate Occupancy; LS: Life Safety; CP: Collapse Prevention; C: Collapse									

Another major result from the analyses is that, shear strength demand capacity ratios of existing columns and RC infill walls also exhibit similar performance against the expected seismic demand for both analyses methods. However, linear analysis method yields to conservative results by means of the shear demand/capacity ratios. In addition, determination of the analyses method for the seismic rehabilitation is not excessively impacted by the story drift limitations as much as the other limit states, especially for

Figure 16. Variation of the elastic and inelastic drift ratios compared to the limits defined in TEC &ASCE/SEI-41

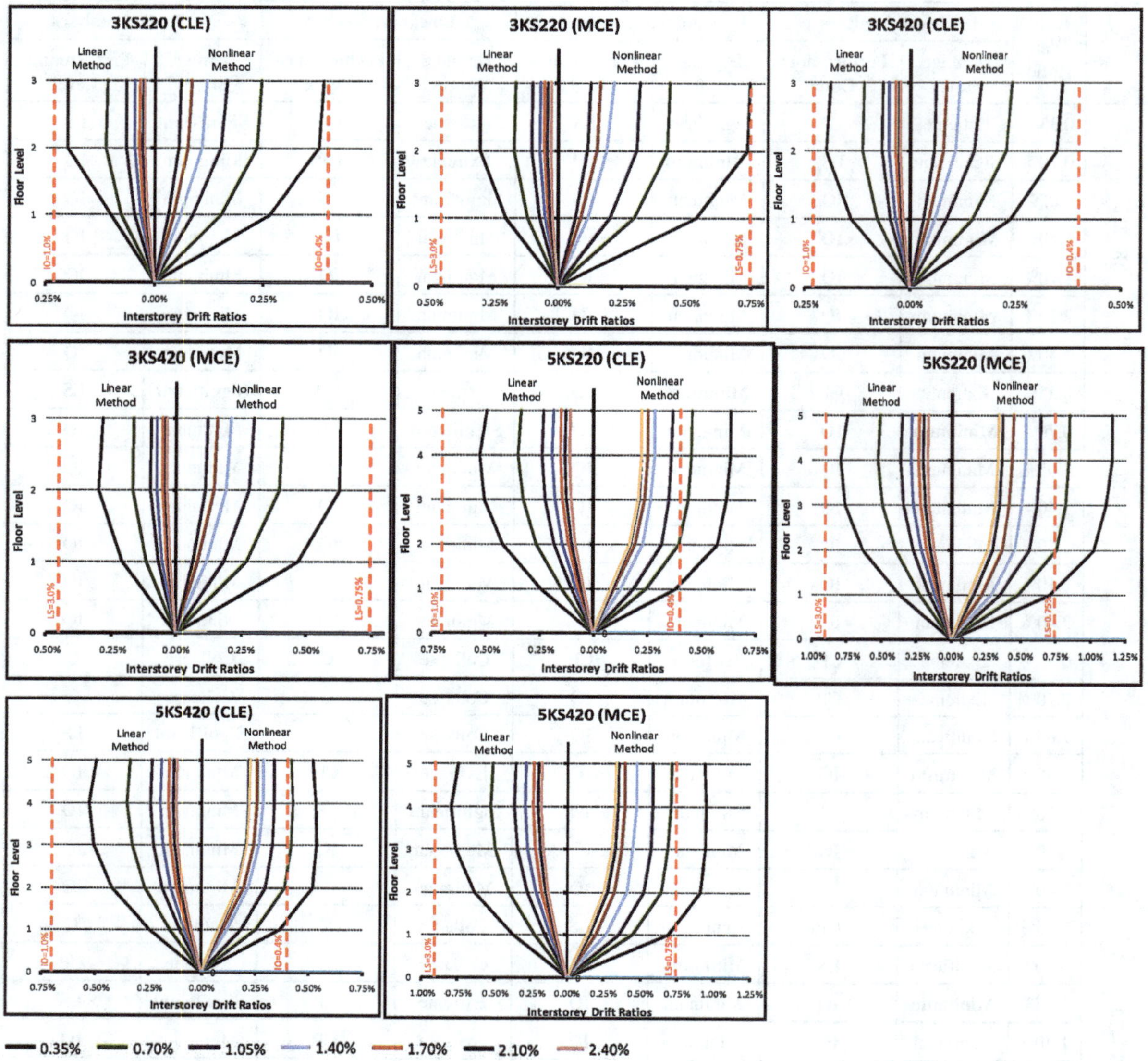

the low rise RC school buildings. On the other hand, school buildings having more than 5 stories should be investigated by using the deformation based nonlinear methods since linear methods provide more favorable results in terms of the RC building performance level.

Consequently, according to the analytically obtained results, all structural elements, except columns, may lead lower performance levels against the target earthquake level if the deformation-based nonlinear analysis is conducted. Therefore, in order to be in safe side, non-linear analyses should be used due to its sensitivity to represent the seismic behavior of the existing RC school buildings. On the other hand, torsional effects related to the RC infill wall location are not taken into consideration due to their symmetrical positioning.

Table 12. Governing analyses method corresponding to the performance limits for entire RC school buildings

RC Building Code	Earthquake Level	Governing Analysis Method				
		Flexural Strength		Shear Strength		Interstorey Drift
		Beam	Column	Column	RC Infill Wall	
3KS220	CLE	NLA	LA	NLA& LA	NLA& LA	NLA
	MCE	NLA	LA	NLA& LA	NLA& LA	NLA
3KS420	CLE	NLA& LA	LA	NLA& LA	NLA	NLA
	MCE	NLA& LA	LA	NLA& LA	NLA& LA	NLA
5KS220	CLE	NLA& LA	LA	LA	NLA& LA	NLA
	MCE	NLA& LA	LA	LA	NLA& LA	NLA
5KS420	CLE	NLA	LA	LA	NLA& LA	NLA
	MCE	LA	LA	LA	NLA& LA	NLA

ACKNOWLEDGMENT

Authors would like to thank Mr. İsmail Hakkı Besler and Mr. Faruk Saka of İdeYAPI Inc. for donating the structural analysis program ideCAD Static. Their support is sincerely acknowledged.

REFERENCES

Antonio, S., & Pinho, R. (2004). Development and verification of a displacement based adaptive pushover procedure. *Journal of Earthquake Engineering*, *8*(5), 643–661. doi:10.1080/13632460409350504

ASCE/SEI 41-06. (2006). *Seismic rehabilitation of existing buildings*. American Society of Civil Engineers.

Canbay, E., Ersoy, U., & Ozcebe, G. (2003). Contribution of reinforced concrete infills to seismic behavior of structural systems. *ACI Structural Journal*, *100*(5), 637–643.

Chopra, A. K., & Goel, R. K. (2002). A modal pushover analysis procedure for estimating seismic demands for buildings. *Earthquake Engineering & Structural Dynamics*, *31*(3), 561–582. doi:10.1002/eqe.144

Elnashai, A. S. (2000). Advanced inelastic (pushover) analysis for seismic design and assessment. *The G.Penelis Symposium.*

Ergun, M., & Sevket, A. (2014). Comparing the effects of scaled and real earthquake records on structural response. *Earthquakes and Structures*, *6*(4), 375–392. doi:10.12989/eas.2014.6.4.375

Fahjan, Y. (2008). Selection and scaling of real earthquake accelerograms to fit the Turkish design spectra.[In Turkish with Extended Summary]. *Technical Journal of Turkish Chamber of Civil Engineers*, *3*(19), 4423–4444.

Ferraioli, M. (2015). Case study of seismic performance assessment of irregular RC buildings: Hospital structure of Avezzano. *Earthquake Engineering and Engineering Vibration, 14*(1), 141–156. doi:10.1007/s11803-015-0012-7

Görgülü, O., & Taskin, B. (2015). Numerical simulation of RC infill walls under cyclic loading and calibration with widely used hysteretic models and experiments. *B EarthqEng, 13*(9), 2591-2610. doi:10.1007/s10518-015-9739-9

Gupta, B., & Kunnath, S. K. (2000). Adaptive spectra-based pushover procedure for seismic evaluation of structures. *Earthquake Spectra, 16*(2), 367–391. doi:10.1193/1.1586117

IdeCAD 3D Integrated analyses, design and detailing software for reinforced concrete structure. (n.d.). IdeYAPI Ltd. Retrieved from www.idecad.com

Jan, T. S., Liu, M. W., & Kao, Y. C. (2004). An upper-bound pushover analysis procedure for estimating seismic demands of high-rise buildings. *Engineering Structures, 26*(1), 117–128. doi:10.1016/j.engstruct.2003.09.003

Kuran, F., Demir, C., Koroglu, O., Kocaman, C., & İlki, A. (2007). Seismic Safety Analysis of an Existing 1502 Type Disaster Building Using New Version of Turkish Seismic Design Code.*Proceedings of the ECCOMAS Thematic Conference on COMPDYN.*

Lopez, O. A., Hernandez, J. J., & Puig, J. (2004). Seismic Risk in Schools: The Venezuelan Project. *Proceedings of the Keeping schools safe in Earthquakes, Ad Hoc Experts' Group Meeting on Earthquake Safety in Schools.*

Perform 3D Components & Elements. (2006). *For Perform-3D and Perform-Collapse*. Computer & Structures Inc.

Postelnicu, T., Popa, V., Cotofana, D., Chesca, B., Ionescu, R., & Vacarenu, R. (2005) Study on seismic performance of existing buildings in Romania. Technical Report, Technical University of Civil Engineering, Bucharest, and Building Research Institute.

Strepelias, T., Fardis, M. N., Bousias, S., Palios, X., & Biskinis, D. (2012). *RC Frames infilled into RC walls for seismic retrofitting: Design, experimental behavior and modeling. Report Series in Structural and Earthquake Engineering*. Patras: University of Patras.

Tasdemir, M. A., & Ozkul, M. (1999). Marmara depremi beton araştırması.[In Turkish]. *Hazır Beton Birligi, 6*(36), 32–35.

Taskın, B., & Tugsal, U. M. (2014) Seismic performance evaluation of school buildings subjected to 2011 Van earthquakes.*Proceedings of the 10th NCEE.*

TEC. (2007). *Specification for buildings to be built in seismic zones. Turkish Standards Institution. Ministry of public works and settlement, Ankara TS-498 (1997) Design loads for buildings. Turkish Standards Institute*. Ankara: Ministry of Public Works and Settlement.

Tehranizadeh, M., & Moshref, A. (2011). Performance-based optimization of steel moment resisting frames. *Scientia Iranica, 18*(2), 198–204. doi:10.1016/j.scient.2011.03.029

Chapter 8
Original and Innovative Structural Concepts for Design, Non-Linear Analysis, and Construction of Multi-Story Base Isolated Buildings

Mikayel G. Melkumyan
Armenian Association for Earthquake Engineering, Armenia

ABSTRACT

Seismic isolation of structures is becoming a more common method of providing protection from earthquake damage. Starting from 1994, 53 buildings and structures have been designed by the author of this chapter with application of base or roof isolation systems. Of these designed buildings, the total number of already constructed and retrofitted buildings has reached 45. The number of seismically isolated buildings per capita in Armenia is one of the highest in the world. Together with that seismic isolation laminated rubber-steel bearings (SILRSBs) different by their shape and dimensions, as well as by damping (low, medium and high) were designed and about 5000 SILRSBs were manufactured in the country, tested locally and applied in construction. since 2003 seismic isolation technologies were designed and then extensively applied in construction of multi-story buildings. these are: base isolated residential complexes, business centers, hotels, schools, and hospital buildings. Original and innovative structural concepts, including the new approach on installation of seismic isolation rubber bearings by clusters, were developed and designed for construction of these base isolated buildings and they are described in the given chapter. The advantages of this approach are listed and illustrated by the examples. It is mentioned that suggested new structural solutions in seismic isolation are bringing to significant savings in construction cost. The earthquake response comparative analyses were carried out for some of the considered buildings in two versions (i.e., when the buildings are base isolated and when they are fixed base). Several histories were used in the analyses and for both cases the buildings were analyzed also according to the requirements of the Armenian Seismic Code. Comparison of the obtained results indicates the high effectiveness of the proposed structural concepts of isolation systems and the need for further improvement of the Seismic Code provisions regarding the values of the reduction factors. A separate section in the chapter dedicated to the design of high damping laminated rubber-steel bearings and to results of their tests.

DOI: 10.4018/978-1-5225-2089-4.ch008

1. INTRODUCTION

The mentioned in the Abstract number of buildings where seismic isolation systems were successfully implemented during a period of the last 21 years is illustrated in Figure 1. Among these buildings there were:

- Bath-houses,
- Private residences,
- School buildings,
- Clinic and hospital buildings,
- Business centers,
- Hotels, and
- Apartment buildings.

Actually more than two buildings per year in average with seismic isolation systems were constructed or retrofitted in Armenia. Also the mentioned above great number of manufactured in the country SILRSBs is illustrated in Figure 2.

Success in implementation of seismic isolation became possible due to applied significant efforts by the author of this chapter, his permanent struggle and overcoming of the invisible barriers caused by envious and conservative community of so called "scientists/engineers".

It should be mentioned that seismic isolation in the country initially (1994-2001) developed mainly through the projects financed by international institutions (World Bank, UNIDO, Huntsman Corpora-

Figure 1. Number of base and roof isolated buildings newly constructed or retrofitted in Armenia by years

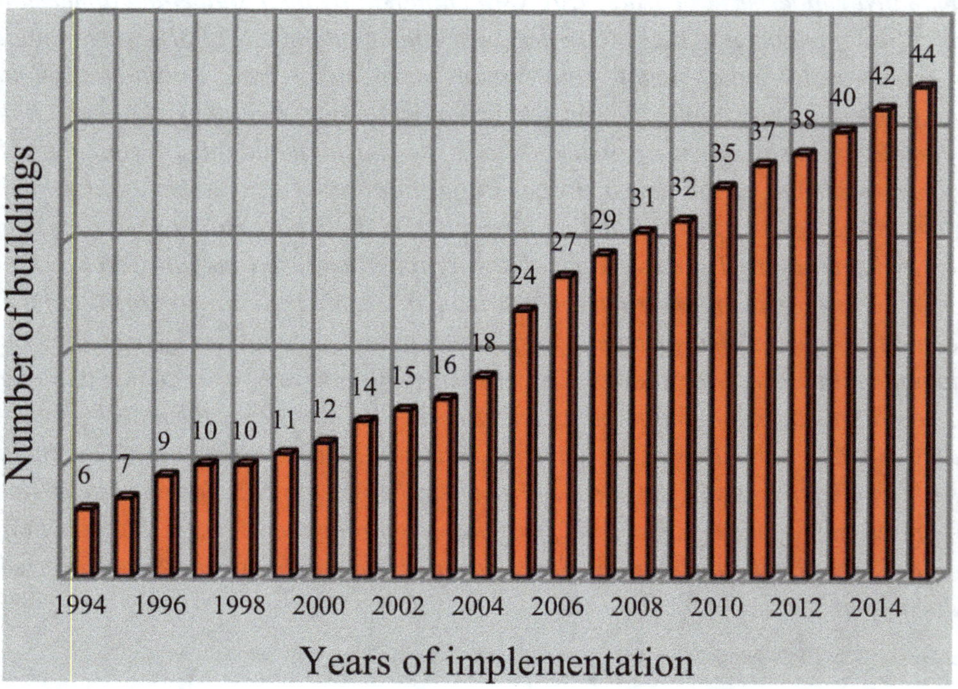

Figure 2. Number of rubber bearings installed in the newly constructed or retrofitted buildings in Armenia by years

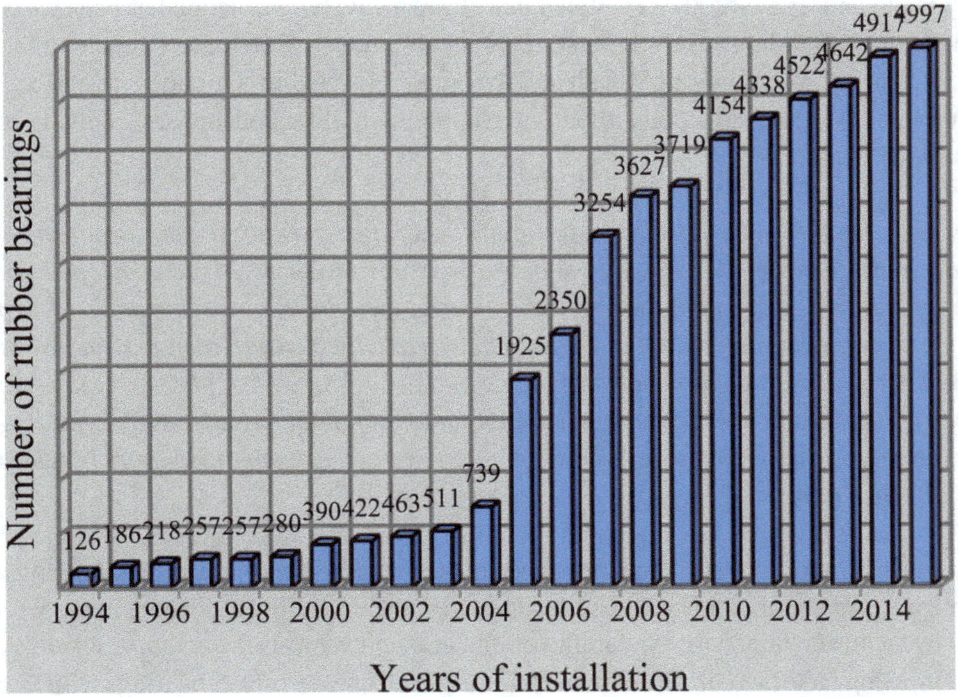

tion, Caritas Switzerland). However, the advantages of seismic isolation were so obvious that in the subsequent years great interest in application of this technology has been shown by private companies and even by the Government of Armenia. Therefore, further development of seismic isolation continued (2002-2016) through the projects financed by the Hayastan All-Armenian Fund, private companies such as PCG International, LLC (USA), Elite Group, CJSC (Armenia), Tufenkian Hospitality, LLC (Armenia), ITARCO Construction CJSC (Armenia) as well as individuals who constructed their own houses and also by the governmental program for providing apartments for young families. Given the local manufacturing of different types of rubber bearings, the developed seismic isolation techniques lead to significant savings in construction costs. Construction of ordinary (apartment) buildings and critical facilities (schools, hospitals, etc.) using seismic isolation costs 30-35% cheaper in comparison with the conventionally designed buildings. These facts attract the attention of the international professional community, of different institutions and private investors (The World Bank Implementation Completion Report 1997). Many experts in the field started to stress and underline the achieved progress. As it is mentioned in Martelli et al.(2001):

...the number of new applications of innovative anti-seismic techniques, especially seismic isolation, is particularly large in Japan, P.R. China and Armenia. Some other countries are beginning to follow the excellent example of Armenia where seismic isolators are locally manufactured also for foreign markets. An existing bank building at Irkutsk-City in Russia was retrofitted by applying the technology invented by Prof. Melkumyan in Armenia.

In Naeim and Kelly (1999), Armenia is mentioned among the few of developing countries where projects that apply low-cost base isolation systems for public housing have been successfully completed. Also in Miyamoto and Gilani (2007) it is stated that "In Armenia base isolation has been used to convert weak and vulnerable buildings to earthquake resistant structures". Reference is made to "… an existing five-story apartment building in Vanadzor, Armenia … located in a highly active seismic zone. It was retrofitted with seismic isolators without interruption to building occupancy". Finally in Garevski (2010) it is stated that:

In the developing countries, base isolation technique has rarely been used due to non-existence of domestic production of bearings and high cost of the bearings produced in the developed countries. In some of these countries, as is Indonesia, Iran and Algeria, there have been some attempts to popularize this technique through development of low-cost bearings and their installation in demonstration structures, but no attempt for production has been made and hence there hasn't been any mass application of such bearings. A greater success in application of base isolation (with isolation of a large number of buildings) was achieved in Armenia where, in addition to placement of isolators in buildings, their production was also adopted.

There are several reasons for the mentioned savings. One of them is that rubber bearings manufactured in Armenia cost significantly cheaper than bearings manufactured elsewhere in the world. This is conditioned by the lower labor cost, availability of rubber components in the country, as well as existence of several competing factories capable of manufacturing high quality rubber bearings with low (LDRB), medium (MDRB) and high (HDRB) damping. Also, the provisions of the Armenian Seismic Code for seismically isolated structures are much more progressive in comparison with, for example, the USA Code in terms of analysis and design of superstructures of base isolated buildings. As a result a huge amount of reinforcement could be reduced in superstructures of R/C base isolated buildings designed in accordance with the Armenian Code. In addition, cross-sections of the bearing structures (columns, beams, shear walls) are smaller, and there is no need to apply high strength concrete for them. Therefore, large amounts of concrete and cement may also be saved in superstructures.

Thus, successful implementation of seismic isolation technologies in the last 21 years, the presence of industry capable of local manufacture of seismic isolators, the presence of capable scientific and engineering brainpower for local development and design of seismic isolation systems, the possibility of retrofitting by seismic isolation without interruption of the use of the facilities (Melkumyan 2011), the low cost of retrofitting and new construction with seismic isolation, and the possibility of speeding up the retrofitting process fully justify further practical application of the advanced seismic isolation technologies in Armenia. Furthermore, worldwide experience proves that seismic isolation is the most reliable technology. Excellent examples demonstrating the effectiveness and high reliability of seismically isolated buildings during the destructive Hanshin-Awaji (Japan) earthquake in 1995 (Fujita 1999), the Great Sichuan (China) earthquake in 2008 (Zhou et al 2009), the Maule (Chile) earthquake in 2010 (Retamales & Boroschek 2014), etc. are well known.

The given chapter is focused on the original approach in applying the clusters of small size rubber bearings under the columns and shear walls instead of single large bearings in isolation systems of seismic isolated buildings. This new approach was in detail elaborated by the author of this chapter and widely implemented in many projects. The chapter also describes the innovative structural concepts and analyses of three buildings which got the great publicity across the world and qualified/characterized by many scientists and engineers of the international professional community as "new concepts of design", "high value of current interest", "quite amazing", "very audacious", "designed, analyzed and presented in a very interesting way".

2. ORIGINAL APPROACH IN APPLYING CLUSTERS OF RUBBER BEARINGS IN ISOLATION SYSTEMS OF SEISMIC ISOLATED BUILDINGS

2.1. The Idea of Applying Clusters of Small Size Rubber Bearings

At the beginning stages of base isolation design and implementation for buildings in Armenia before 2001, the usual way was to install a single rubber bearing under each of the columns of a frame building or to place rubber bearings one by one with certain spacing under the load-bearing walls of the building (Whittaker and Robinson 2009, Garevski 2010, Martelli and Forni 2011). However, having analyzed and designed an increasing number of base isolated buildings and with transition from low-story small buildings to large multistory buildings, as well as via thorough observations after the process of creating seismic isolation systems on the construction sites, the author of this chapter gradually came to a conclusion that the better way is to use a cluster of small bearings instead of a single large bearing. This seemed beneficial also from the viewpoint of local manufacturing of rubber bearings.

What observations brought to this new idea? In designing base isolated structures very often the engineers have to deal with buildings that are irregular in their plan and along their height. Such asymmetry causes rotation to buildings as their center of mass does not coincide with the isolation system's center of rigidity. In this case if the isolation system consists of rubber bearings installed one by one under the columns or load-bearing walls, then it is difficult to avoid rotation, although, generally speaking, it is possible to do by using bearings with different horizontal stiffness (or different geometrical dimensions). But this is inconvenient in practical terms both for manufacturers and constructors. Also, the bearings installed one by one will not be uniformly loaded by the vertical forces. The range of the vertical loads on bearings could be quite wide (Fuller et al., 2000) and again this will require application of bearings different by their rubber compounds and physical/mechanical characteristics. Another important factor which speaks in favor of installations of small size rubber bearings in clusters is that such bearings can be installed or replaced manually without using any mechanisms or expensive equipment. This is especially important during the execution of retrofitting works. Obviously, the space under the existing buildings to be retrofitted is usually narrow and machineries cannot operate inside such buildings to carry and install large and heavy isolators. Actually the same is true for newly constructed buildings if for some reason a need arises to replace the bearings.

It should be specially emphasized that installation of rubber bearings by clusters increases the seismic stability of base isolated buildings. Figure 3 shows how using clusters of bearings can enable increasing the distance between the edge bearings by 2 m (1 m from axis "8" to the left and 1 m from axis "11" to the right), which will significantly improve the overall performance of the superstructure (Melkumyan, 2011). Moreover, it is both apparent and confirmed by comparative response analyses that in case of clusters of small rubber bearings the stresses and deformations from seismic impact would be distributed more evenly in the structural elements below and above the bearings without any significant concentration in one joint, as it is the case for one large bearing.

Figure 3. Vertical elevation of the isolation system of 14-story building in "Arami" complex with the increased distance between the edge bearings

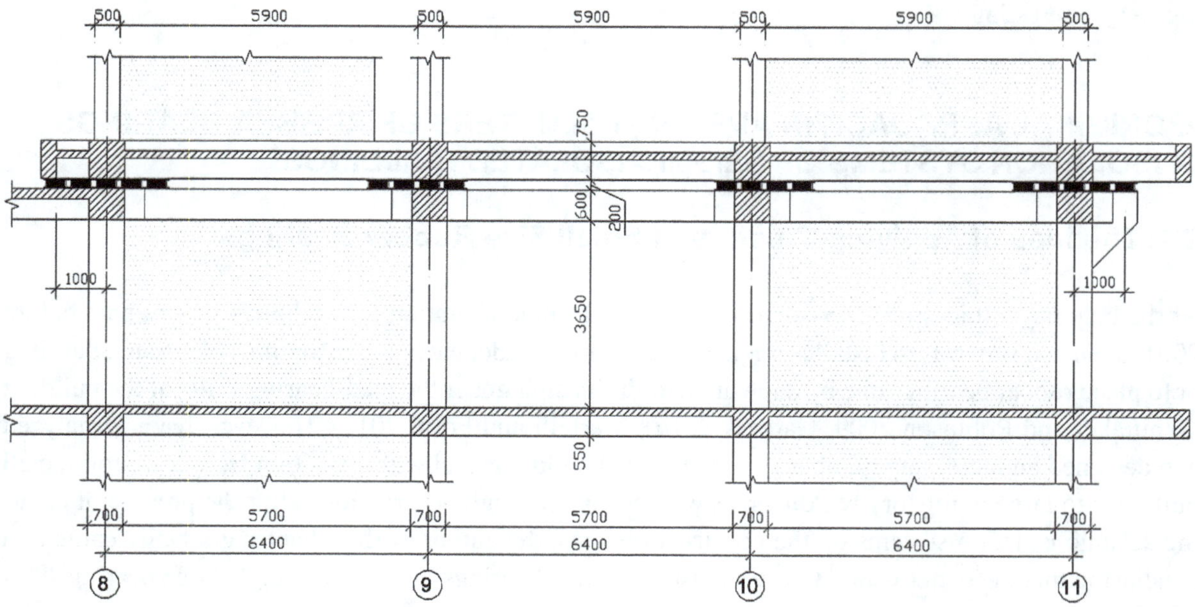

Also, the use of small size bearings simplifies such construction processes as precise installation of isolator's sockets, casting concrete under them and fixing them in the design position (Figure 4, 5, 6). All these processes are much easier to implement at the construction site with a socket of small diameter than with a large one.

Thus, after 2001 a new approach in creation of seismic isolation systems by installing clusters of small rubber bearings instead of one large bearing was proposed (Melkumyan, 2004; Melkumyan et al. 2005; Melkumyan & Hovhannisyan, 2006) and implemented in the design and construction of base isolated structures. Advantages of the proposed new approach are listed and discussed below. Original and in-

Figure 4. Precise fixing of the isolator's sockets in the design position using special conductors

Figure 5. The concrete under the sockets of small size isolators is casted without difficulties and with high quality

Figure 6. Construction process on installation of small size rubber bearings clusters

novative structural concepts for residential complexes, commercial/business centers, hotels, hospitals and schools were developed. Different quantities of seismic isolation rubber bearings have to be used under different columns of R/C frames and different shear walls of these buildings. Some examples are given in Figure 7.

2.2. Examples Illustrating Advantages of the New Approach on Installation of Clusters of Small Size Rubber Bearings Instead of a Single Large Size Bearing

Based on the above observations, as well as on the analysis and design of a number of base isolated buildings, the following advantages of the new approach on installation of clusters of small size rubber bearings instead of a single large size bearing can be summarized (Foti & Mongelli, 2011; Melkumyan, 2011):

Figure 7. Examples on installation of clusters with different numbers of rubber bearings

- Increased seismic stability of the buildings;
- More uniform distribution of the vertical dead and live loads, as well as additional vertical seismic loads on the rubber bearings;
- Small bearings can be installed manually, without using any mechanisms;
- Easy replacement of small bearings, if necessary, without using any expensive equipment;
- Easy casting of concrete under the steel plates with anchors and recess rings of small diameter for installation of bearings;
- Neutralization of rotation of buildings by manipulation of the number of bearings in the seismic isolation plane.

One more advantage was pointed out by Prof. Kelly during the 11th World Conference on Seismic Isolation in Guangzhou, China. Positively evaluating the suggested approach he mentioned that in the course of decades the stiffness of neoprene bearings may increase, and in order to keep the initial dynamic properties of the isolated buildings the needed number of rubber bearings can be dismantled from the relevant clusters.

In order to illustrate the advantages of the suggested approach two base isolated buildings, namely "Sayat-Nova" and "Arami", designed for construction in the city of Yerevan, were examined. These buildings were analyzed by several versions of installation of rubber bearings clusters. There is no need to bring here the results of all analyses, so only those of the last two versions for each of the considered buildings are presented and discussed below. Figure 8 shows the plan of the isolators' location in "Sayat-Nova" building according to a version conditionally referred to as "preceding version." There are 250 isolators planned in this version. Some results of analysis for the preceding version are given in Table 1. The final version of the isolators' location in "Sayat-Nova" building is shown in Figure 9 and the results of analysis by this version are also presented in Table 1.

The example of the "Arami" building is similar to the above described one. Figure 10 displays the plan of isolators' location for this building in the preceding version of design and analysis. The final version is illustrated in Figure 11. Some results of the analyses for both versions are given in Table 2.

The obtained results indicate that in the preceding version for "Sayat-Nova" building the difference in horizontal displacements along the axes "K" and "A" is 7.5 mm, which means rotation will occur in the isolation system for this version of isolators' location. With the final version, the rotation is neutralized by changing the number and location of the isolators. The difference in horizontal displacements in this case is equal to 0.7 mm. Also, in the final version periods of the first mode of vibrations increase, whereas total shear forces and story drifts somewhat decrease. This means the proposed approach enables improving the overall effectiveness of the isolation system and achieving a more rational solution by manipulating the number and location of isolators.

For the preceding version of "Arami" building the difference in horizontal displacement along the axes "I" and "A" is 19 mm, while for the final version this difference is only 2.3 mm. This was also achieved by manipulating with the number of isolators and changing their location in plan of the isolation system.

3. NON-LINEAR ANALYSES OF SOME BASE ISOLATED BUILDINGS BY CODE REQUIREMENTS AND BY THE TIME HISTORIES

Earthquake response analyses of the buildings where the approach on installation of clusters of rubber bearings was used in their isolation systems were carried out by SAP2000 non-linear program. Calculations were carried out considering the non-linear behavior of SILRSBs with the application of the following input parameters based on the historical tests data of seismic isolation rubber bearings (Melkumyan, 2001; Melkumyan & Hakobyan, 2005): yield strength for SILRSBs is equal to 56 kN and their yield displacement is equal to 19 mm.

The non-linearity was considered only for seismic isolators because due to application of base isolation the stories' drifts appeared to be much smaller than allowable values. Therefore, the structural elements of superstructures will actually work in elastic stage and there is no need to apply non-linearity to the superstructures. For non-linear model the isolators have initial stiffness equal to 3 kN/mm and post yield stiffness – 0.81 kN/mm.

The earthquake response analyses carried out for different buildings have shown that in comparison with the fixed base buildings, seismic isolation significantly reduces the maximum spectral acceleration, also proving to be cost effective for the isolated structures and ensuring high reliability of their behavior under seismic impacts. Comparison of the Code based analyses results with those obtained by the time history analyses indicates that the shear forces at the level of isolation systems, the maximum

Figure 8. Preceding version of the isolators' location plan in "Sayat-Nova" building

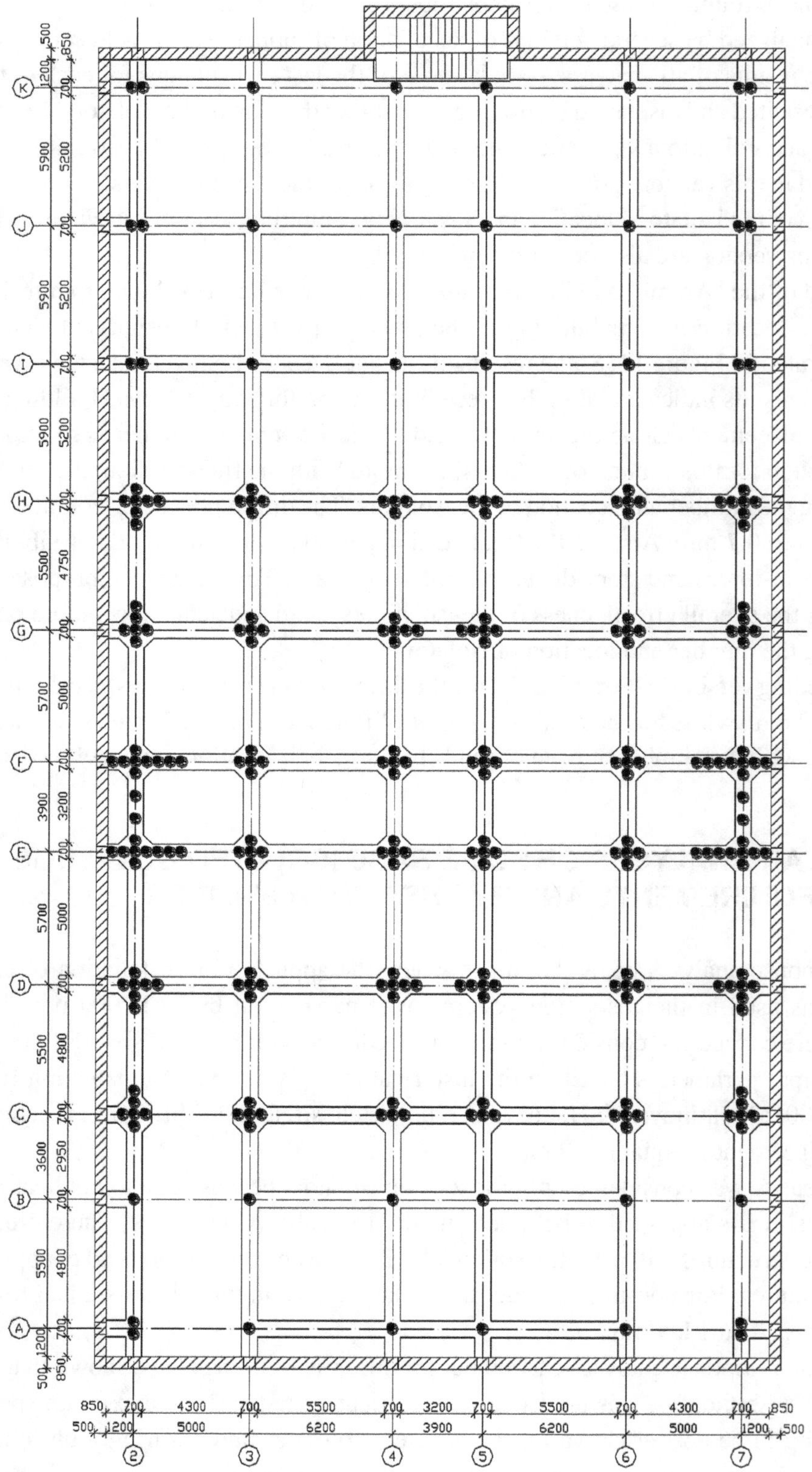

Figure 9. Final version of the isolators' location plan in "Sayat-Nova" building

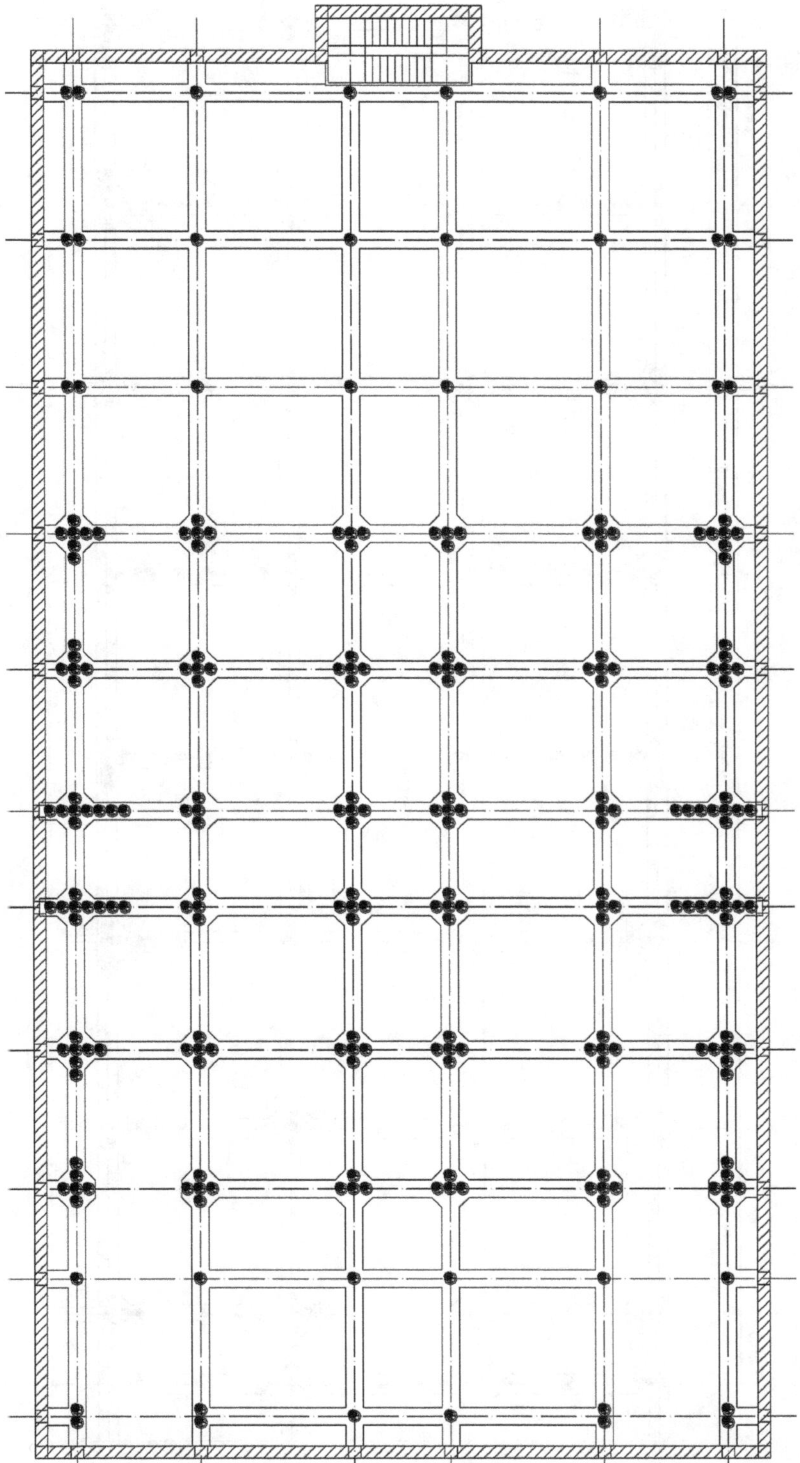

Figure 10. The preceding version of the isolators' location plan in "Arami" building

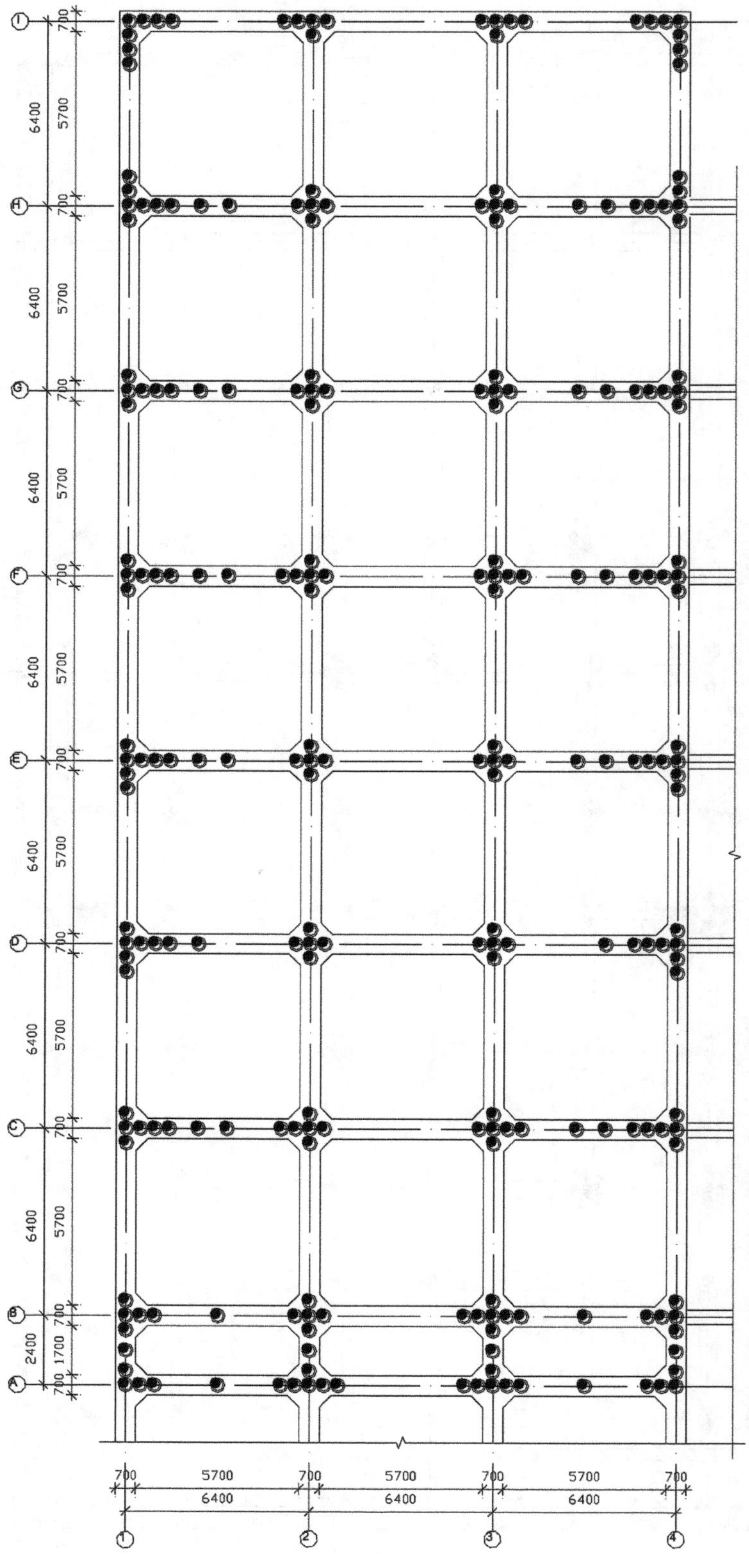

Figure 11. The final version of the isolators' location plan in "Arami" building

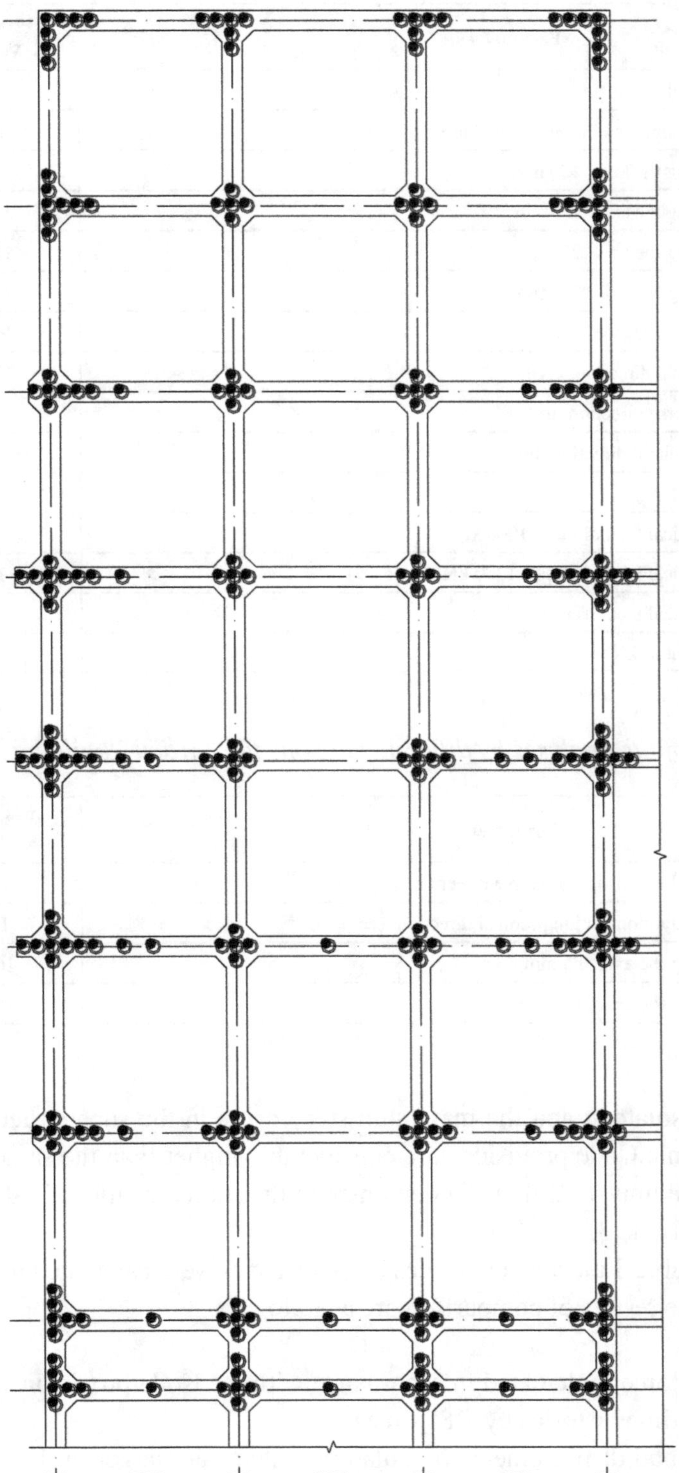

Table 1. Some results of the analyses by different versions for "Sayat-Nova" building

Parameters	Preceding Version	Final Version
Periods of the first mode of vibrations in transverse direction, sec	1.87	1.91
Periods of the first mode of vibrations in longitudinal direction, sec	1.90	1.95
Horizontal displacement along the axis "K", mm	136	135.2
Horizontal displacement along the axis "A", mm	128.5	134.5
Horizontal displacement along the axis "2", mm	134.8	138.8
Horizontal displacement along the axis "7", mm	133.8	137.8
Total shear force in transverse direction, kN	26760	26350
Total shear force in longitudinal direction, kN	26190	25830
Maximal story drift in transverse direction, mm	4.60	3.97
Maximal story drift in longitudinal direction, mm	6.40	6.07
Total number of seismic isolators	250	240
Number of isolators with vertical load of up to 1000 kN	109	97
Number of isolators with vertical load of 1000 - 1250 kN	126	105
Number of isolators with vertical load of over 1250 kN	15	38
Average vertical load per isolator, kN	548	571

Table 2. Some results of the analyses by different versions for "Arami" building

Parameters	Preceding Version	Final Version
Periods of the first mode of vibrations in transverse direction, sec	1.98	2.00
Periods of the first mode of vibrations in longitudinal direction, sec	1.95	1.96
Horizontal displacement along the axis "I", mm	169.4	153.2
Horizontal displacement along the axis "A", mm	150.4	155.5

displacements of the isolators, and the maximum story drifts in the superstructures calculated based on the Armenian Seismic Code provisions are considerably higher than the same values calculated by the time histories (Melkumyan, 2013). To demonstrate this, some results of calculations for different buildings are given in Table 3.

Using the data of Table 3 the average values of Q, D and Δ were calculated for both cases and compared to each other. The results of comparison are as follows:

- Horizontal shear forces calculated in accordance with the Code provisions are greater than those calculated by the time histories by 1.85 times,
- Maximum horizontal displacements of isolation systems are larger by 1.89 times and maximum story drifts in superstructures – by 2.03 times in average.

Table 3. Some results of calculations for different base isolated buildings by the Armenian Seismic Code and by the time histories

Name of Building	By the Armenian Seismic Code			By the Time Histories		
	Q, kN	D, mm	Δ, mm	Q, kN	D, mm	Δ, mm
10-story "Our Yard" building, T=2.04 sec	20950	220	3.1	13037	133	2.1
11-story "Cascade" building, T=1.91 sec	21386	188	3.6	12583	112	2.3
14-story "Arami" building, T=2.00 sec	17860	218	4.3	11303	132	2.9
13-story "Dzorap" building, T=2.00 sec	12831	217	4.0	6970	130	2.0
17-story "Baghramian" building, T=2.46 sec	51810	259	5.7	25943	138	3.0
15-story "Avan" building, T=2.03 sec	44341	222	2.3	24068	106	0.9
17-story "Sevak" building, T=1.98 sec	32092	215	3.0	19838	100	1.4

Q - horizontal shear force at the level of isolation system;
D - maximum horizontal displacement of isolation system;
Δ - maximum story drift in superstructure

Obviously, the differences should have not been so large. This means some further steps should be taken to more realistically reflect the characteristics of seismically isolated buildings (including the reduction factors for isolation systems) in the design models for the calculations based on the Code (Melkumyan, 2005; Saito, 2006). For zone 3, where the expected maximum acceleration is equal to a=400 cm/sec^2 there are different permissible damage coefficients stipulated in the Code for base isolated structures. It is required to apply the permissible damage coefficient (reduction factor) of k_1=0.4 for superstructure and k_1=0.8 for seismic isolators and structures below the isolation plane.

Actually, the Code requires that any base isolated building should be analyzed twice: first, by applying k_1=0.8 and the obtained results will serve as a basis to design the isolation system and structures below it, and then the second analysis should be carried out by applying k_1=0.4 and the derived results will serve as a basis to design the superstructure. However, the data regarding the analyses of multistory buildings indicate that the displacements of isolation systems, inter-story drifts and horizontal shear forces obtained by calculations of the base isolated buildings by the Armenian Seismic Code are close to the same values obtained by the time history analyses when the permissible damage coefficient of k_1=0.4 is applied. In case if k_1=0.8, the Code based results are higher by a factor of about 2 in average. Therefore, the Code needs a more accurate designation of reduction factors for seismic isolation systems. At this stage it is suggested to accept k_1=0.6 for zone 3 in the next edition of the Code, as a compromise solution.

4. STRUCTURAL CONCEPT AND ANALYSIS OF THE 20-STORY BUSINESS CENTER "ELITE PLAZA" WITH AND WITHOUT BASE ISOLATION SYSTEM

4.1. Structural Concept of the 20-Story Base Isolated Building

Analysis and design of 20-story base isolated building of the "Elite Plaza" business center is one of the recent projects and its construction has already been completed in the city of Yerevan (Figure 12). This is one of the tallest base isolated buildings in Europe.

Figure 12. Architectural design view of the 20-story base isolated building of the "Elite Plaza" business center and its current view

Similarly to the many base isolated buildings implemented in Armenia (Melkumyan, 2005, 2006, 2007, 2009, 2011, 2014; Melkumyan & Gevorgyan, 2008, 2009), the considered building also has parking floors below the isolation plane designed using strong and rigid R/C structural elements. The cross sections of columns vary from 700×700 mm to columns with the round cross sections of 700 mm in diameter and beams from 700×600(h) mm to 700×650(h) mm. The thickness of shear walls is equal to 300-400 mm. The cross section of the foundation strips is equal to 1000×1200(h) mm. It is worth noting that elevators' shafts together with staircases in the building were also designed as the rigid cores with the thickness of their walls equal to 300 mm. Thus, the accepted structural solution allowed obtaining a rigid system below the isolation plane, which provides a good basis for effective and reliable behavior of isolators during the seismic impacts. Of course, the superstructure should have substantial rigidity for the same purpose. This was achieved by using R/C columns with cross section of 500×500 mm along the interior axes, round cross section columns with diameter of 700 mm along the exterior axes, and 160 mm thick shear walls between them. Along all exterior axes the beams were designed with cross sections of 700×350(h) mm and along the interior axes the beams have cross sections 500×350(h) mm. The thickness of R/C slabs was set at 150 mm for all floors. The elevators' shafts and staircases in superstructure were designed in the same way as for the part of the building below the seismic isolation plane.

From the drawing in Figure 13 it can be seen that there are two low-story parts on the left and right sides of the designed building, which are rigidly connected to the main building. This was dictated by the necessity to increase the overall stability of the building in lateral direction and reliably resist the overturning. Plan of location of SILRSBs is shown in Figure 14. All of them are of the same size (di-

Figure 13. Vertical elevations of the 20-story base isolated building of the "Elite Plaza" business center in transverse (a) and longitudinal (b) directions

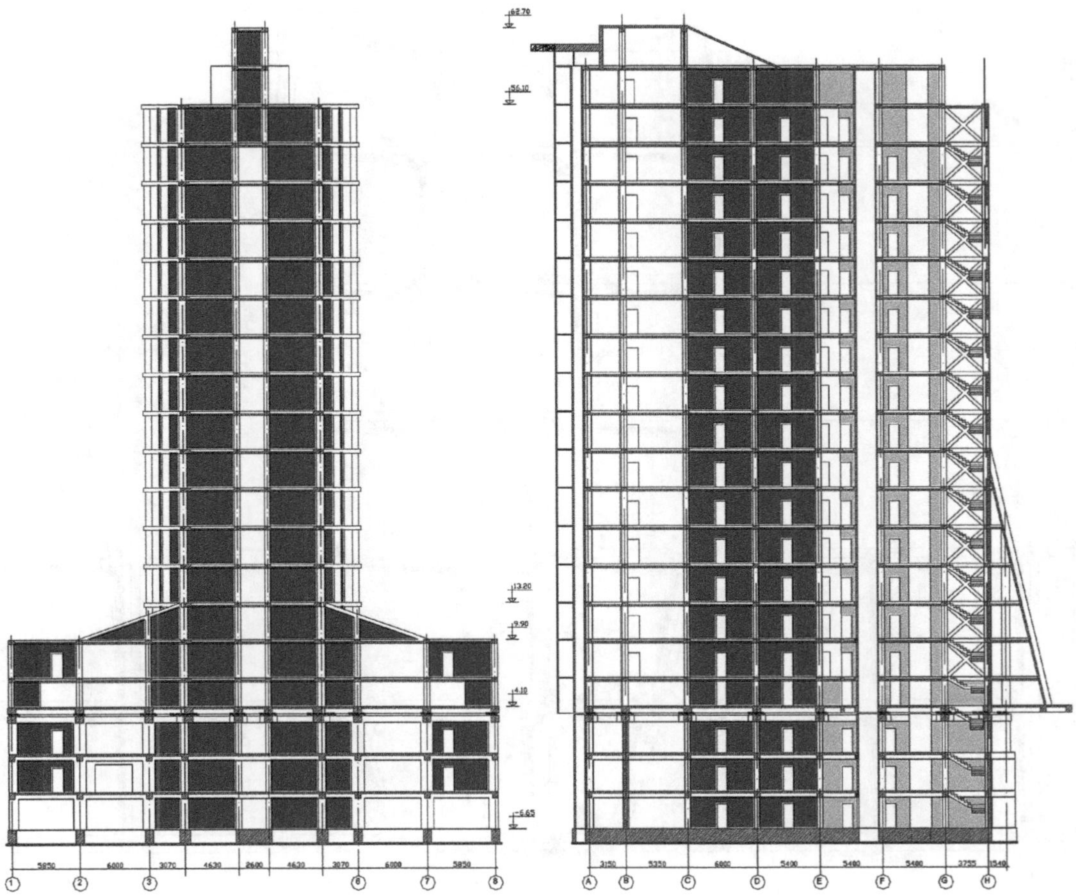

ameter – 380 mm, and height – 202 mm) and characteristics. They have a horizontal post yield stiffness of 0.81 kN/mm, damping factor of about 9-10%, and can develop a horizontal displacement of up to 280 mm (about 220% of shear strain). They are made of neoprene and were designed and tested locally (Melkumyan, 2001; Melkumyan & Hakobyan, 2005).

4.2. Results of Analysis for the 20-Story Building With and Without Seismic Isolation System

Earthquake response analysis of the considered building was carried out using SAP 2000 non-linear program. The design model (Figure 15) was developed using shell finite elements (FE) for shear walls and floor slabs, and frame FE for columns and beams. Special FE was used for seismic isolators. Nine time histories were selected for calculations and were scaled to 0.4g acceleration (see paragraph 5.2 below). Also the building was analyzed based on the provisions of the Armenian Seismic Code. The soil conditions of the construction site for the considered building correspond to category II, for which the soil conditions coefficient is $k_0=1$. The site is located in zone 3, where the expected maximum acceleration

Figure 14. Plan of location of SILRSBs with color-coded indication of vertical forces on rubber bearings

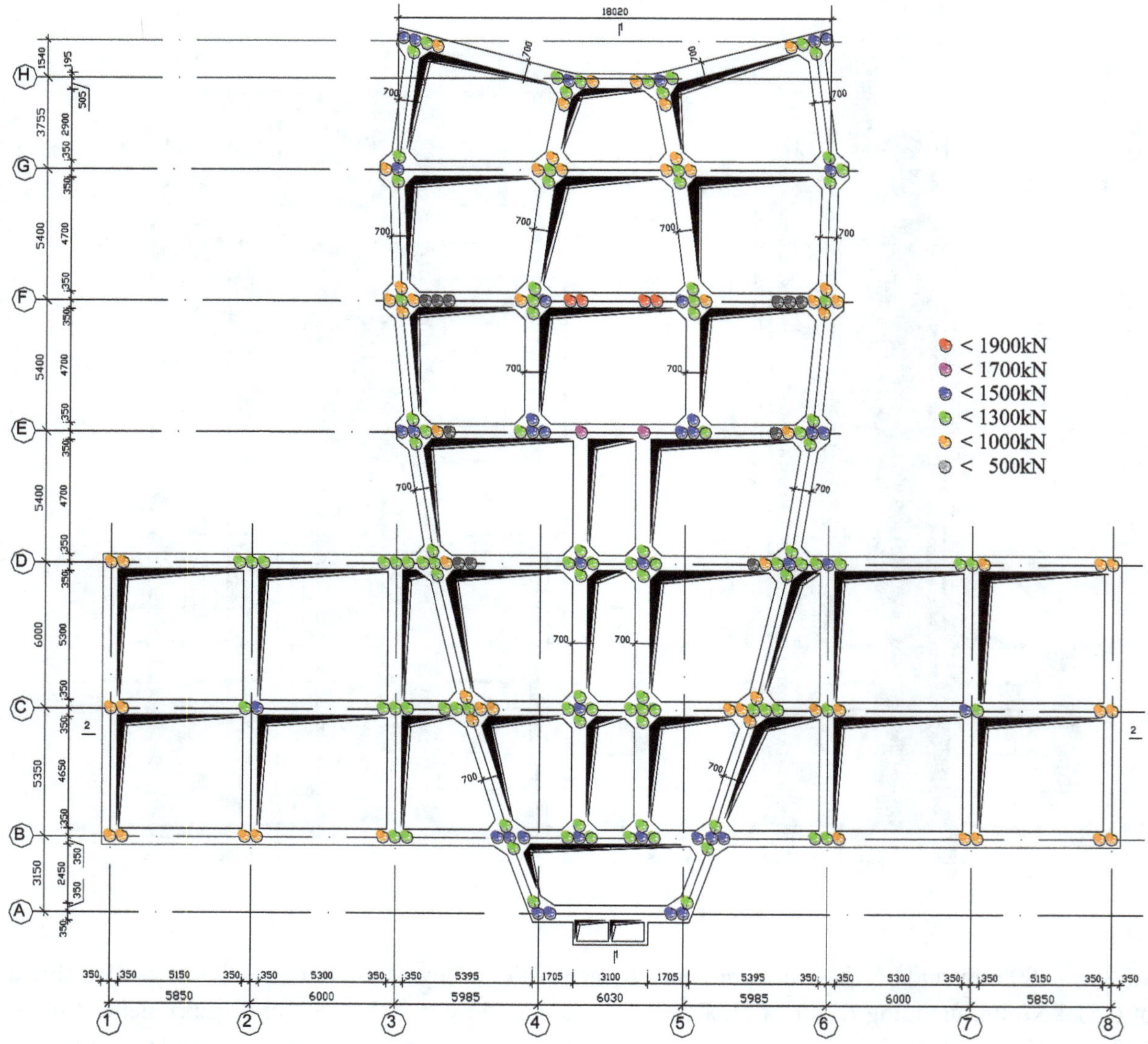

is equal to a=400 cm/sec^2. As it is mentioned in Section 3 there are different allowable damage coefficients stipulated in the Code. For this particular case of R/C building with shear walls it is required to apply allowable damage coefficient (reduction factor) of k_1=0.4 for superstructure and k_1=0.8 for seismic isolators and the structures below the isolation plane. The changes of the input acceleration at the top of the building in transverse (X) and in longitudinal (Y) directions for both cases with and without seismic isolation are shown in Table 4. Some results of calculations are given in Table 5.

It follows from the obtained results that the first mode vibrations' period of base isolated building is longer than that for the fixed base building by a factor of 2.4 in transverse direction and by a factor of 3.5 in longitudinal direction. This means seismic isolation has reduced the maximum spectral acceleration by a factor of 2.3 in average, which can be also told from Table 43. Conversely, for fixed base building the input acceleration amplifies along the height of the building. At the top of the fixed base building the

Table 4. Changes of input acceleration (1988 Spitak Earthquake accelerogram recorded at Ashotsk) at the top of the 20-story base isolated and fixed base buildings in transverse (X) and in longitudinal (Y) directions

Response acceleration	
Top of the isolated building, X direction, Max. acceleration = 0.17g	Top of the fixed base building, X direction, Max. acceleration = 1.10g
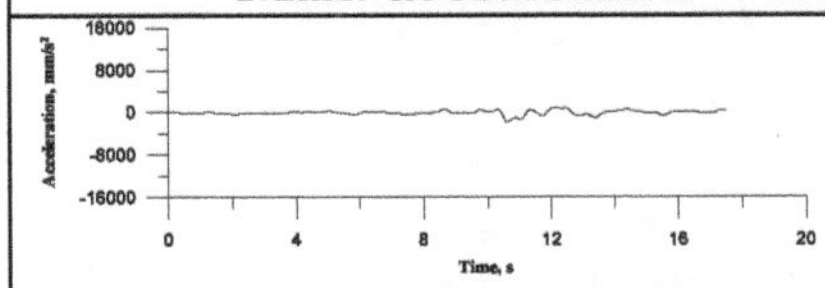	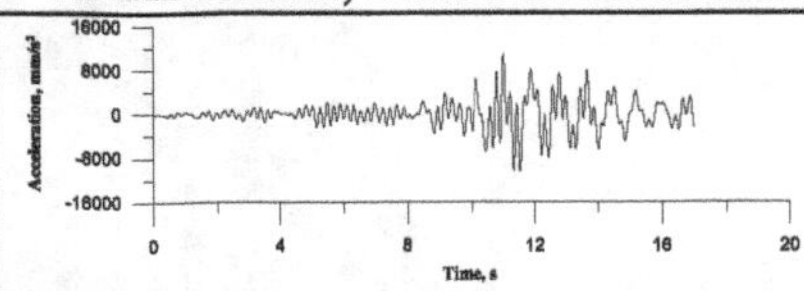
Top of the isolated building, Y direction, Max. acceleration = 0.18g	Top of the fixed base building, Y direction, Max. acceleration = 1.51g
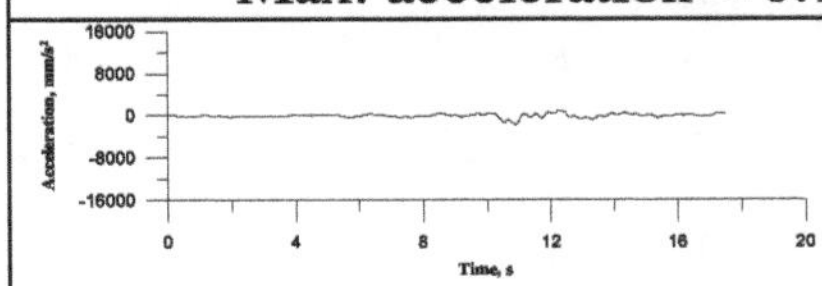	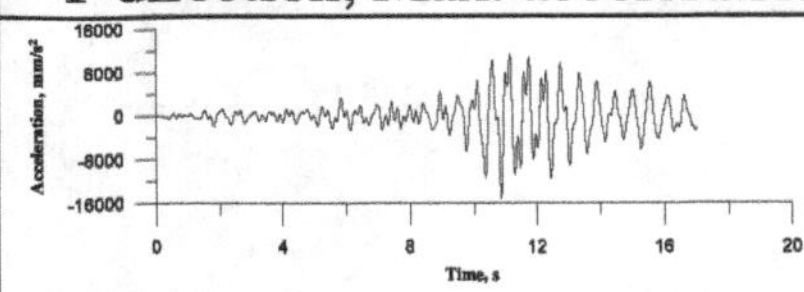
Input acceleration at foundation, 0.4g	
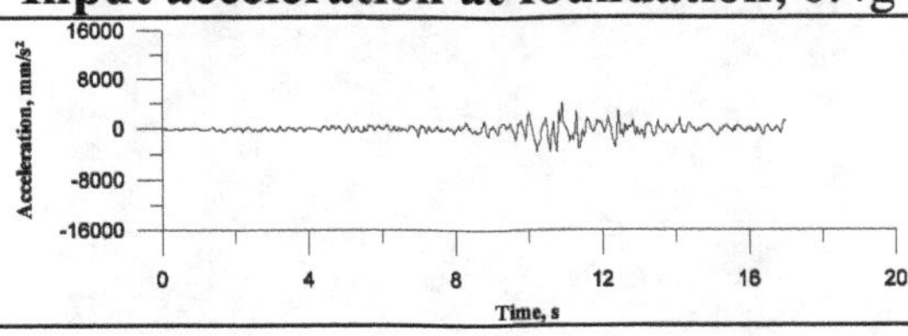	

acceleration in transverse and in longitudinal directions exceeds the input acceleration by a factor of 3.3 in average. Direct comparison of accelerations at the top level of the building indicates that acceleration in the base isolated building is in average about 7.4 times smaller than that in fixed base building. Results of calculations also show that inter-story drifts in base isolated building are in average 3.1 times smaller than in fixed base building and horizontal shear forces are smaller by 3.8 times in average.

It can be noticed that the displacements of isolation system, inter-story drifts and horizontal shear forces for the base isolated building obtained by calculations based on the Armenian Seismic Code are close to the same values obtained by the time history analysis when the allowable damage coefficient (reduction factor) of k_1=0.4 is applied. However, in case if k_1=0.8, the Code based results are higher by a factor of 2.6 in average.

The conducted study confirms that base isolation brings to simultaneous reduction of floor accelerations and inter-story drifts and to significant reduction of shear forces in comparison with those in the fixed base buildings. The suggested structural concept of the 20-story base isolated building and the new approach of installing clusters of seismic isolation rubber bearings brings to rational solution for the whole bearing structure, increases overall stability of the building and effectiveness of the isolation system. The obtained results also indicate the high reliability of the proposed concept of base isolation system and the need for further improvement of the Armenian Seismic Code provisions regarding the values of the reduction factors.

Figure 15. Design model of the 20-story base isolated building of the "Elite Plaza" business center

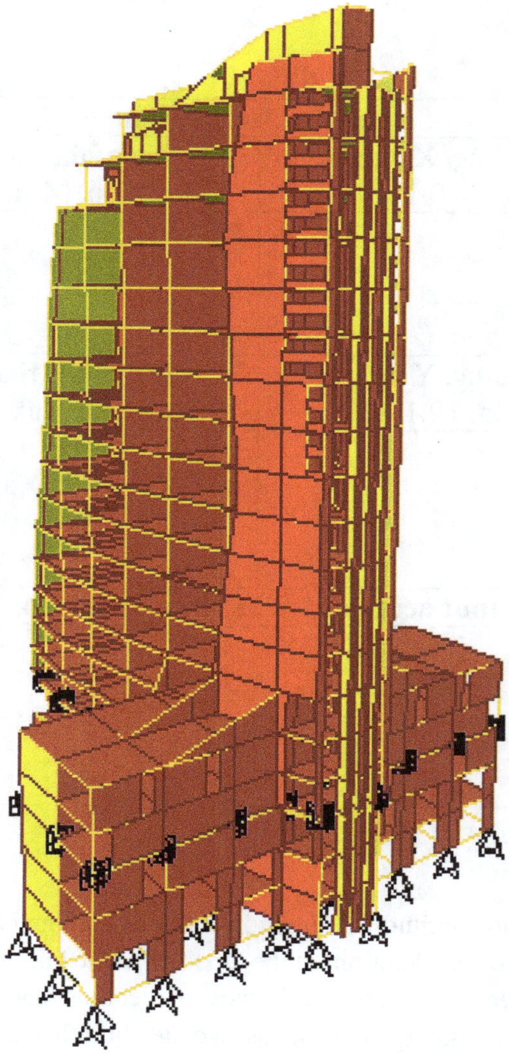

5. STRUCTURAL CONCEPT AND ANALYSIS OF THE 18-STORY RESIDENTIAL COMPLEX "NORTHERN RAY" WITH AND WITHOUT BASE ISOLATION SYSTEM

5.1. Structural Concept of the 18-Story Base Isolated Building

Analysis and design of 18-story base isolated building of the multifunctional residential complex "Northern Ray" is also one of the recent projects and its construction has already been completed as well in the city of Yerevan (Figure 16).

The building is designed in an unusual shape. This was dictated by architectural solution as the complex crowns the Northern Ray Street (Figure 17) and serves as a great gate, opening a magnificent and majestic view of Mount Ararat from the North.

Table 5. Some results of analysis for the 20-story building of the "Elite Plaza" business center with and without base isolation system

Design Parameters	By the Armenian Seismic Code for:			
	Base Isolated Building		**Fixed Base Building**	
Period of vibrations (sec)	T_x=1.99	T_y=1.92	T_x=0.82	T_y=0.55
Inter-story drift (mm)	4.2 (k_1=0.4) 8.1 (k_1=0.8)	2.2 (k_1=0.4) 4.5 (k_1=0.8)	16.5 (k_1=0.4)	5.0 (k_1=0.4)
Horizontal shear force on the level of foundation (kN)	24141 (k_1=0.4) 48281 (k_1=0.8)	23157 (k_1=0.4) 46313 (k_1=0.8)	80107 (k_1=0.4)	88287 (k_1=0.4)
Displacement of the isolation system (mm)	139.8 (k_1=0.4) 278.6 (k_1=0.8)	111.5 (k_1=0.4) 223.0 (k_1=0.8)	-	-
	Average by the time histories for:			
	Base Isolated Building		**Fixed Base Building**	
Inter-story drift (mm)	2.4	1.4	6.7	4.5
Horizontal shear force on the level of foundation (kN)	19322	19857	77373	83399
Displacement of the isolation system (mm)	113.7	114.9	-	-

Figure 16. Architectural design view of the 18-story base isolated buildings/twins of the multifunctional residential complex "Northern Ray" from the Northern side (left) and the current view of the constructed base isolated buildings/twins from the Southern side (right)

The considered building has one parking floor below the isolation plane (Figure 18a) designed using R/C strong and rigid structural elements. The cross sections of columns vary from 700×700 mm, 700×1000 mm to 700×1900 mm, and those of beams – from 700×600(h) mm to 700×650(h) mm. The thickness of shear walls is equal to 300-400 mm. The cross section of the foundation strips is 1000×1200(h) mm. The buildings are located on a very complicated terrain. The ground surface on the Northern side is about 9 m higher than on the Southern side. Therefore, a deep retaining wall was designed in order to provide free horizontal movement of the structure (Figure 18b). Both the elevators' shafts and staircases in the building were also designed as the rigid cores with the thickness of their walls equal to 300 mm.

Figure 17. Northern Ray Street in Yerevan with the view from South on the base isolated 18-story buildings at its end

The accepted structural solution allowed obtaining a rigid system below the isolation plane and also the substantial rigidity was provided to the superstructure. This was achieved by using R/C columns with cross section of 500×500 mm and 600×600 mm and shear walls between them with the thickness of 160 mm. Along all exterior axes strong beams were designed with a cross section of 500×650(h) mm and along the interior axes the beams have cross section of 500×350(h) mm. The thickness of R/C slabs was set at 150 mm for all floors. The elevators' shafts and staircases in superstructure were designed in the same way as for the part of the building below the seismic isolation plane.

Starting from the level of 17.45 m the building has a cantilever part the span of which increases towards the top of the building (Figure 19). Such a solution would have brought to significant complications if this building were to be designed with conventional foundations. Actually, nobody in the country agreed to the design this building. Only the structural concept suggested by the author of this chapter, along with application of base isolation technology, made it possible to design and erect this structure, which is very interesting from the engineering point of view and quite unusual. In the considered building also the approach to install clusters of small rubber bearings was used. It can be seen in Figures 18a and 19 that different numbers of rubber bearings are installed under different columns and shear walls. All rubber bearings are of the same size and characteristics as mentioned above.

5.2. Results of Analysis for the 18-Story Building With and Without Seismic Isolation System

Earthquake response analysis of the considered building was carried again using SAP 2000 non-linear program based on the developed design model shown in Figure 20. Calculations were carried out taking into account the non-linear behavior of seismic isolation rubber bearings with the same as mentioned above input parameters: yield strength – 56 kN; yield displacement – 19 mm; effective horizontal post yield stiffness – 0.81 kN/mm.

Figure 18a. Plan of location of seismic isolation rubber bearings and the retaining wall, and distribution of vertical forces indicated by color-coded rubber bearings

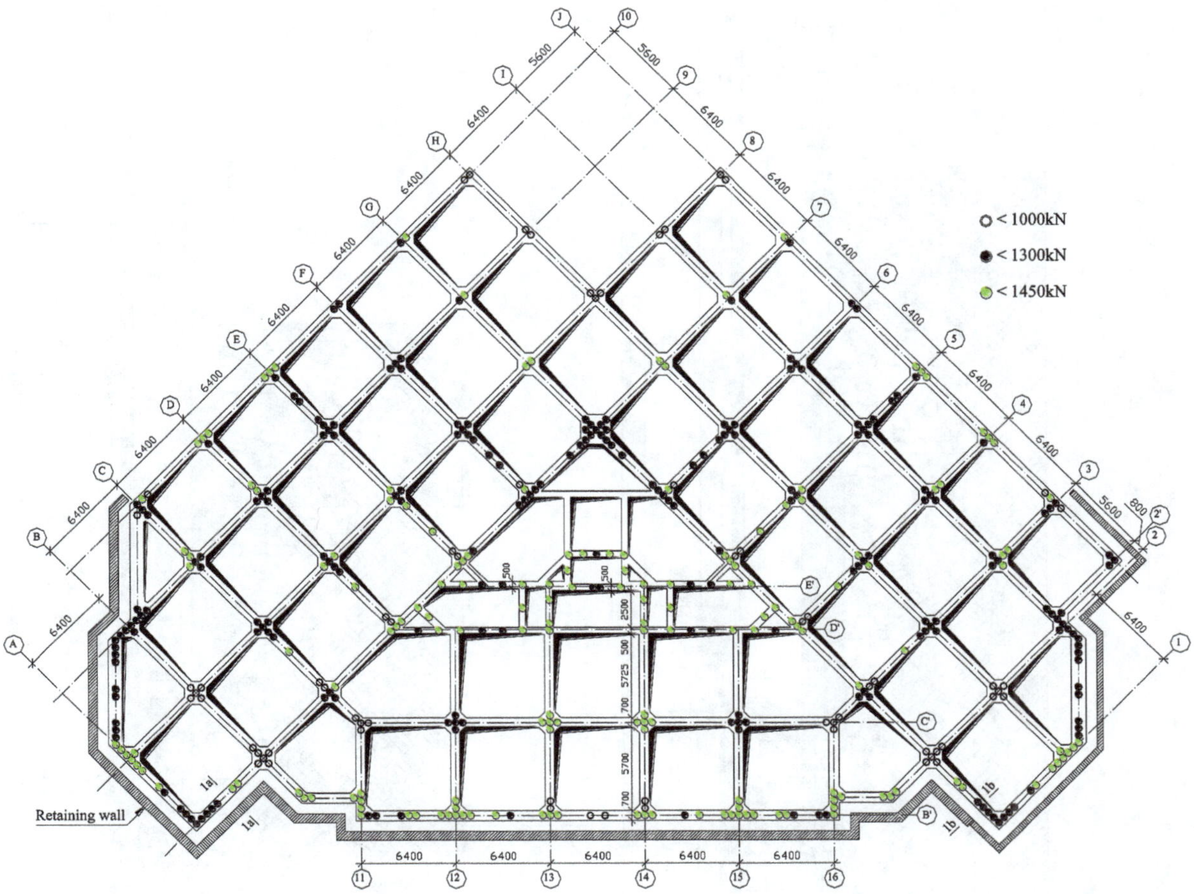

For the time history non-linear earthquake response analysis a group of accelerograms was used, including a synthesized accelerogram (Table 6). They were chosen in a manner that the predominant periods of the Fourier spectra do not exceed 0.5-0.6 sec as the soils in this construction site are classified based on the provisions of the Armenian Seismic Code as category II with the predominant period of vibrations of not more than 0.6 sec. Also the building was analyzed considering that it located in seismic zone 3 with the expected maximum acceleration of 0.4g. Some results of calculations are given in Table 7 and Figure 21.

It follows from the obtained results that the first mode vibrations' period of the base isolated building is longer than that for the fixed base building by a factor of 2.4 in transverse (X) direction and by a factor of 1.9 in longitudinal (Y) direction. Direct comparison of accelerations at the top level of the building shows that accelerations in the base isolated building are about 5 times smaller in average than in fixed base building. Results of calculations also show that inter-story drifts in base isolated building are in average 2.6 times smaller than those in fixed base building and horizontal shear forces are smaller by 2.3 times in average.

Figure 18b. Plan of location of seismic isolation rubber bearings and the retaining wall, and distribution of vertical forces indicated by vertical elevations of the retaining wall

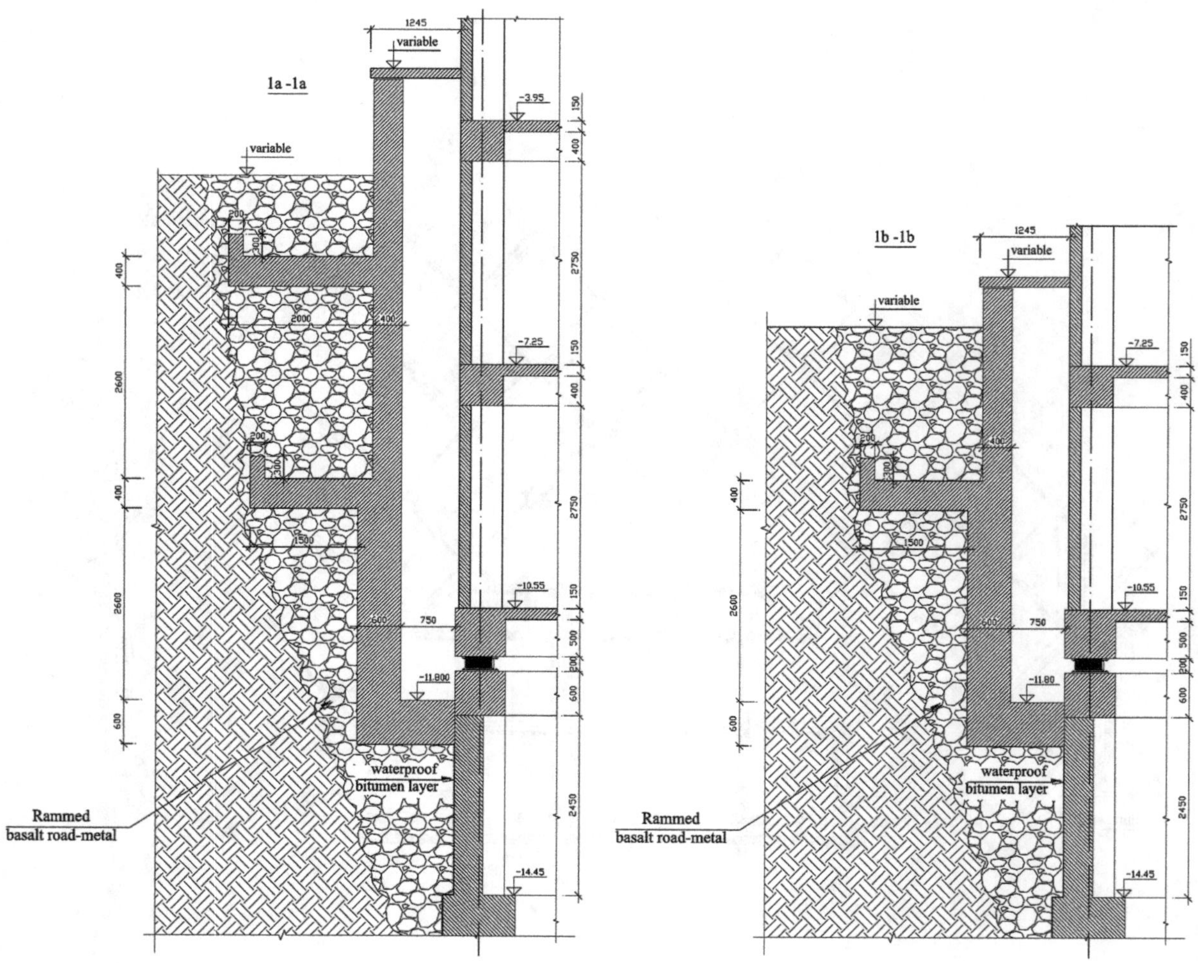

It also can be noticed that the displacements of isolation system, inter-story drifts and horizontal shear forces for the base isolated building obtained by calculations in accordance with the Armenian Seismic Code are close to the same values obtained by the time history analysis when the allowable damage coefficient (reduction factor) of $k_1=0.4$ is applied. However, in case if $k_1=0.8$ the Code based results are in average higher by a factor of 2.4. It is also necessary to state that in none of the isolators the vertical force exceeds 1500 kN. Figure 18a shows that thanks to the proposed approach of location of rubber bearings in the seismic isolation system, a more or less uniform distribution of the vertical loads was achieved. The differences in vertical loads for different isolators do not exceed the factor of 1.5 as required by the Code. Also rotation of the building in the horizontal plane is neutralized.

Figure 19. Vertical elevation of the exterior frame by axis "9" of the 18-story base isolated building of the multifunctional residential complex "Northern Ray"

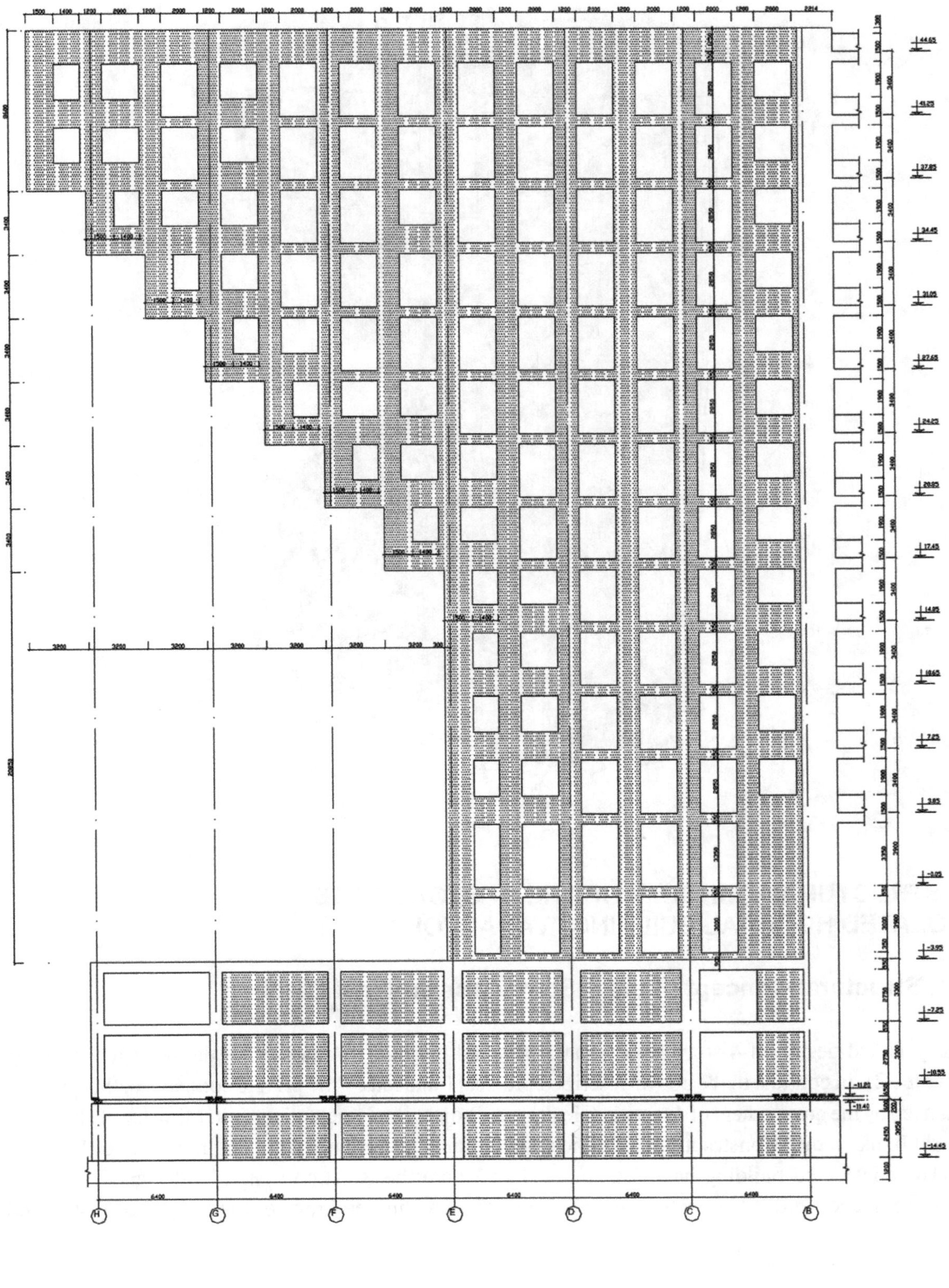

Figure 20. Design model of the 18-story base isolated building of the multifunctional residential complex "Northern Ray"

6. STRUCTURAL CONCEPT OF THE 4 - STORY BASE ISOLATED HOSPITAL BUILDING "VANADZOR"

6.1. Structural Concept of the 4-Story Base Isolated Building

Analysis and design of 4-story base isolated hospital building "Vanadzor" is the most recent project (Figure 22). In contrast to the above described projects financed by the private companies, this project is financed by the government of Armenia (Healthcare Project Implementation Unit, Ministry of Health, the World Bank credit). Construction of this hospital building is currently going on in the city of Vanadzor.

The considered building has a basement floor where the isolation system is designed. There are also strong and rigid R/C structural elements designed within the limits of the basement. However, in

Table 6. Acceleration time histories selected for earthquake response analysis of the 18-story base isolated building, scaled to 0.4g

Earthquake and Record Component	Predominant period, sec	Duration, sec	View of Accelerogram
09.03.49 Hollister (USA)	0.30	9	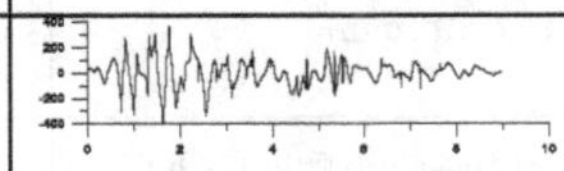
15.04.79 Bar (former Yugoslavia) in horizontal EW direction	0.55	15	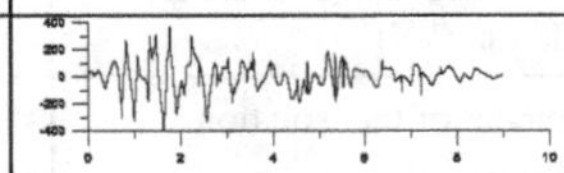
20.12.54 Eureka (USA) horizontal NE direction	0.44	26	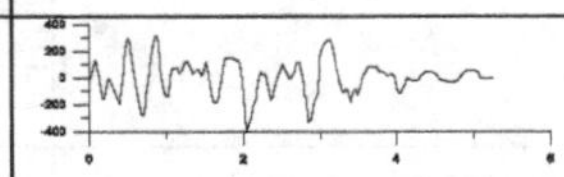
17.12.87 Chiba (Japan) in horizontal NS direction	0.35	39	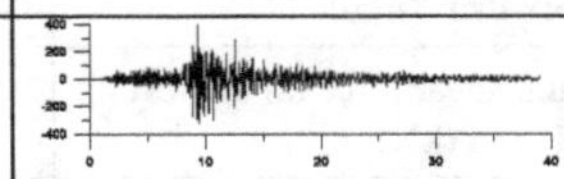
20.06.90 Manjil (Iran) in horizontal NE direction	0.49	20	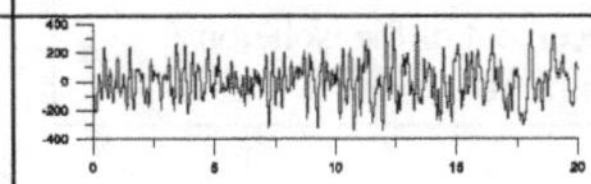
7.12.88 Spitak (Armenia) in horizontal EW direction at Ashotsk station	0.43	18	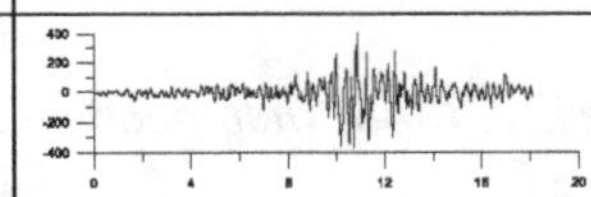
7.12.88 Spitak (Armenia) in horizontal NS direction at Ashotsk station	0.47	18	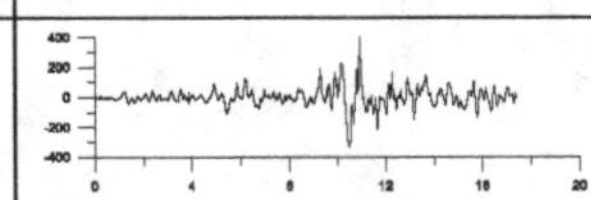
17.10.89 Loma Prieta (USA)	0.34	10	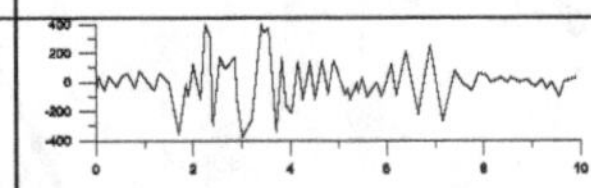
Synthesized acceleration time history (obtained by E. Gevorgyan in 2003)	0.26	18	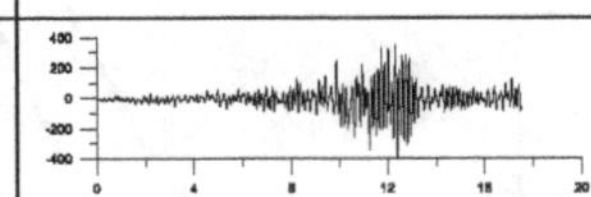

the above mentioned buildings the top parts of all columns right under the seismic isolation plane are connected by the beams. In this case, because of the accepted architectural solution, columns are not connected by the beams and, therefore, they have rather big cross sections in order to be able to reliably carry horizontal seismic shear forces (Figure23).

The cross section of the foundation strips (Figure 24) is T - shaped with the area of its bottom flange equal to 1400×400 (h) mm and of its web – 800×700 (h) mm. The cross section of most of the columns is also T - shaped so that the area of its flange is equal to 1800×600 mm and of its web – 900×600 mm. Some of the columns are designed with the L - shaped cross section. Consequently, the central core of all columns has a cross section equal to 600×600 mm. The height of the columns is equal to 2850 mm

Table 7. Some results of analysis for the 18-story building of the multifunctional residential complex "Northern Ray" with and without seismic isolation

Design Parameters	By the Armenian Seismic Code for:			
	Base Isolated Building		Fixed Base Building	
Period of vibrations (sec)	T_x=2.06	T_y=2.17	T_x=0.85	T_y=1.13
Inter-story drift (mm)	2.5 (k_1=0.4) 4.9 (k_1=0.8)	4.0 (k_1=0.4) 7.2 (k_1=0.8)	6.1 (k_1=0.4)	8.7 (k_1=0.4)
Horizontal shear force on the level of foundation (kN)	38786 (k_1=0.4) 96964(k_1=0.8)	41336 (k_1=0.4) 103339 (k_1=0.8)	93151 (k_1=0.4)	81452 (k_1=0.4)
Displacement of the isolation system (mm)	133 (k_1=0.4) 265 (k_1=0.8)	141 (k_1=0.4) 282 (k_1=0.8)	-	-
	Average by the time histories for:			
	Base Isolated Building		Fixed Base Building	
Inter-story drift (mm)	1.4	3.4	3.8	9.9
Horizontal shear force on the level of foundation (kN)	47667	49240	86042	152352
Displacement of the isolation system (mm)	123	118	-	-

Figure 21. Floor displacements of the 18-story residential building with and without seismic isolation (k_1=0.4)

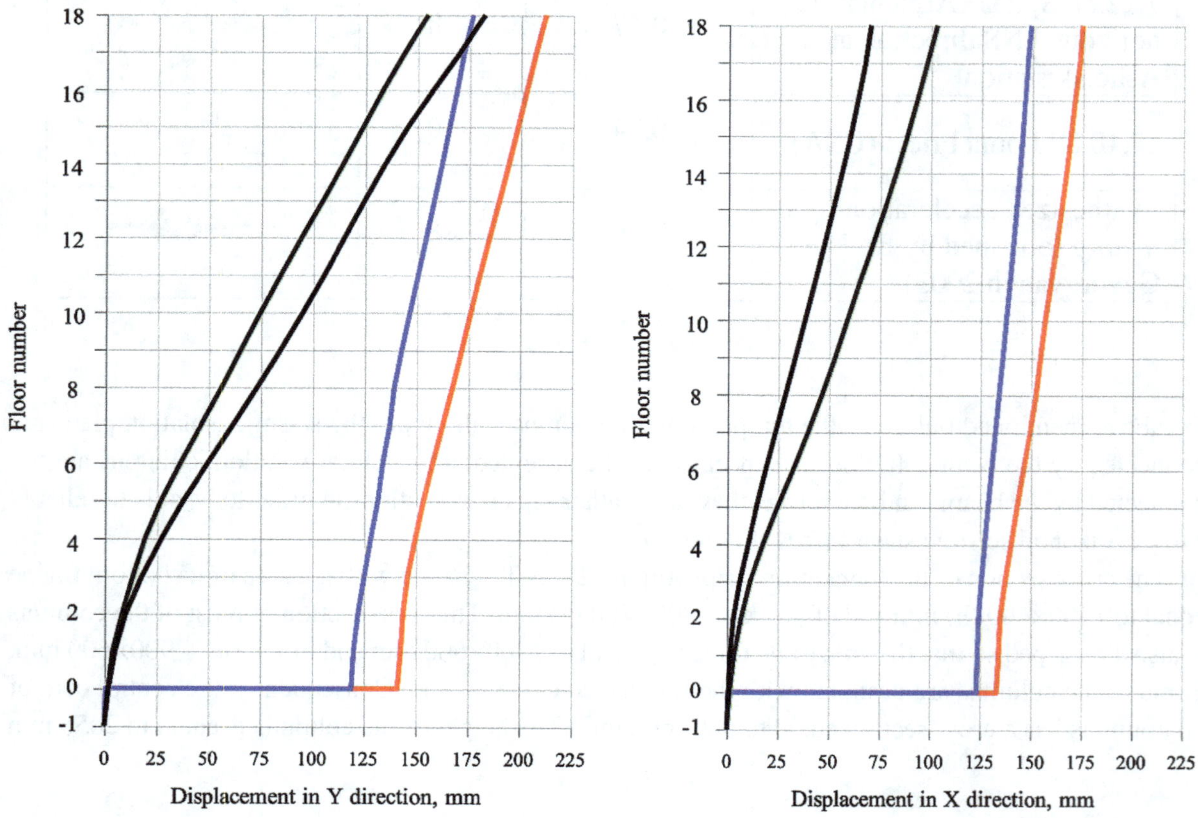

Figure 22. Design view of the 4-story base isolated hospital building "Vanadzor"

Figure 23. Types of the columns and examples on installation of rubber bearings' clusters in the basement of the 4-story base isolated hospital building "Vanadzor"

and the seismic isolators are installed just on top of them. The cross section of beams above the seismic isolators is equal to 600×650(h) mm. The shear walls are designed with the thickness equal to 300 mm and they are located along the all exterior axes of the basement (see Figure 24). Some of the shear walls are solid and others have door or window openings.

As for the above considered buildings in the given case the accepted structural solution also allowed obtaining a rigid system below the isolation plane and providing effective and reliable behavior of isolators during the seismic impacts. The superstructure (the part of building above the isolation plane, which consists of 3 floors) possesses substantial rigidity as well and this was achieved by using R/C columns

Figure 24. Plan of foundation strips and location of columns and shear walls at the mark of –3.70 in the 4-story base isolated hospital building "Vanadzor"

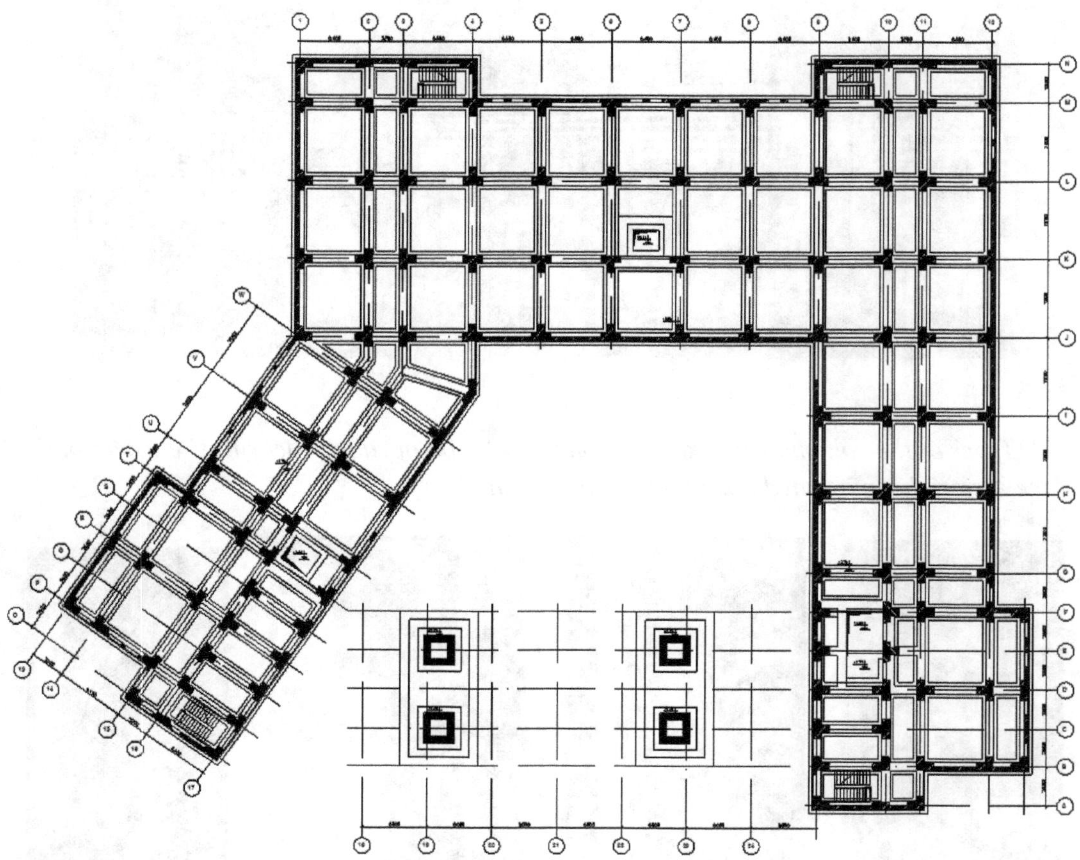

with cross section of 400×400 mm and 200 mm thick shear walls around the staircases and elevators' shafts. The thickness of R/C slabs was set at 150 mm for all floors. Plan of location of seismic isolators is shown in Figure 25.

In the considered building again (see Figure 25) the approach suggested earlier (Melkumyan & Hovhannisyan, 2006; Melkumyan, 2007, 2009) on installation of the clusters of small rubber bearings instead of a single large bearing under the columns was used. Corresponding examples of installed isolators are shown in Figure 23. Thus, different numbers of SILRSBs are installed under the different columns but in contrast to the above mentioned characteristics of isolators their damping factor for this particular project was increased and equal to 13 - 15%. All other parameters of SILRSBs were the same as described above.

6.2. Results of Analysis of the 4-Story Base Isolated Hospital Building "Vanadzor"

Earthquake response analysis of the considered building was carried out using LIRA - SAPR2013 R2 software as well as SAP 2000 non-linear program and the same selected time histories as given in Table 6. Also the building was analyzed based on the provisions of the Armenian Seismic Code for zone 3.

Figure 25. Plan of location of seismic isolation rubber bearings at different marks in the 4-story base isolated hospital building "Vanadzor"

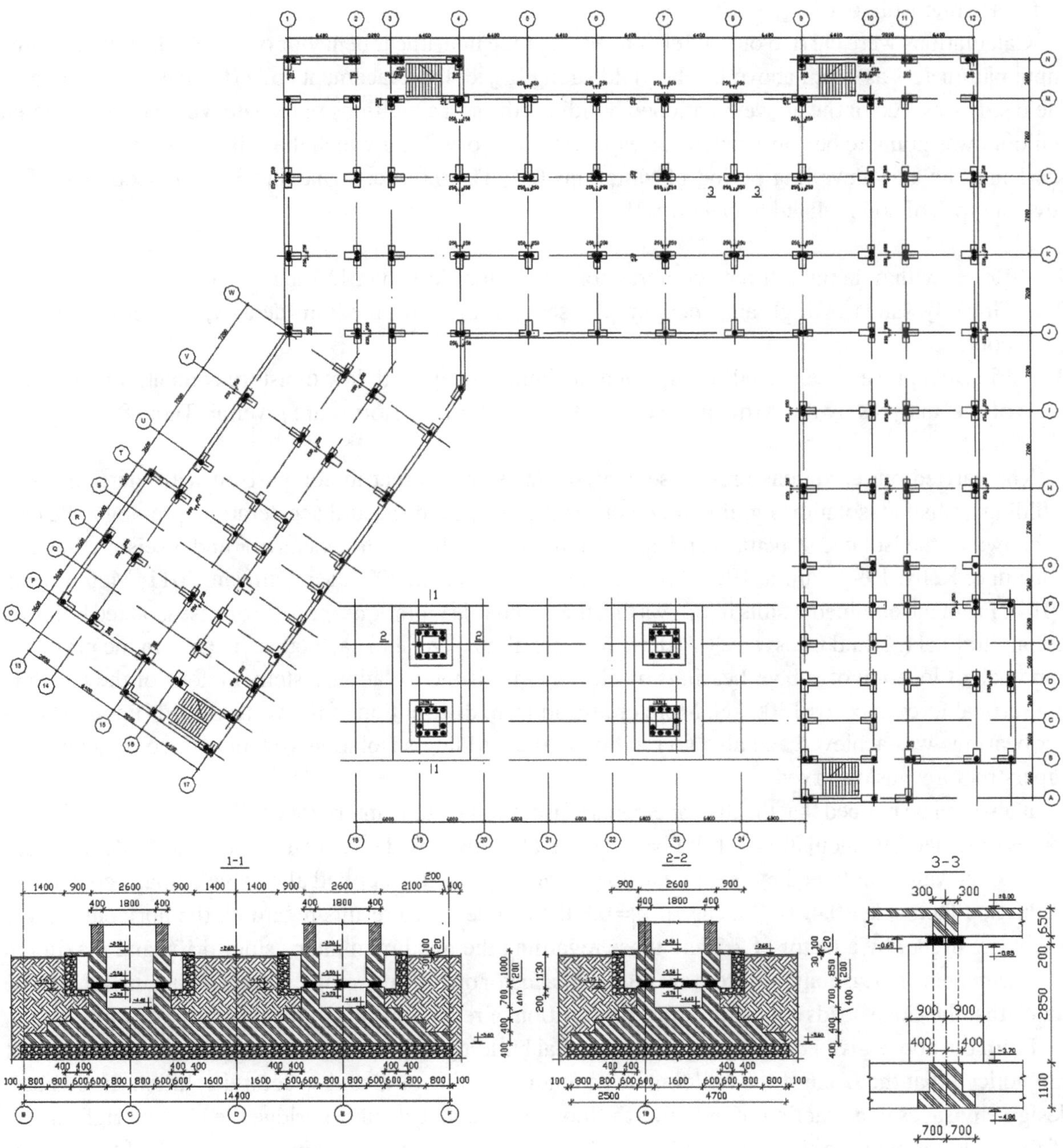

In this project the selected time histories were scaled to 0.52g acceleration. The matter is that Code stipulates the importance coefficient k_2=1.3 for the buildings like schools, hospitals and other critical facilities. Taking this into account the hospital building "Vanadzor" was analyzed under the seismic impact of 0.52g (0.4g×1.3). Also for this hospital R/C frame building with shear walls the same allowable damage coefficients (reduction factors) were applied: k_1=0.4 for superstructure and k_1=0.8 for seismic isolators and the structures below the isolation plane. The design model (Figure 26) was developed by application of different types of finite elements for shear walls, floor slabs, columns and beams, as well as for seismic isolators.

Calculations were carried out taking into account the non-linear behavior of SILRSBs with the same input parameters as given above for the yield strength, yield displacement and effective horizontal post yield stiffness. As for the above mentioned buildings the soil conditions of the site where the considered building was going to be constructed correspond to category II, for which the soil conditions coefficient k_0=1 and the site prevailing period of vibrations 0.3≤T0≤0.6 sec. Armenian Seismic Code specifies several types of soils related to category II:

1. Rocks with uniaxial ultimate compression strength of less than 15 mpa;
2. Gravelly sands of high and medium coarseness, high and medium density, with low moisture content;
3. Fine and pulverescent sands of high and medium density, with low moisture content. Some results of the analyses by the Armenian Seismic Code and time histories are given in Table 8.

The carried out earthquake response analyses have shown that in comparison with the fixed base buildings, seismic isolation significantly reduces the maximum spectral acceleration, proving to be cost effective for the isolated structures and ensuring high reliability of their behavior under seismic impacts (Naeim & Kelly, 1999; Fujita, 1999; Saito, 2006; Martelli et al., 2008; Melkumyan, 2011; Melkumyan, 2014). From the obtained results it follows that the first mode vibrations' periods of base isolated building in longitudinal (X) and transverse (Y) directions are almost equal to each other. Thanks to the proposed approach of location of rubber bearings by clusters in seismic isolation system, in none of the isolators the vertical force exceeds 1500 kN. More or less uniform distribution of the vertical loads upon the rubber bearings was achieved and also no rotation in the building's isolation system and, consequently, in superstructure was observed.

It also can be noticed that the displacements of isolation system, inter-story drifts and horizontal shear forces obtained by calculations of the base isolated building by the Armenian Seismic Code are close to the same values obtained by the time history analysis when the applied allowable damage coefficient (reduction factor) k_1=0.4. But in case if k_1=0.8 then Code based results regarding the horizontal shear forces are higher by a factor of 2.2 and those regarding the maximum inter - story drifts and maximum displacements of the isolation system are higher by a factor of 1.75 in average. Therefore, again it should be stated that the Code needs a more accurate designation of reduction factors for seismic isolation systems.

From the above given description of the hospital building "Vanadzor" and from Figures 22 and 26 it can noticed that the front of the building has a unique architectural and structural solution. In this part design envisages construction of only fourth floor which connects like a bridge the left and right wings of the building between the axes "17" and "9". Within the limits of its span this so called "bridge" is supported by four hollow columns with outside dimensions equal to 2600×2600 mm and inside dimensions – 1800×1800 mm (see section 1 - 1 in Figure 25). Under each of such columns ten seismic isolation rubber bearings are installed with different location as it is shown in Figure 25. Obviously, that the differences in location of rubber bearings were dictated by the results of analysis. The deformed view of this part of the building under the action of the vertical static design loads is shown in Figure 27, from where it is seen that the maximum vertical displacement is not more than 33 mm, which is within the allowable range.

Figure 26. Design model of the 4 - story base isolated hospital building "Vanadzor". The model is given in two versions for better understanding

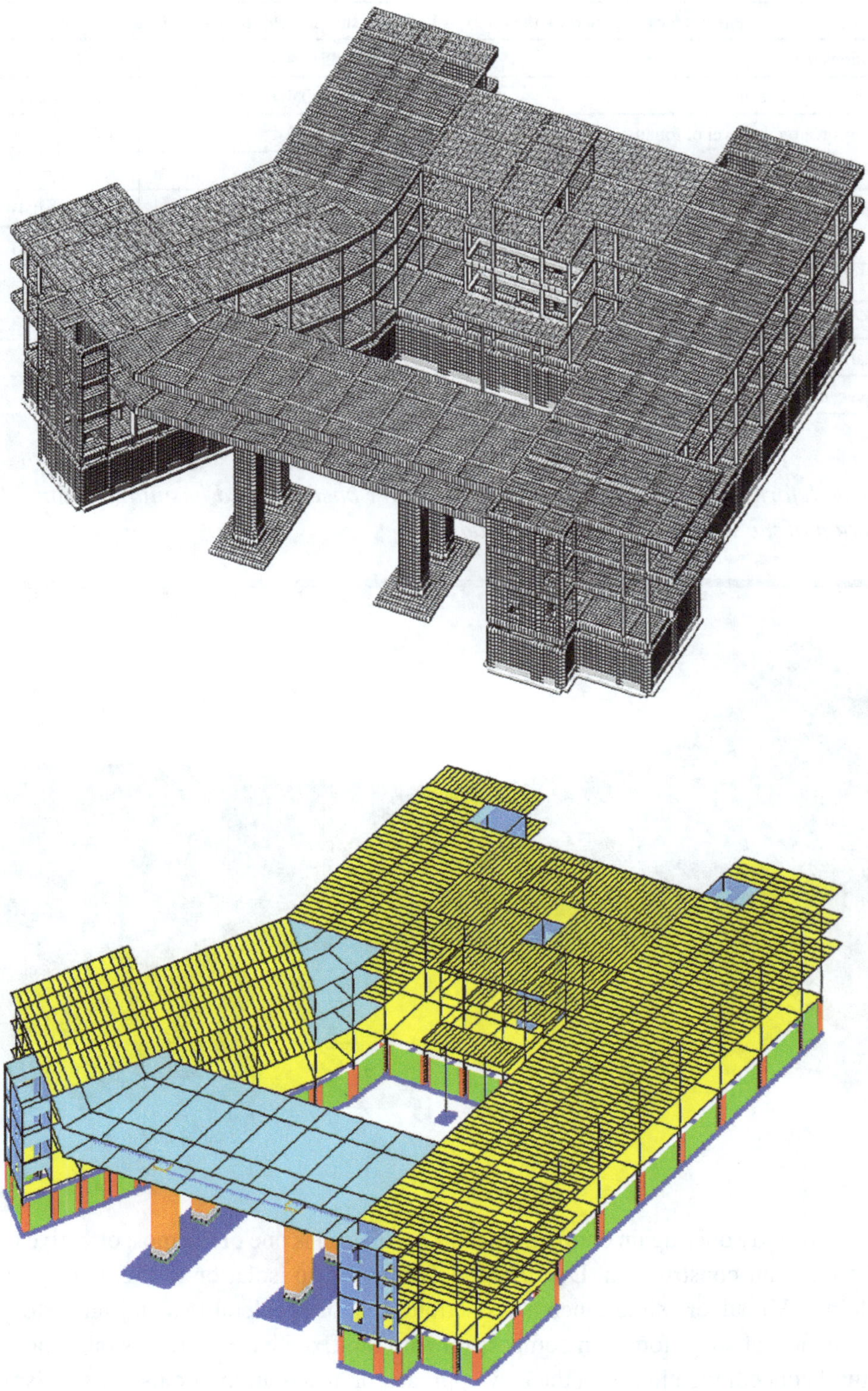

Table 8. Some results of the analyses of 4 - story base isolated hospital building "Vanadzor" by the Armenian Seismic Code and time histories

Parameters obtained by the analysis based on the Armenian Seismic Code		
Period of vibrations (sec)	T_x=1.85*	T_y=1.81**
Maximum inter - story drift (mm)	11.0 (k_1=0.4)	14.2 (k_1=0.4)
Horizontal shear force on the level of foundation (kN)	22640 (k_1=0.4)	24630 (k_1=0.4)
	40970 (k_1=0.8)	43530 (k_1=0.8)
Maximum displacement of the isolation system (mm)	102.0 (k_1=0.4)	111.0 (k_1=0.4)
	178.0 (k_1=0.8)	195.0 (k_1=0.8)
Average parameters obtained by the time histories analyses		
Maximum inter - story drift (mm)	6.5	8.3
Horizontal shear force on the level of foundation (kN)	18360	19780
Maximum displacement of the isolation system (mm)	105.0	114.0

Figure 27. The deformed state of the front part of 4-story base isolated hospital building "Vanadzor" under the action of the vertical static design loads

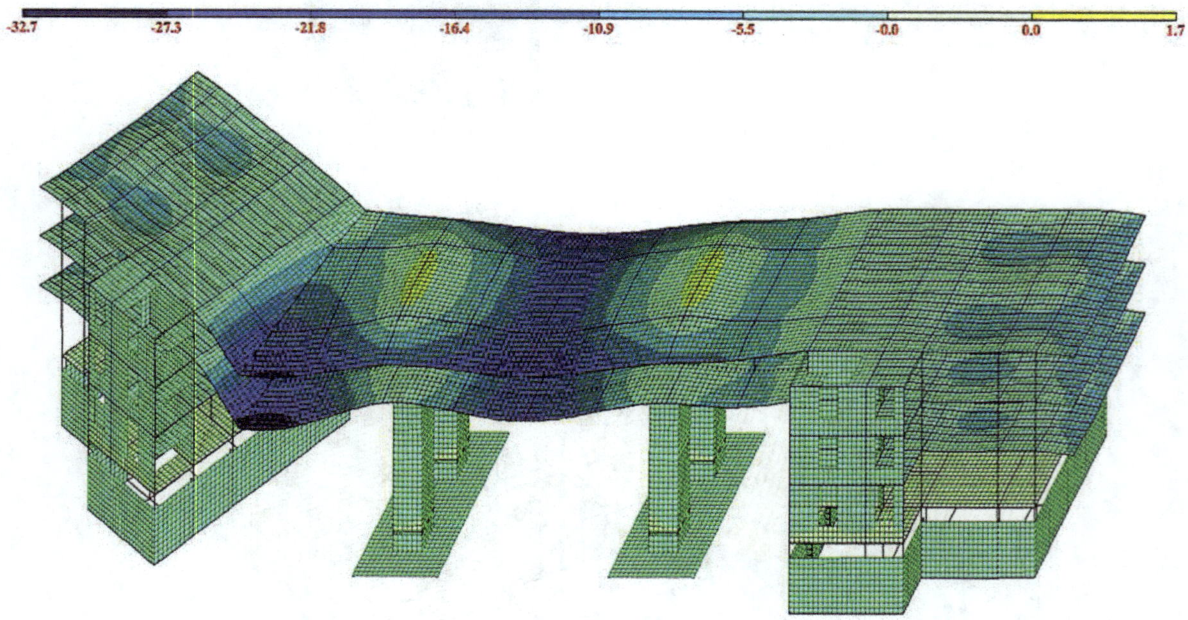

The conducted study once again confirms that base isolation is one of the most effective technologies in earthquake resistant construction. Due to application of base isolation to the 4-story base isolated hospital building "Vanadzor" simultaneous reduction of floor accelerations and inter-story drifts and significant reduction of shear forces in comparison with the fixed base buildings takes place. The suggested structural concept together with the new approach on installation of clusters of seismic isolation rubber bearings provides high reliability to the whole bearing structure and effectiveness of the isolation system, neutralizing the rotation in the building's isolation system and, consequently, in its superstructure. In this case almost uniform distribution of the vertical loads upon the rubber bearings was achieved. The obtained results also indicate that first mode vibrations' periods of base isolated building in longitudinal and transverse directions are almost equal to each other.

7. DESIGN AND TESTING OF SEISMIC ISOLATION RUBBER BEARINGS

7.1 Design of the Bearings

Based on the carried out analyses for different buildings the high damping seismic isolation laminated rubber-steel bearings (HDSILRSBs) to be installed in clusters were designed. Table 9 summarizes the design parameters of the bearings. In all base isolated buildings constructed in Armenia the simple recess connection detail to fix the bearings was chosen. Such option necessitates a check that the bearings are safe against roll-out at the maximum horizontal displacement, with a due regard to the reduction in vertical load on some of the bearings attributable to the overturning of the building at large displacements. Geometrical dimensions of the designed HDSILRSBs and their location in upper and lower recesses are shown in Figure 28.

Table 9. Design parameters of HDSILRSBs

Parameters	Values
Number of rubber layers	14
Number of internal metal plates	13
Thickness of rubber layers, mm	9
Thickness of internal metal plates, mm	2.5
Radius of internal metal plates, mm	180
Thickness of side cover layer, mm	10
Thickness of steel end-plate, mm	20
Thickness of end cover layer, mm	2
Overall height, mm	202.5
Overall diameter, mm	380
Rubber shear modulus	0.97
Static compressive stress (max), MPa	8.7
Critical load, kN	4500
Design vertical load	1500
Load for internal plate yield, kN	4800
Horizontal stiffness, kN/mm	0.81
Horizontal displacement at onset of roll-out, mm	
for design vertical load	300
for min. vertical load	260
Nominal vertical stiffness, kN/mm	400

7.2. Bearing Tests

The performance of the bearing design was checked by carrying out further tests. The bearings were subjected to QC tests using a single bearing testing facility soon after manufacture as well as two years later, during which time the bearings had stiffened very slightly. The degree of stiffening (2-3%) over two years suggests a figure of 10-15% over a 50 year service life on the basis that stiffening proceeds approximately as $\sqrt{\text{time}^9}$; this would represent good long-term stability. The shear dynamic stiffness of bearings was measured over a range of rubber strains at a frequency of 0.1 Hz under a constant vertical load equal to 500 kN. These tests were carried out using a biaxial load-cell directly under the bearing. The magnitude of the vertical load applied was limited by the capacity of the biaxial load-cell, and thus the bearings could not be tested under the design vertical load; this should not have influenced the result significantly. The horizontal capacity of the dynamic test facility limited the rubber strain amplitudes to 50% (Fuller et al., 2000).

The sixth cycle force-displacement loops are given in Figure 29. The corresponding stiffness and damping values are listed in Table 10 as functions of the displacement amplitude for all the tests. The bearing stiffness is seen to decrease with increasing displacement. The stiffness at 50% of the design displacement (D) is 2.53 times smaller than at 0.05D. The increased stiffness at small displacements reduces the movement of the building under wind loading without the need for additional wind restrain devices.

The observed dynamic stiffness of the bearings appears somewhat lower than expected. Using the test data for the variation of rubber shear modulus with strain to estimate a bearing stiffness at 100% rubber strain from the observed value at 50% strain, and making an allowance for the effect of test frequency and the reduced vertical load of 500 kN gives a dynamic stiffness of 0.73 kN/mm – 12% below the design value. The modulus of the rubber batch for bearings was 0.96 MPa (very close to the design requirement of 0.97 MPa). It therefore appears that calculation of the bearing horizontal stiffness from the rubber shear modulus and using the standard design equations worked less well. The reason may be that the modulus data from the rubber tests pieces correlates less well with the properties of the rubber within the bearing. The dynamic bearing tests confirmed the advisability of using a stiffer compound for bearings.

The damping of the bearings (after allowing for the effect of frequency and the reduction between 50 and 100% strain amplitude indicated by the measurements on rubber samples) is slightly greater than that obtained directly from tests at 100% strain on rubber samples. The bearings were finally tested quasistatically in shear under the vertical load of 820 kN up to the maximum horizontal displacement of 195 mm (Figure 30a). The corresponding force-deflection plot (Figure 30b) shows a slight stiffening at large deflection; there is no sign of an approach to the displacement capacity of the isolator.

Thus, a rubber compound suited to sites with severe winter temperatures has been developed. Dynamic tests on the bearings showed the performance of the design to be satisfactory. The bearing test result confirmed that their stiffness and damping is predicted reasonably well from the design equations and rubber properties, as measured on small tests pieces.

Figure 28. Geometrical dimensions of the designed HDSILRSBs and their location in upper and lower recesses provided by annular steel rings bolted to the outer steel plates connected to reinforcement in the upper and lower continuous beams

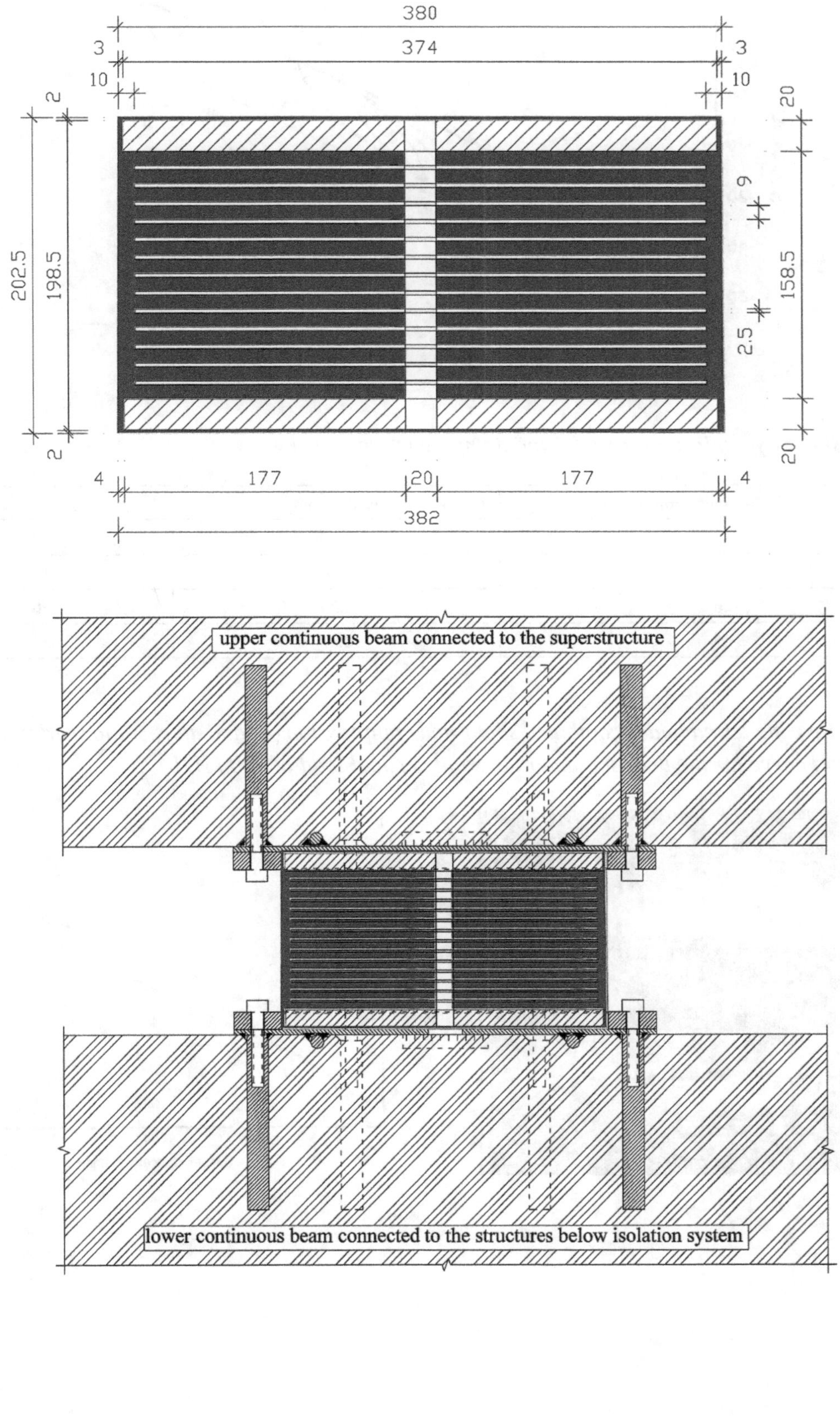

Figure 29. Shear force-deflection loops for designed bearing
Test frequency 0.1 Hz. Nominal vertical load 500 kN

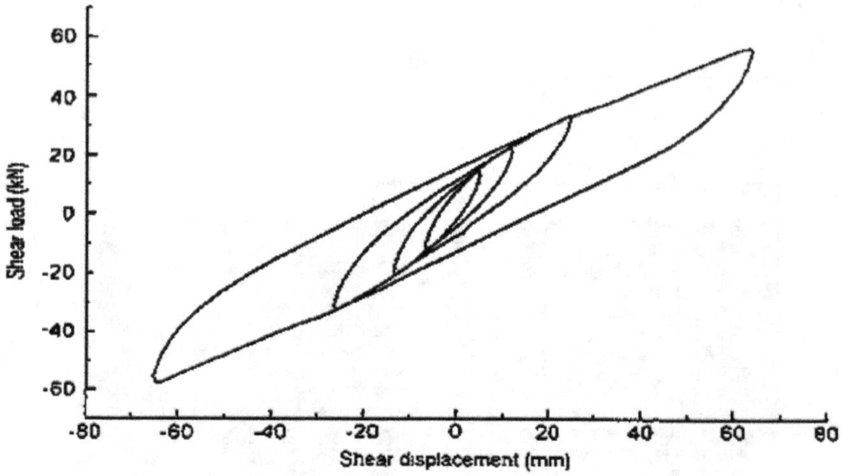

Table 10. Shear dynamic stiffness (K_s) and damping (d) of bearings

Displacement, mm	K_s, kN/mm	d%
6.5	2.20	19
13	1.60	17
26	1.30	16
65	0.87	14

Figure 30. Bearing tested under combined shear and compression (a) and quasistatic shear force-shear displacement curve for the bearing under constant vertical load of 820 kN (b)

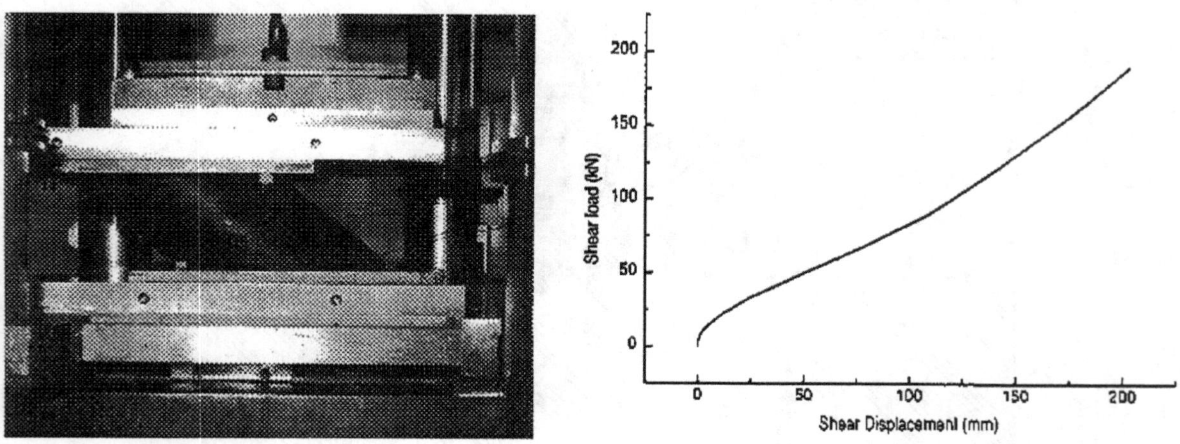

8. CONCLUSION

Diagrams which illustrate increase of the number of seismic isolated buildings and corresponding number of rubber bearings manufactured, tested and installed in Armenia by years are given. It is stated that more than two buildings per year in average with seismic isolation systems were constructed or retrofitted in Armenia during the period of the last 21 years. Currently the number of seismically isolated buildings per capita in Armenia is one of the highest in the world.

Original approach in design of seismic isolation systems of base isolated buildings by installation of clusters of small size rubber bearings under the columns and/or shear walls is presented and the advantages of this approach are listed and illustrated by the examples. The obtained results indicate that due to the proposed approach the rotation in the buildings is neutralized, periods of the first mode of vibrations increase, whereas total shear forces and story drifts, as well as the number of isolators somewhat decrease. This means the proposed approach enables improving the overall effectiveness of the isolation system and achieving a more rational solution by manipulating the number and location of isolators.

Results of analyses of some buildings where the approach on installation of clusters of rubber bearings was used in their isolation systems are given for two cases:

1. When the analyses are carried out based on the provisions of the Armenian Seismic Code, and
2. When the time history analyses are carried out.

It was obtained that the shear forces at the level of isolation systems, the maximum displacements of the isolators, and the maximum story drifts in the superstructures calculated based on the Armenian Seismic Code provisions are considerably higher (by a factor of about 2 in average) than the same values calculated by the time histories. Therefore, it is suggested that some further steps should be taken to more realistically reflect the characteristics of seismically isolated buildings (including the reduction factors for isolation systems) in the design models for the calculations based on the Code.

Proposed innovative structural concepts, details of analysis and design of three unique base isolated buildings, namely,

- The 20-story business center “Elite Plaza”,
- The 18-story residential building of the multifunctional complex “Northern Ray”, and
- The 4-story hospital building “Vanadzor” are described.

Using 9 selected time histories earthquake response analyses of these buildings were carried out in two versions, i.e.

- When the buildings are base isolated and
- When they are fixed base.

Comparison of the obtained average results indicates the high effectiveness of the proposed structural concepts for the isolation systems, the substructures and superstructures.

Some information is given on the design parameters of the high damping laminated rubber-steel bearings and the results of their tests. The bearings were tested in shear under the constant vertical load up to the maximum horizontal displacement. The corresponding force-deflection plot shows a slight stiffening at large deflection; there was no sign of an approach to the displacement capacity of the isolator. Tests on the bearings showed their performance to be quite satisfactory and confirmed that stiffness and damping is predicted reasonably well.

REFERENCES

Foti, D., & Mongelli, M. (2011). *Isolatori sismici per edifice esistenti e di nuova costruzione*. Dario Flaccovio Editore. (in Italian)

Fujita, T. (1999), *Demonstration of Effectiveness of Seismic Isolation in the Hanshin-Awaji Earthquake and Progress of Applications of Base-Isolated Buildings*. Report on 1995 Kobe Earthquake by INCEDE, ERC and KOBEnet.

Fuller, K., Lim, C., Loo, S., Melkumyan, M., & Muniandy, K. (2000). *Design and Testing of High Damping Rubber Earthquake Bearings for Retrofit Project in Armenia.* In S. Balassanian, A. Cisternas, & M. Melkumyan (Eds.), *Earthquake Hazard and Seismic Risk Reduction* (pp. 379–385). Kluwer Academic Publishers. doi:10.1007/978-94-015-9544-5_40

Garevski, M. (2010). Development, Production and Implementation of Low Cost Rubber Bearings. Springer. doi:10.1007/978-90-481-9544-2_17

Martelli, A., & Forni, M. (2011). Seismic retrofit of existing buildings by means of seismic isolation: some remarks on the Italian experience and new projects. *Proceedings of the 3rd International Conference on Computational Methods in Structural Dynamics and Earthquake Engineering, COMPDYN 2011*.

Martelli, A., Forni, M., Arato, G.-B., & Spadoni, B. (2001). Overview and Summary of the 7th International Seminar on Seismic Isolation, Passive Energy Dissipation and Active Control of Vibrations of Structures. *Proceedings of the 7th Int. Seminar*, i-xxxvii.

Martelli, A., Forni, M., & Rizzo, S. (2008). Seismic isolation: present application and perspectives. *Proceedings of the ASSISi International Workshop on Base Isolated High-Rise Buildings*, 1 - 26.

Melkumyan, M. (2001). The state of the art in development of testing facilities and execution of tests on isolation and bridge bearings in Armenia. *Proceedings 5th World Congress on Joints, Bearings and Seismic Systems for Concrete Structures*.

Melkumyan, M. (2004). Recent Applications of Seismic Isolation in Civil Buildings in Armenia. *Proceedings of the 13th World Conference on Earthquake Engineering*.

Melkumyan, M. (2005). Current Situation in Application of Seismic Isolation Technologies in Armenia. *Proceedings of the International Conference dedicated to the 250th anniversary of the 1755 Lisbon Earthquake*, 493-500.

Melkumyan, M. (2006). Armenia is one of the world leaders in development and application of base isolation technologies. *MENSHIN Journal of the Japan Society of Seismic Isolation*, *54*, 38 - 41.

Melkumyan, M. (2007). Seismic isolation of civil structures in Armenia - development and application of innovative structural concepts. *Proceedings ECCOMAS Thematic Conference on Computational Methods in Structural Dynamics and Earthquake Engineering.*

Melkumyan, M. (2009). Armenian seismic isolation technologies for civil structures - example on application of innovative structural concepts, R&D and design rules for developing countries. *Proceedings 11th World Conference on Seismic Isolation, Energy Dissipation and Active Vibration Control of Structures.*

Melkumyan, M. (2011a). Assessment of the Seismic Risk in the City of Yerevan and its Mitigation by Application of Innovative Seismic Isolation Technologies. *J. Pure and Applied Geophysics, 168*(3-4), 695–730. doi:10.1007/s00024-010-0110-4

Melkumyan, M. (2011b). *New Solutions in Seismic Isolation.* Yerevan: LUSABATS.

Melkumyan, M. (2013). Comparison of the analyses results of seismic isolated buildings by the design code and time histories *Journal of Civil Engineering and Science*, *2*(3), 184–192.

Melkumyan, M. (2014a). Structural concept and analysis of the 15-story base isolated apartment building "Avan". *International Journal of Engineering Research and Management*, *1*(7), 157–161.

Melkumyan, M. (2014b). Structural concept and analysis of the 17-story base isolated apartment building "Sevak". *International Journal of Engineering and Applied Sciences*, *1*(3), 13–17.

Melkumyan, M., & Gevorgyan, E. (2008). Structural concept and analysis of a 17-story multifunctional residential complex with and without seismic isolation system. *Proceedings 2008 Seismic Engineering Conference Commemorating the 1908 Messina and Reggio Calabria Earthquake*, 1425 - 1432. doi:10.1063/1.2963766

Melkumyan, M., & Gevorgyan, E. (2010). Structural concept and analysis of 18-story residential complex "Northern Ray" with and without base isolation system. *Proceedings of the 14th European Conference on Earthquake Engineering.*

Melkumyan, M., Gevorgyan, E., & Hovhannisyan, H. (2005). Application of Base Isolation to Multifunctional Multistory Buildings in Yerevan, Armenia. *Proceedings of the 9th World Seminar on Seismic Isolation, Energy Dissipation and Active Vibration Control of Structures*, 2, 119-127.

Melkumyan, M., & Hakobyan, A. (2005). Testing of Seismic Isolation Rubber Bearings for Different Structures in Armenia. *Proceedings of the 9th World Seminar on Seismic Isolation, Energy Dissipation and Active Vibration Control of Structures*, 2, 439-446.

Melkumyan, M., & Hovhannisyan, H. (2006). New approaches in analysis and design of base isolated multistory multifunctional buildings. *Proceedings 1st European Conference on Earthquake Engineering and Seismology (a joint event of the 13th ECEE & 30th General Assembly of the ESC).*

Miyamoto, K., & Gilani, A. (2007). Base Isolation for Seismic Retrofit of Structures: Application to a Historic Building in Romania. *Seismic Risk Reduction,Proceedings of the International Symposium*, 585-592.

Naeim, F., & Kelly, J. (1999). *Design of Seismic Isolated Structures. From Theory to Practice.* John Wiley & Sons Inc. doi:10.1002/9780470172742

Retamales, R., & Boroschek, R. (2014). State-of -the-Art of Seismic Isolation and Energy Dissipation Applications in Chile. *MENSHIN Journal of the Japan Society of Seismic Isolation*, *84*, 55-69.

Saito, T. (2006). In M. Higashino & S. Okamoto (Eds.), *Observed Response of Seismically Isolated Buildings. In Response Control and Seismic Isolation of Buildings* (pp. 63–88). Taylor & Francis.

The World Bank Implementation Completion Report No. 17255. (1997). *Armenia Earthquake Reconstruction Project*. World Bank.

Whittaker, D., & Robinson, W. (2009). Progress of Application and R&D for Seismic Isolation and Passive Energy Dissipation for Civil and Industrial Structures in New Zealand. *Proceedings of the 11th World Conference on Seismic Isolation, Energy Dissipation and Active Vibration Control of Structures*.

Zhou, F., Tan, P., Heisha, W., & Huang, X. (2009). Recent Development and Application on Seismic Isolation in China. *Proceedings of the JSSI 15th Anniversary International Symposium on Seismic Response Controlled Buildings for Sustainable Society*.

Chapter 9
Fuzzy Logic Applications for Performance-Based Design

Ferhat Pakdamar
Gebze Technical University, Turkey

ABSTRACT

In this chapter, it is expressed how performance based design criteria are modeled with fuzzy logic and why it needs to be modeled with fuzzy logic. Firstly, a brief information is given about current theories and techniques such as Fuzzy Logic, Chaos, Fractal Geometry, Artificial Neural Networks; and it is mentioned about advantages of applying these techniques to the problems. Then graphical inference deduction technique is expressed by giving a summary info about Fuzzy Set Theory, membership functions, clustering, rule-based systems and defuzzification. Then the divergence of performance based design criteria is set forth by making curvature evaluations for various regulations on a column section as a structure element. Finally, it is shown in detail how to use the fuzzy set theory for solution of this disharmony by using the clustering technique on a univariate sample first and then on a multivariate sample.

INTRODUCTION

Today, there are various techniques and theories that most of scientist are not familiar with and these are being developed to simplify operations and/or easing the life by effectively applying on all of the science branches. Names of some of these techniques can be defined according to their relevant science branch: Physics – Quantum, Mathematics – Chaos, Geometry – Fractal, Classical Logic – Fuzzy Logic; Neural Networks and Genetic Algorithm must be also added.

For example; progression of Physics towards quantum was in early of 20th century. It was specified that laws described for macro objects were not functional in micro sizes. Even if concept of quantum occurred with the studies made upon blackbody radiation by Max Planck (1900), the theory was grounded by Max Planck, Albert Einstein, Niels Bohr, Werner Heisenberg and Erwin Schrödinger.

Meeting of Mathematics with Chaos became by Edward Norton Lorenz. Edward Lorenz, who was a meteorologist, was working upon weather forecasting with his computer in 1963. In one of his studies again, Lorenz entered the number of 0,506127 as starting data in his calculation. In the next study,

DOI: 10.4018/978-1-5225-2089-4.ch009

Lorenz considered that entering the number of 0,506127 as 0,506 would not affect the system, because he thought that the change he made was a small change. In mathematically it was so… But Lorenz shocked with the obtained results. This small change he made caused the system to give hugely different results. Lorenz thought that his computer was broken but calculations he made again and again were telling the opposite. So indeed, this change he made in his computer which was as unimportant as a flapping of a bird led system to totally differentiate. In March of 1963, Lorenz published this phenomenon he discovered on Atmospheric Sciences Magazine and he used the term of "Butterfly Effect" for the first time. Thus, "Butterfly Effect" took its place in public and scientific language.

The term of Fractal was derived from the word of "fractus" in Latin which means fragmented or broken. Fractal is the common name of complex geometrical shapes that show the characteristic of auto-simulation mostly. Even if first mathematical studies done by Karl Weierstrass in 1861, the concept of fractal proposed for the first time by Polish mathematician Benoit B. Mandelbrot in 1975 led a new geometry system creating important effect not only on mathematics but also on various areas such as physical chemistry, physiology and fluid mechanics (Mandelbrot, 1983). Expressing objects about nature with this geometry system became easier compared to Euclidean geometry.

John Holland, a psychology and computer science expert in Michigan University, is the person who made first studies about Genetic algorithms (Holland, 1975). Holland working for machine learning, considered to carry out genetic process of creatures in computer environment by being impressed by evolution theory of Darwin. He saw that successful (able to learn) new individuals can be created by undergoing a community in genetic processes such as reproduction, pairing and mutation, instead of developing the learning ability of only one mechanical structure. He did his studies by looking at natural selection and genetic evolution to find the search and optimum. During the process, he modelled computer software in problems of finding optimum and machine learning by taking example that individual adapts to environment and becomes more suitable. Genetic Algorithms (or shortly GA) is the name of the method that he developed after publishing of his book in 1975 that he explained the results of his studies.

Artificial Neural Networks (ANN) is a data processing technology which works to try to copy the data processing system of neurons and brain in human nervous system. It is based on a basis of processing data in parallel and creating a network just like the human nervous system. This parallel data processing technique provides scientist very important opportunities such as able to derive data, learn, discover and a speed. Warren McCulloch and Pitts (1943) created a computational model for neural networks based on mathematics and algorithms called threshold logic.

The philosophy which predicts everything in nature can be expressed with two values such as correct-wrong, good-bad, black-white, open-close, beautiful-ugly, full-empty etc., can be referred to as Dialectic Logic or Syllogistic Logic. According to Aristoteles, a proposition is correct or wrong, and the data can be established by propositions. In fact, there were studies about logic before Aristoteles (B.C. 384-322) but categorizing and defining was proposed by Aristoteles for the first time in B.C. 310's. For this reason, he is called as the founder of logic (Smith, 2016). Indeed, science couldn't develop so much if binary logic did not exist because the relevant stable stops namely threshold values which are the main points of problems and propositions in Syllogistic Logic, were easily expressed and understood. However, unfortunately Syllogistic Logic remained incapable to define these problems and propositions for the requirement of human life and technological improvements of 20th century trying to meet these requirements. Because now humanity and technology wanted to use and apply to their lives not only threshold points in data but also the data in intermediate values correspond to margins of these threshold points. How would the philosophy of using these intermediate values was integrated to the data network patterned by binary logic within about 2350 years. Professor Lutfi Asker ZADEH found out the answer of this question in 1965 and he gave the name of "Fuzzy" to this new logic conception, unfortunately. I'm telling "unfortunately" because if the names such as "relativist", "relative", "proportional", "conditional" would be given, then there might not get a lot of objections and reaction. It is impossible that even refuser and deliberate scientists cannot be unconcerned with this technology. Because Fuzzy Logic is located as an extremely normal mechanism at devices in every point of our lives. In fact, fuzzy is in a lot of areas of life from robotics, unmanned air vehicles, driverless vehicles to electronical household appliances such as air conditioning, washing machine, dish washer etc.

Sadly, stated that fuzzy systems reaching a lot of points of life cannot get well-deserved place in construction science and technology. It will be wrong to tell that there is no study because it can be seen a lot of studies in indexes but there is a lack in point of transferring the results of these studies to life. Some of the articles can be given as examples for the performance based design limits fuzzification or non-linear account fuzzification studies or other important studies made in a similar manner (Doran, Yetilmezsoy, & Murtazaoglu, 2015; Guclu & Yazici, 2007; Mistakidis & Georgiou, 2003; Ozkul, Ayoub, & Altunkaynak, 2014; Pakdamar & Guler, 2012; Sgambi, 2004; Tesfamaraim & Saatcioglu, 2008).

As Hamburger (1996) and Fasan et al. (2015) stated: Nowadays, almost all seismic building codes are based on principles of Performance Based Design (PBD). The main contribution to the development of this philosophy of design has been given by the Vision 2000 report (Committee, 1995). This document and the following papers and codes, define a series of performances (in terms of acceptable damage) that a building should reach during earthquakes of different strength. These performance levels are usually defined as: Operational Limit (OL), Immediate Occupancy (IO), Life Safety (LS) and Collapse Prevention (CP).

Subjective equals of IO, LS, CP performance levels for a general structure are given inclusively under the title of "Performance objectives and seismic hazards" in Section 2 in ASCE 41-13 (Pekelnicky & Poland, 2013) Expressions related to performance matter are available in many sections of this book. In order to explain the relation between Fuzzy and performance better, performance matter is not explained in detail in this chapter. General performance expressions for a similar structure are available in regulations such as Eurocode 8 (Fardis, 2005) and Turkish Seismic Code (TSC, 2007), even if the structure differs according to its intended use. For example, even if structure of a hospital or education institution are not the same as structure a house or workplace or differ according to earthquake magnitudes, performances can be defined in the coarsest way as follows: some of main performance objectives given in ASCE41-13 and Eurocode8 are given in Table 1:

- **Immediate Occupancy:** It has high probability to be in structure economical life (50% probability in 50 years) but it remains standing to continue its structure activities in minor earthquakes which release low energy.
- **Life Safety:** It has intermediate probability to be in structure economical life (about 10% in 50 years) and not damaging the carrier systems of structure in moderate earthquakes which release midsize energy but damaging the structure in general but being able to continue its activity in a manner of not damaging the life safety of structure.
- **Collapse Prevention:** It has low probability to be in structure economical life (about 2% in 50 years) and damaging the structural systems in major earthquakes which release very high energy but allowing the discharge of structure in a manner of not damaging the life safety of structure.

Table 1. Some main performance objectives according to ASCE41-13 and Eurocode8

	ASCE 41-13	Eurocode8
Collapse Prevention (severe)	Little residual stiffness and strength to resist lateral loads, but gravity loadbearing columns and walls function. Large permanent drifts. Some exits blocked. Building is near collapse in aftershocks and should not continue to be occupied. Significantly more damage and greater life safety risk.	Limit states of near collapse (NC). The structure is heavily damaged, with small residual strength and stiffness, although vertical elements are still capable of sustaining vertical loads. Most non-structural components have collapsed. Large permanent drifts are present. The structure is near collapse and would not survive another earthquake, even of moderate intensity.
Life Safety (Moderate)	Some residual strength and stiffness left in all stories. Gravity-load-bearing elements function. No out-of plane failure of walls. Some permanent drift. Damage to partitions. Continued occupancy might not be likely before repair. Building might not be economical to repair. Somewhat more damage and slightly higher life safety risk.	Limit states of significant damage (SD). The structure is significantly damaged, with some residual strength and stiffness, and vertical elements are capable of sustaining vertical loads. Non-structural components are damaged, although partitions and infills have not failed out-of-plane. Moderate permanent drifts are present. The structure is likely to be uneconomic to repair.
Immediate Occupancy (Light)	No permanent drift. Structure substantially retains original strength and stiffness. Continued occupancy likely. Less damage and low life safety risk.	Limit states of damage limitation (DL). The structure is only lightly damaged, with structural elements prevented from significant yielding and retaining their strength and stiffness properties. Non-structural components, such as partitions and infills, may show a diffused state of cracking that could however be economically repaired. No permanent drifts are present. The structure does not need any repair measures.

As is seen, these definitions and most of performance definitions in regulations include linguistic elements. Namely the concepts such as little, some, middle, much, large. It is difficult to express these concepts with mathematical models. Thus, a method is needed to process and express these concepts with classical mathematics and Fuzzy Logic is a very suitable method for this. However, acceptance criteria for per structural element (Concrete, Steel, Wood etc.) are given in regulations because these subjectively expressed concepts need to have objective corresponds on structure materials. These criteria include crisp values showing the limits of behaviors of each structural element under what kind of pressure it is. Engineers decide the performances of structure elements and then the performances of structures by using tables specified by these crisp limits.

Whereas, if history of performance based design studies is analyzed, it is seen that a lot of acceptance criteria have changed over time. For example, a beam rotation is accepted as 0.015 rad today, while it was accepted as 0.01 rad at past. Even, the same element is accepted as 0.01 rad in America, 0.02 rad in Europe and 0.015 rad in Turkey. Unfortunately, a fuzzy logic implementation to consider these differences and elasticize acceptance criteria couldn't be seen clearly in performance based design until the study of Pakdamar and Guler (2012). Differently from the things expressed in this chapter, clustering technique is not used in that study.

DISHARMONY AT PERFORMANCE BASED DESIGN CRITERIA

A consensus is out of the question about element acceptance criteria even if performance based design took the shape of law and regulation in most countries. In order to show this disharmony, a sample analysis is made in following example. This analysis is performed for FEMA356 (FEMA356, 2000),

ASCE41-13 (Pekelnicky & Poland, 2013), Eurocode 8 (Fardis, 2005) and Turkish Seismic Code (TSC, 2007), and expressed on a single graphic. Sample analysis can be extended for other regulations.

A general moment-curvature curve is given in Figure 1. Bending moment – curvature diagram of a concrete section of Figure 2 equipped symmetrically with square section, is showed in this analysis. Curvature of this section can easily be obtained by utilizing the basic reinforced concrete knowledge. Geometry and material characteristics belong to relevant section are given in Table 2.

Various models and softwares can be used for a more certain calculation of section, of course; but the purpose here is to investigate the limits on moment – curvature relation, not to find the clearest behavior of section.

In all regulations, calculation methods for deformation rate, ε_y is same and Hook's Law is valid until the bottom reinforcement bars begin to yield.

$$\varepsilon_{s3} = \varepsilon_y = \frac{f_y}{E_s} = \frac{420}{200000} = 0.0021$$

Figure 1. Three deformation states on section under the effect of changed curvature related to moment

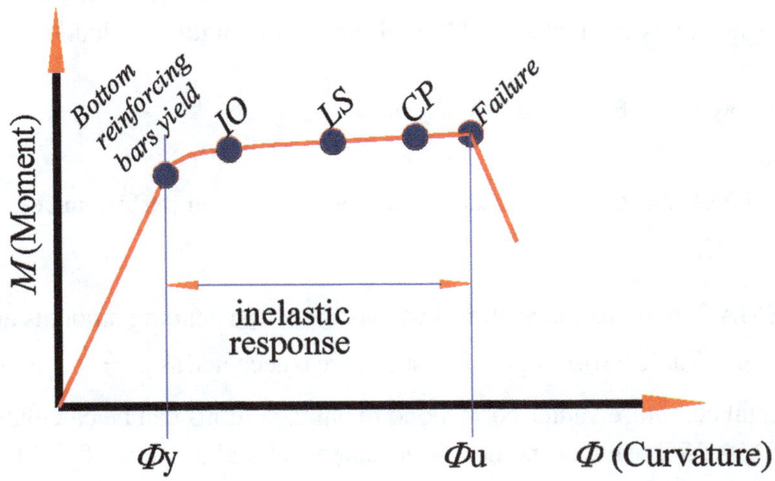

Figure 2. Three deformation states on section under the effect of changed curvature related to moment

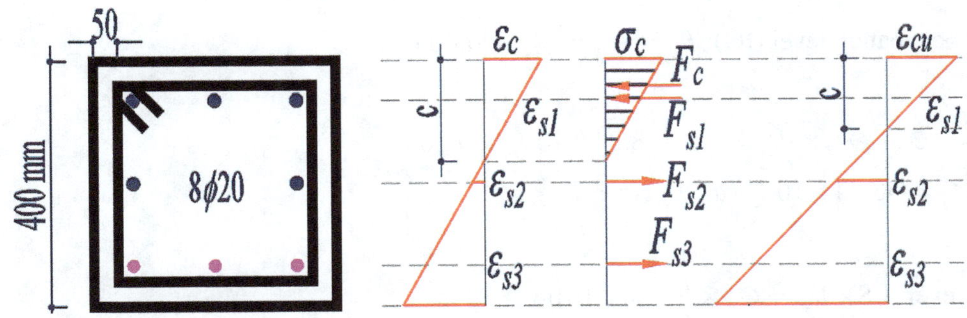

Table 2. Section geometry and material properties

Description	Quantity
Section base/height:	b/h=400 mm/400 mm
Compressive strength of concrete:	$f_c = 25\ MPa$
Yield strength of reinforcing steel:	$f_y = 420\ MPa$
Modulus of elasticity for concrete (Young's modulus):	$E_c = 30000\ MPa$
Modulus of elasticity for reinforcing steel (Young's modulus):	$E_c = 200000\ MPa$
Ultimate compressive strain of concrete:	$\varepsilon_{cu} = 0.004$
Rebars in section:	8 $\varnothing$ 20
Concrete cover:	d' = 50 mm

According to yield deformation state, if equation of equilibrium on section is written and solved, bending moment on section is calculated as M_y and section curvature is calculated as $\varnothing_y$.

$M_y = 134.4\ kNm$ and $\varnothing_y = 8.24 \times 10^{-3}\ rad / m$

- According to FEMA356, damage limits, in other words section performances in relevant section are as follows:

According to FEMA 356, performance limits of plastic section bending amounts are given as plastic rotations, θ_p (FEMA356, Table 6-6). Length of plastic hinge is accepted as $l_p = h/2 = 0.40/2 = 0.20\ m$. According to this, total curvature values correspond to damage limits can be calculated as follows:

According to FEMA 356, bending moment is accepted as fixed for state of yield and failure.

$M_y = 134.4\ kNm$

Immediate occupancy level (IO): $\theta_p = \varnothing_p \times h/2 = 0.005$

$\varnothing_p \times h/2 = (\varnothing - \varnothing_y) \times h/2 = \left(\varnothing - 8.24 \times 10^{-3}\right) \times 0.20 =$
$0.005 \rightarrow \varnothing_y = 33.24 \times 10^{-3}\ rad / m$

Life safety level (LS): $\theta_p = \varnothing_p \times h/2 = 0.010$

$$\varnothing_p \times h/2 = (\varnothing - \varnothing_y) \times h/2 = \left(\varnothing - 8.24 \times 10^{-3}\right) \times 0.20 = 0.010 \rightarrow \varnothing_y = 58.24 \times 10^{-3}\ rad/m$$

Collapse prevention level (CP): $\theta_p = \varnothing_p \times h/2 = 0.020$

$$\varnothing_p \times h/2 = (\varnothing - \varnothing_y) \times h/2 = \left(\varnothing - 8.24 \times 10^{-3}\right) \times 0.20 = 0.020 \rightarrow \varnothing_y = 108.24 \times 10^{-3}\ rad/m$$

- According to ASCE41-13, damage limits, in other words section performances in relevant section are as follows:

According to ASCE41-13, performance limits of plastic section bending amounts are given as plastic rotations, θ_p (ASCE41-13 Table 10-7). Length of plastic hinge is accepted as $l_p = h/2 = 0.40/2 = 0.20\ m$. According to this, total curvature values correspond to damage limits can be calculated as follows:

According to ASCE41-13, bending moment is accepted as fixed for state of yield and failure.

$M_y = 134.4\ kNm$

Immediate occupancy level (IO): $\theta_p = \varnothing_p \times h/2 = 0.005$

$$\varnothing_p \times h/2 = (\varnothing - \varnothing_y) \times h/2 = \left(\varnothing - 8.24 \times 10^{-3}\right) \times 0.20 = 0.005 \rightarrow \varnothing_y = 33.24 \times 10^{-3}\ rad/m$$

Life safety level (LS): $\theta_p = \varnothing_p \times h/2 = 0.020$

$$\varnothing_p \times h/2 = (\varnothing - \varnothing_y) \times h/2 = \left(\varnothing - 8.24 \times 10^{-3}\right) \times 0.20 = 0.020 \rightarrow \varnothing_y = 108.24 \times 10^{-3}\ rad/m$$

Collapse prevention level (CP): $\theta_p = \varnothing_p \times h/2 = 0.040$

$$\varnothing_p \times h/2 = (\varnothing - \varnothing_y) \times h/2 = \left(\varnothing - 8.24 \times 10^{-3}\right) \times 0.20 = 0.040 \rightarrow \varnothing_y = 208.24 \times 10^{-3}\ rad/m$$

- According to Eurocode 8, damage limits, in other words section performances in relevant section are as follows:

According to Eurocode 8, a frame model is needed for section performance limits. If this column is considered as a basic portal frame at $3.00\ m$ high consists of two columns and a beam, normal force of column is foreseen as very low, limit values can be calculated as follows:

Immediate occupancy level (IO):

$$M_y = 134.4\ kNm\ \varnothing_y = 8.24 \times 10^{-3}\ rad / m$$

Life safety level (LS): $\varnothing = \varnothing_{total} \times {}^{3}\!/_{4} = 92.38 \times 10^{-3} \times {}^{3}\!/_{4} = 69.29 \times 10^{-3}\ rad / m$

Collapse prevention level (CP): $\varnothing_{total} = 92.38 \times 10^{-3}\ rad / m$

- According to TSC2007, damage limits, in other words section performances in relevant section are as follows:

According to TSC2007, performance limits can be calculated with regard to steel and concrete strain limits given in Table 3.

Immediate occupancy level (IO): if it is accepted as $\varepsilon_{s3} = \varepsilon_{su} = 0.010$

$$M_u = 167.8\ kNm\ \varnothing = \varepsilon_{s3} / \left(d - c\right) = 35.84 \times 10^{-3}\ rad / m$$

$$\varepsilon_c = \varepsilon_{s3} \times c / \left(d - c\right) = 0.0025 < \varepsilon_{cu} = 0.0035$$

Life safety level (LS): if it is accepted as $\varepsilon_c = \varepsilon_{cg} = 0.0060$

Table 3. Strain limits for concrete and reinforcement steel in TSC2007

Deformation Limits	Concrete Strain	Reinforcement Strain
Minimum Damage Limit	0.0035	0.010
Safety Limit	$\varepsilon_{cg} = \min[0.0035+0.010\, {}^{\rho_s}\!/_{\rho_{sm}};0.0135]$	0.040
Collapse Limit	$\varepsilon_{cg} = \min[0.0040+0.014\, {}^{\rho_s}\!/_{\rho_{sm}};0.0180]$	0.060

$M_u = 169.8\ kNm\ \varnothing = \varepsilon_c / c = 0.0060 / 0.006 = 100.00 \times 10^{-3}\ rad / m$

$\varepsilon_{s3} = 0.038 < \varepsilon_{su} = 0.040$

Collapse prevention level (CP): if it is accepted as $\varepsilon_c = \varepsilon_{cg} = 0.0085$

$M_u = 168.8\ kNm\ \varnothing_u = \varepsilon_{cu} / c = 0.0085 / 0.0059 = 144.07 \times 10^{-3}\ rad / m$

$\varepsilon_{s3} = \varepsilon_{cu} \times \left(d - c\right) / c = 0.0151 < \varepsilon_{su} = 0.060$

The plot which shows performance limits on section taken as sample, calculated according to FEMA356, ASCE41-13, Eurocode 8 and TSC2007 is given in Figure 3.

In fact, it is a conflict that limit values to decide the performance of a structure element change in various regulations as is seen in Figure 3. Furthermore, this state of disharmony can change in regulation even with regard to time, as is seen in FEMA356 and ASCE41-13 results.

This controversy is one of the biggest problems of performance based design method which needs to be solved. Even if a lot of experimental and theoretical studies are made, it is seen that a consensus will not be ensured soon. In this point, to reveal the existence of Fuzzy Logic which is a calculation method includes all regulations and able to consider all of limit values, and to show the its applicability for the problems of performance based design are realized in this chapter.

FUZZY SET THEORY

Even though Fuzzy Logic is a method known what it is now, having an implementation technique and chance to reach any kind of information easily about it, in this chapter it is found suitable to give a place for briefly theoretical information and general implementation details.

Figure 3. Performance limits for bending moment-curvature of the section

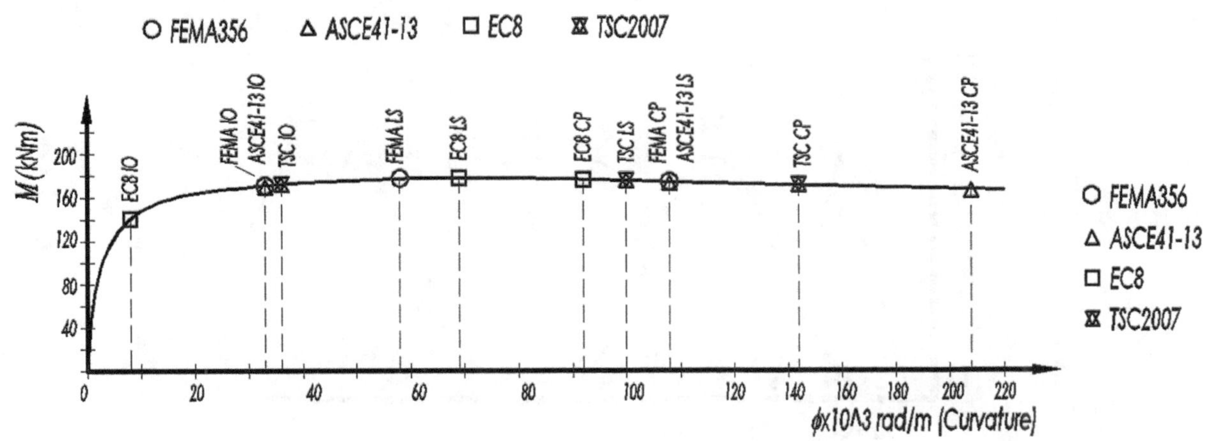

Fuzzification with Sets

Set A represented like $\tilde{A}$ is a fuzzy set. A few sample membership functions (MF) for fuzzy sets (triangular, Gaussian, bell type) are given in Figure 4. However, triangular type membership function is commonly used due to its simplicity.

If an element in the universe, say x, is a member of fuzzy set $\tilde{A}$, then this mapping is given as, $\mu_{\tilde{A}}(x) \in [0,1]$ shown in Figure 4.

A notation convention for fuzzy sets when the universe of discourse, *X*, is discrete and finite, is as follows for a fuzzy set $\tilde{A}$ (1) (Hong, Hong, & Chang, 2003)

$$\tilde{A} = \left\{ \frac{\mu_{\tilde{A}}(x_1)}{x_1} + \frac{\mu_{\tilde{A}}(x_2)}{x_2} + \ldots \right\} = \left\{ \sum_i \frac{\mu_{\tilde{A}}(x_i)}{x_i} \right\} \quad (1)$$

where summation symbol denotes the collection or aggregation of each element, $\frac{\mu_{\tilde{A}}(x_i)}{x_i}$. When the universe, X, is continuous and infinite, the fuzzy set $\tilde{A}$ is denoted by (2)

$$\tilde{A} = \left\{ \int \frac{\mu_{\tilde{A}}(x)}{x} \right\} \quad (2)$$

Except the excluded middle (3) and the contradiction (4) laws,

$$\tilde{A} \cup \overline{\tilde{A}} \neq X \quad (3)$$

$$\tilde{A} \cap \overline{\tilde{A}} \neq \varnothing \quad (4)$$

Figure 4. Sample membership functions

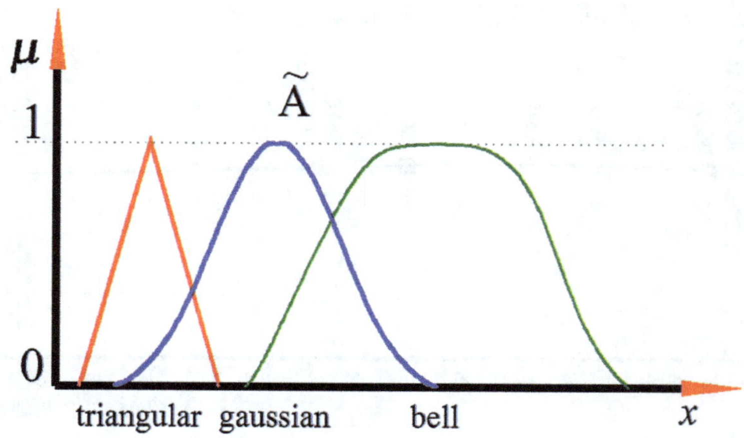

all the set operations as Union, Intersection, Complement and the set properties as Associativity, Distributivity, Idempotency, Identity, Transitivity, Involution, De Morgan's law are exactly the same with classical sets. For the details of set operations and properties elementary logic are given mathematic sources (Zadeh, 1965).

Rule Based System of Fuzzy

In the fuzzy rule based systems rules are defined by conditional sentences with IF, THEN, AND, OR logical conjunctions. AND and OR represent the minimum and maximum operations, respectively.

A single fuzzy IF-THEN rule assumes the following form as (5),

IF x is A THEN y is C (5)

where A and C are linguistic values defined by fuzzy sets on the ranges (universes of discourse) X and Y, respectively. The IF part of the rule "x is A" is called the antecedent or premise, while THEN part of the rule "y is C" is called the consequent or conclusion.

If x1 is $\widetilde{A_1^k}$ and x_2 is $\widetilde{A_2^k}$ THEN y_k is $\widetilde{C^k}$ for k=1, 2, …., r (6)

where $\widetilde{A_1^k}$ and $\widetilde{A_2^k}$ are the fuzzy sets representing the kth-antecedent pairs, and $\widetilde{C^k}$ are the fuzzy sets representing the kth-consequent (6).

For max-min inference method, membership for the crisp inputs x_1 and x_2, based on Mamdani's implication, will be described by (7, 8) (Mamdani, 1976)

$$\mu\left(x_1\right) = \delta\left(x_1 - input\left(i\right)\right) = \begin{Bmatrix} 1, x_1 = input\left(i\right) \\ 0, otherwise \end{Bmatrix} \tag{7}$$

$$\mu\left(x_2\right) = \delta\left(x_2 - input\left(j\right)\right) = \begin{Bmatrix} 1, x_2 = input\left(j\right) \\ 0, otherwise \end{Bmatrix} \tag{8}$$

For a set of disjunctive rules, the aggregated output for the r rules will be given by (9, 10) (Zadeh, 1973)

$$\mu_{\widetilde{C^k}}\left(y\right) = max_k\left[min\left[\mu_{\widetilde{A_1^k}}\left(input\left(i\right)\right), \mu_{\widetilde{A_2^k}}\left(input\left(j\right)\right)\right]\right] \text{k=1, 2, …., r} \tag{9}$$

$$\mu_{\widetilde{C^k}}\left(y\right) = max_k\left[max\left[\mu_{\widetilde{A_1^k}}\left(input\left(i\right)\right), \mu_{\widetilde{A_2^k}}\left(input\left(j\right)\right)\right]\right] \text{k=1, 2, …., r} \tag{10}$$

Graphical Technique of Fuzzy Inference

The most popular and practical way to create fuzzy model is graphical technique. Symbolic presentation of the graphical technique is already given in Figure 5 illustratively. Processes of entering inputs, intersecting the sets by inputs, applying fuzzy operators (AND, OR …etc), implicating the outputs, applying the aggregation method and defuzzifying the general output between the rules are clearly presented in the Figure 5. There is a very simple graphical interpolation of Mamdani's implication for two rules in Figure 5, where the symbols $\widetilde{A_1^1}$ and $\widetilde{A_2^1}$ refer to the first and second fuzzy antecedents of the first rule, respectively, and the symbol $\widetilde{C^1}$ refers to the fuzzy consequent of the first rule; the symbols $\widetilde{A_1^2}$ and $\widetilde{A_2^2}$ refer to the first and second fuzzy antecedents, respectively, of the second rule, and the symbol $\widetilde{C^2}$ refers to the fuzzy consequent of the second rule.

Figure 5. Graphical presentation of the max-min inference method for two rules

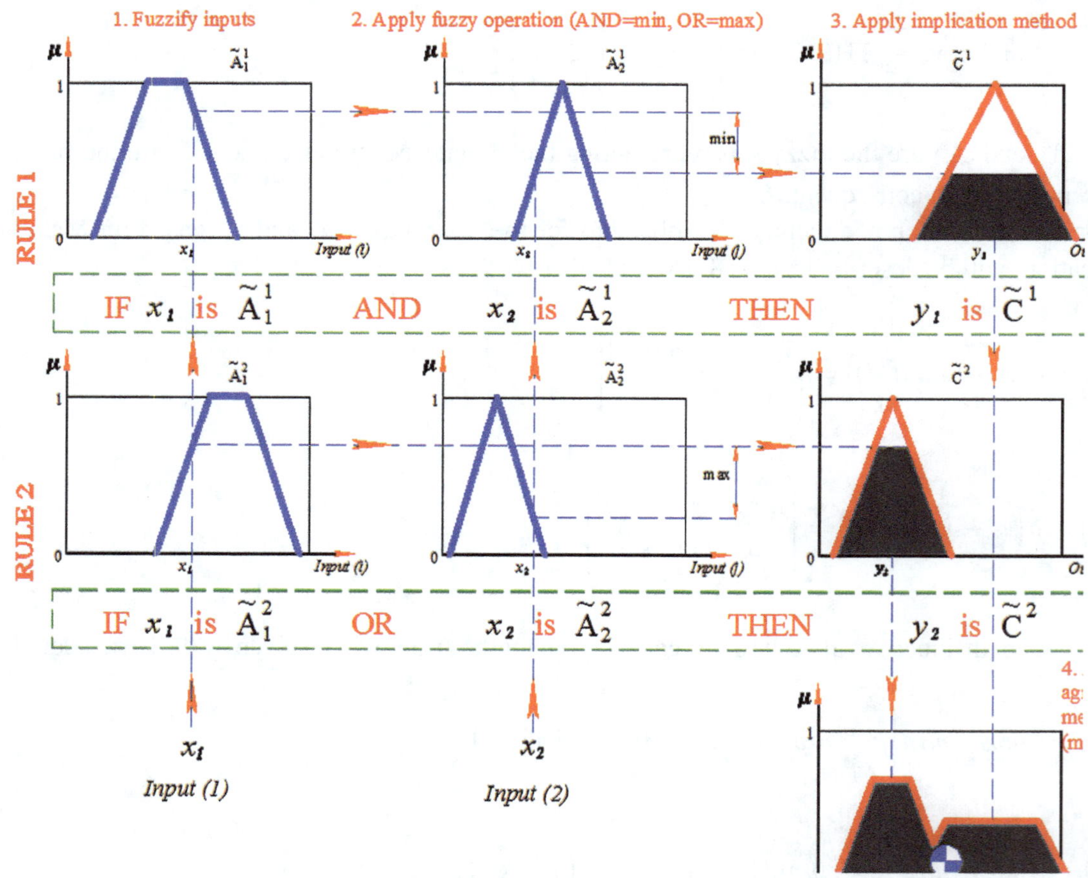

Defuzzification

The input for the defuzzification process is a fuzzy set (the aggregate output fuzzy set) and the output is a single number. As much as fuzziness helps the rule evaluation during the intermediate steps, the final desired output for each variable is generally a single number. However, the aggregate of a fuzzy set encompasses a range of output values which must be defuzzified in order to resolve a single output value from the set.

There is not a specific way to define a defuzzification method for fuzzy logic applications (Zadeh, 1968). Probably the most popular defuzzification method is the centroid calculation, which returns the center of area under the curve. There are a few more methods, such as Bisector, Middle of maximum (the average of the maximum value of the output set), Largest of maximum and Smallest of maximum. Middle of maximum is also called mean-max membership. The centroid method is also called the method of the center of gravity or the center of area. The mathematical expressions and graphical representations of the Centroid (11) and the Middle of Maximum (MOM) (12) methods are given in Figure 6.

$$y^{*} = \frac{\int \mu_{\tilde{C}}\left(y\right) \rightarrow y dy}{\mu_{\tilde{C}}\left(y\right) dy} \tag{11}$$

$$y^{**} = \frac{a+b}{2} \tag{12}$$

Clustering and Fuzzy C-Means Clustering

Clustering is a method which is used for classification of data to create convenience in interpreting the graphical representations of variances in any problem. A sample is given in Figure 7a for data in graphical representation and classification of them. A sample is given in 7b for fuzzyfying of these sets.

Fuzzy c-means is a method of clustering which allows one piece of data to belong to two or more clusters. This method developed by Dunn (1973) and improved by Bezdek, Ehrlich, and Full (1984).

Figure 6. (a) The centroid and (b) the middle of maximum defuzzification method

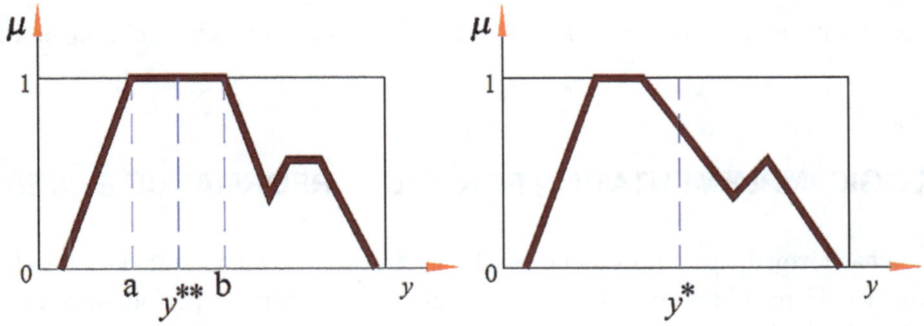

Figure 7. (a) Data clustering (b) Fuzzy c-means clustering

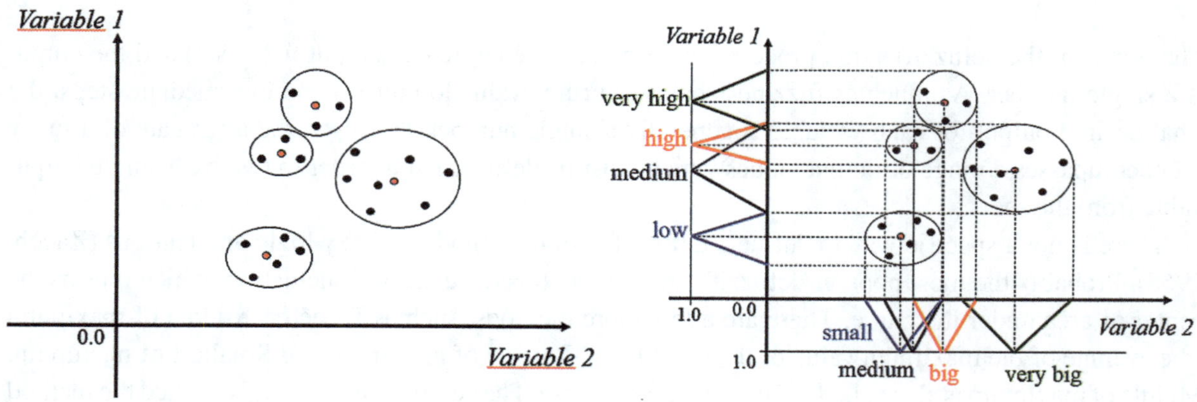

It is based on minimization of the following objective function:

$$J_m = \sum_{i=1}^{N}\sum_{j=1}^{C} u_{ij}^{m} x_i - c_j^{2} \qquad , \qquad 1 \leq m < \infty$$

where m is any real number greater than 1, u_{ij} is the degree of membership of x_i in the cluster j, x_i is the ith of d-dimensional measured data, c_j is the d-dimension center of the cluster, and ||*|| is any norm expressing the similarity between any measured data and the center.

Fuzzy partitioning is carried out through an iterative optimization of the objective function shown above, with the update of membership u_{ij} and the cluster centers c_j by:

$$u_{ij} = \frac{1}{\sum_{k=1}^{C}\left(\frac{x_i - c_j}{x_i - c_k}\right)^{\frac{2}{m-1}}} \qquad , \qquad c_j = \frac{\sum_{i=1}^{N} u_{ij}^{m} \rightarrow x_i}{\sum_{i=1}^{N} u_{ij}^{m}}$$

This iteration will stop when $max_{ij}\left\{\left|u_{ij}^{(k+1)} - u_{ij}^{(k)}\right|\right\} < \varepsilon$, where ε is a termination criterion between 0 and 1, whereas k are the iteration steps. This procedure converges to a local minimum or a saddle point of J_m.

A FUZZY LOGIC IMPLEMENTATION FOR THE PERFORMANCE BASED DESIGN

Each regulation has formed algorithms and flow charts include its own analysis and details for evaluation of performance. Here, a flow chart includes basic elements for seismic performance evaluation, was formed by taking basis of ASCE41-13. It is showed in Figure 8 which part of this algorithm is modeled with fuzzy system.

Figure 8. Flow chart of PBD for retrofit process using fuzzy logic

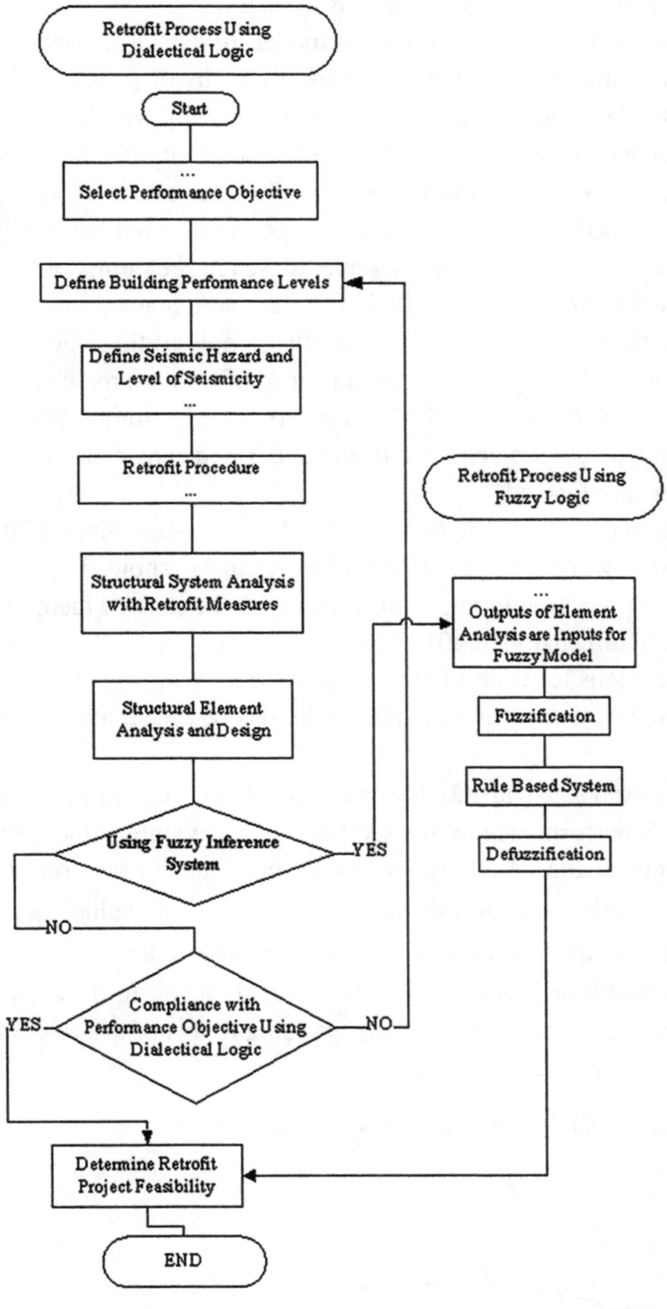

Fuzzification of the Model Which Has One Variable

The disharmony of performance based design criteria is inclusively expressed in title of "DISHARMONY AT PERFORMANCE BASED DESIGN CRITERIA." It is expressed in title of "FUZZY SET THEORY" that the method of implementing a problem to a fuzzy logic is fuzzification, rules and defuzzification in general. Fuzzy clustering method for fuzzyfying the obtained values is expressed in title of "Clustering and Fuzzy c-means Clustering". In this chapter, it is shown practically in Figure 3, clustering first and then fuzzyfying the Moment – Curvature graphic as is seen in Figure 9.

There is no obligation to follow a mathematical model for clustering operation because there is no certainty in system. It is not a logical effort to ensure the limits of an uncertain system. As is seen in Figure 9, a setting was made for IO, LS and CP. A center of gravity was determined for each defined set and marked with red points inside of sets. These centers of gravity can be determined by using fuzzy c-means but the essence of process is as follows: It determines the center of gravity according to distribution of elements in sets defined in shape of ellipse. This point refers to the value having 1.0 membership value according to Syllogistic Logic. As an example, these are the points of this section being in state of %100 IO, %100 LS and %100 CP. It is needed to express here that a point having %100 membership value is needed because triangle function is used for this sample. If these points cannot be determined, then trapezoidal membership function can be used instead of triangle membership function; in such as case, a conditions releases which includes %100 membership value for a certain margin, not a point. As is seen in Figure 10, triangle feet are deducted from the defined center of gravities to the edges of sets, and outlines of fuzzy sets are formed.

Even though the system here is formed by two variables as of Moment and Curvature, as the variable of moment values are so low, only the fuzziness of curvature is showed by foreseeing fixed. But the systems having more than one variable can be fuzzyfied with the showed methodology. Fuzzy curvature set of a concrete section in dimension of 40x40 can be found in Figure 11, with expressing the names of triangle membership functions to fit into fuzzy notation, improving the representation of membership values by removing variable on vertical axis, adding the sets of Linear and Collapse (C) expressing the edge conditions.

In this way, the triangle expressed with IO is trying to explain that: If curvature value is $30 \times 10^{-3}\ rad / m$, there is no share from LS performance in this section. This curvature value shows that the section has 100% of IO performance value. As long as curvature value on section starts to increase from $30 \times 10^{-3}\ rad / m$ value and closes towards the $80 \times 10^{-3}\ rad / m$ value; membership value of being on IO performance level of section decreases towards 0% and when curvature reaches $80 \times 10^{-3}\ rad / m$ value, then LS performance level has the 100% membership level. In these two curvatures, there are transitions between IO and LS at the level of corresponding to feet of triangle sets. Here, fuzzy logic is

Figure 9. Clustering of performance limits for the sample section

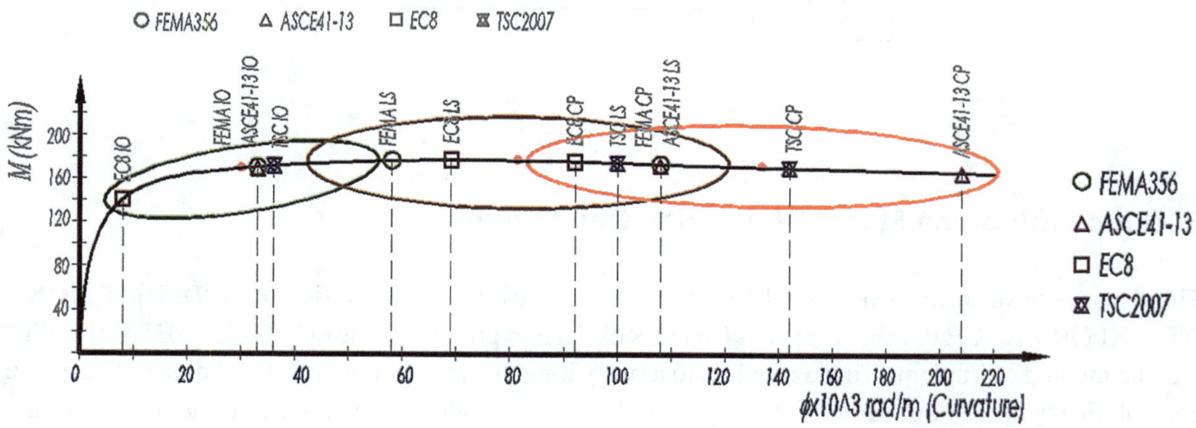

Figure 10. Transformations of performance limits of the sample to fuzzy sets via c-means

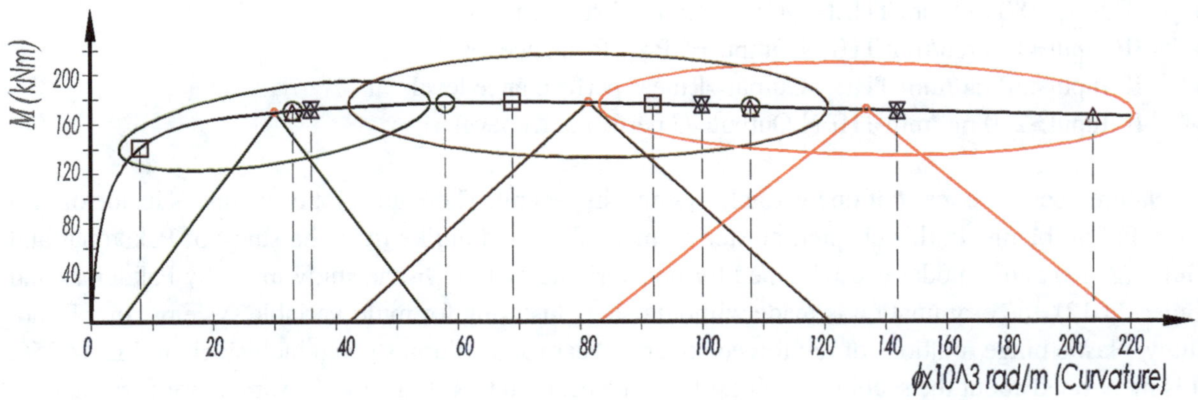

Figure 11. Fuzzy set of performance criteria of the sample section depending as curvature

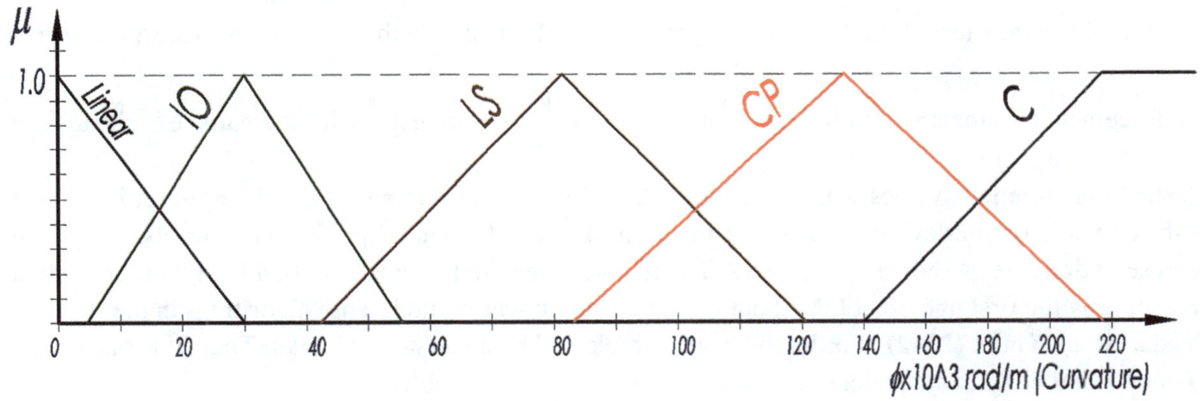

that transition between variables is relative. It can be another study subject that the variables have the biggest membership value 1.0; sensitivity of 30, 80 and 130 rad/mm curvature values. In this book chapter, only the formation of system and methodology is expressed.

Fuzzification of a Model Which Has More Than One Variable

Fuzzification of a single variable which concern to structural engineering and performance based design is tried to be explained inclusively by using clustering method. It can be found that curvature values given in that problem have which one and which rate among IO, LS and CP performance levels. Additionally, C performance Level and Linear performance Level are described for the out of limit situations. As explained previously, that problem was carried out for a single input and output. It is tried to be explained through an example how fuzzy set theory works. In single-input and single-output systems, the rules can be established in a matter of interrelation between input and output sets and subsets. The rules for relevant example are as follows:

- IF Input=30 rad/mm THEN Output=IO performance level
- IF Input=80 rad/mm THEN Output=LS performance level
- IF Input=130 rad/mm THEN Output=CP performance level
- IF Input<30 rad/mm THEN Output=Linear performance level
- IF Input>220 rad/mm THEN Output=C performance level

Nature does not consist of one variable systems in general. There are a lot of variables inside of systems and problems. In this chapter, clustering method is used unlike from the study of Pakdamar and Guler (2012) and a model is established for one variable system. In the study made by Pakdamar and Guler (2012), the examination is made without using clustering for multi variable systems. In relevant study, plastic hinge rotations of reinforced concrete beam and columns given inclusively in FEMA356. It is mentioned about the section which explains the beam in this study, as an example for fuzzification of multi variable problems.

In a part of the study made by Pakdamar and Guler (2012), a fuzzy model is formed for following Table 4 taken from FEMA356 for a beam as a structure element.

From the parameters of Table 4, $\frac{\rho - \rho'}{\rho_{bal}}$ parameter is defined with the name of "balance", transverse reinforcement parameters with the name of "trans", $\frac{V}{b_w d \sqrt{f_c'}}$ parameter with the name of "shear" and plastic hinge rotations values with the name of "rotationradian" as input sets, and output set is defined with the name of "perlev" in a matter of including IO, LS, CP and C performance levels. Graphical representations are as shown in Figure 12. The fuzzy made with the name of "first fuzzy rotationradian set" for only the first line of table without giving the complicated and intense algorithms in the study of Pakdamar and Guler (2012) in order to be an example and be understood simpler. Thus, it is made easy to understand how fuzzy model of a multi variable problem is established.

Table 4. Numerical boundaries of the performance levels for beams (FEMA356)

Primary Beams, Controlled by Flexure					
Conditions			**Plastic Hinge Rotations, in Radians**		
$\frac{\rho - \rho'}{\rho_{bal}}$	**Transverse Reinforcement**	$\frac{V}{b_w d \sqrt{f_C'}}$	**IO**	**LS**	**CP**
≤ 0.0	C	≤ 0.25	0.010	0.020	0.025
≤ 0.0	C	≥ 0.50	0.005	0.010	0.020
≥ 0.5	C	≤ 0.25	0.005	0.010	0.020
≥ 0.5	C	≥ 0.50	0.005	0.005	0.015
≤ 0.0	NC	≤ 0.25	0.005	0.010	0.020
≤ 0.0	NC	≥ 0.50	0.0015	0.005	0.010
≥ 0.5	NC	≤ 0.25	0.0050	0.010	0.010
≥ 0.5	NC	≥ 0.50	0.0015	0.005	0.005

Figure 12. Fuzzy sets and subsets of the beam model

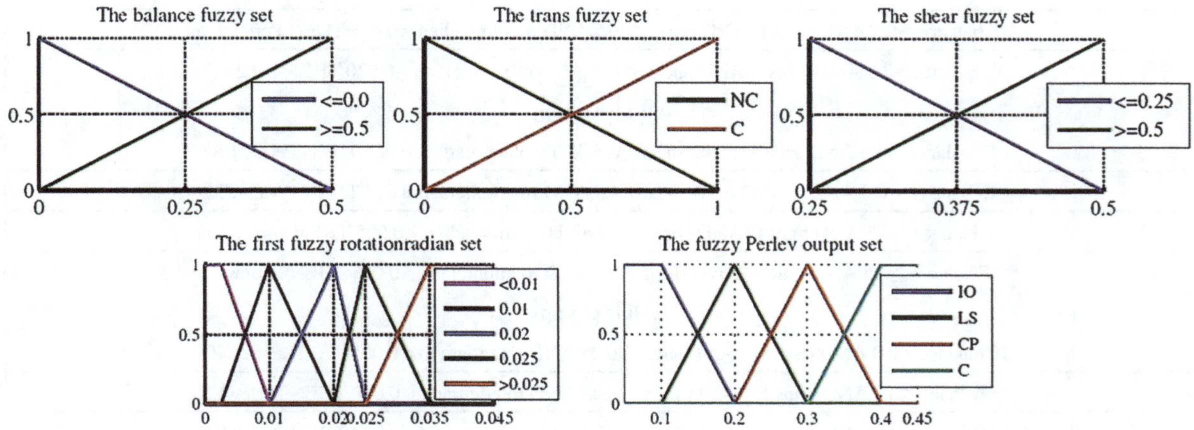

The relationships between the input and output sets are described by rules with the logical operators IF, THEN, AND, OR and NOT. These rules are described using numerical acceptance criteria from Table 4. Some additional rules, such as greater than (>) or less than (<), which are not included in the table, are added to make the system more stable. 37 rules, in Table 5 are used to describe the flexible performance model. Every rule group corresponds to each row in Table 4. As an example, the first group in Table 5 corresponds to the first row in Table 4. It is recommended that the rules in the Table 5 are followed according to Table 4 for beams. Technically for the first row of Table 4 can be described as:

Table 5. Fuzzy model rules for beams of flexible performance model

1. Rule Group
IF balance ≤0.0 AND trans C AND shear ≤0.25 AND rotationradian 0.010 THEN Perlev IO
IF balance ≤0.0 AND trans C AND shear ≤0.25 AND rotationradian 0.020 THEN Perlev LS
IF balance ≤0.0 AND trans C AND shear ≤0.25 AND rotationradian 0.025 THEN Perlev CP
IF balance ≤0.0 AND trans C AND shear ≤0.25 AND rotationradian <0.010 THEN Perlev IO
IF balance ≤0.0 AND trans C AND shear ≤0.25 AND rotationradian >0.025 THEN Perlev C
2. Rule Group
IF balance ≤0.0 AND trans C AND shear ≥0.5 AND rotationradian 0.005 THEN Perlev IO
IF balance ≤0.0 AND trans C AND shear ≥0.5 AND rotationradian 0.010 THEN Perlev LS
IF balance ≤0.0 AND trans C AND shear ≥0.5 AND rotationradian 0.020 THEN Perlev CP
IF balance ≤0.0 AND trans C AND shear ≥0.5 AND rotationradian <0.005 THEN Perlev IO
IF balance ≤0.0 AND trans C AND shear ≥0.5 AND rotationradian >0.020 THEN Perlev C
3. Rule Group
IF balance ≥0.5 AND trans C AND shear ≤0.25 AND rotationradian 0.005 THEN Perlev IO
IF balance ≥0.5 AND trans C AND shear ≤0.25 AND rotationradian 0.010 THEN Perlev LS
IF balance ≥0.5 AND trans C AND shear ≤0.25 AND rotationradian 0.020 THEN Perlev CP

continued on following page

Table 5. Continued

IF balance ≥0.5 AND trans C AND shear ≤0.25 AND rotationradian <0.005 THEN Perlev IO
IF balance ≥0.5 AND trans C AND shear ≤0.25 AND rotationradian >0.020 THEN Perlev C
4. Rule Group
IF balance ≥0.5 AND trans C AND shear ≥0.5 AND rotationradian 0.005 THEN Perlev LS
IF balance ≥0.5 AND trans C AND shear ≥0.5 AND rotationradian 0.015 THEN Perlev CP
IF balance ≥0.5 AND trans C AND shear ≥0.5 AND rotationradian <0.005 THEN Perlev IO
IF balance ≥0.5 AND trans C AND shear ≥0.5 AND rotationradian >0.015 THEN Perlev C
5. Rule Group
IF balance ≤0.0 AND trans NC AND shear ≤0.25 AND rotationradian 0.005 THEN Perlev IO
IF balance ≤0.0 AND trans NC AND shear ≤0.25 AND rotationradian 0.010 THEN Perlev LS
IF balance ≤0.0 AND trans NC AND shear ≤0.25 AND rotationradian 0.020 THEN Perlev CP
IF balance ≤0.0 AND trans NC AND shear ≤0.25 AND rotationradian <0.005 THEN Perlev IO
IF balance ≤0.0 AND trans NC AND shear ≤0.25 AND rotationradian >0.020 THEN Perlev C
6. Rule Group
IF balance ≤0.0 AND trans NC AND shear ≥0.5 AND rotationradian 0.0015 THEN Perlev IO
IF balance ≤0.0 AND trans NC AND shear ≥0.5 AND rotationradian 0.005 THEN Perlev LS
IF balance ≤0.0 AND trans NC AND shear ≥0.5 AND rotationradian 0.010 THEN Perlev CP
IF balance ≤0.0 AND trans NC AND shear ≥0.5 AND rotationradian <0.0015 THEN Perlev IO
IF balance ≤0.0 AND trans NC AND shear ≥0.5 AND rotationradian >0.010 THEN Perlev C
7. Rule Group
IF balance ≥0.5 AND trans NC AND shear ≤0.25 AND rotationradian 0.005 THEN Perlev IO
IF balance ≥0.5 AND trans NC AND shear ≤0.25 AND rotationradian 0.010 THEN Perlev CP
IF balance ≥0.5 AND trans NC AND shear ≤0.25 AND rotationradian <0.005 THEN Perlev IO
IF balance ≥0.5 AND trans NC AND shear ≤0.25 AND rotationradian >0.010 THEN Perlev C
8. Rule Group
IF balance ≥0.5 AND trans NC AND shear ≥0.5 AND rotationradian 0.0015 THEN Perlev IO
IF balance ≥0.5 AND trans NC AND shear ≥0.5 AND rotationradian 0.005 THEN Perlev CP
IF balance ≥0.5 AND trans NC AND shear ≥0.5 AND rotationradian <0.0015 THEN Perlev IO
IF balance ≥0.5 AND trans NC AND shear ≥0.5 AND rotationradian >0.005 THEN Perlev C

If the balance set is less than or equal to "0.0" and trans set is equal to "C" and the shear set is less than or equal to "0.25" and the rotationradian set is 0.010 rad then perlev set is "IO".

As it is given before, there is not a specific way to define a defuzzification method for fuzzy logic applications (Zadeh, 1968). Generally, engineering judgements and/or heuristic approaches are used for the definition. In this example, MOM (middle of maximum, mean-max method) or CENTROID methods are used to defuzzify the fuzzy inference systems. The accuracy of the chosen defuzzification method is tested by using particular samples. Expected outputs are taken via the MOM or CENTROID methods.

REFERENCES

Bezdek, J. C., Ehrlich, R., & Full, W. (1984). FCM: The fuzzy c-means clustering algorithm. *Computers & Geosciences*, *10*(2), 191–203. doi:10.1016/0098-3004(84)90020-7

Committee, S. V. (1995). *Performance-based seismic engineering*. Sacramento, CA: Structural Engineers Association of California.

Doran, B., Yetilmezsoy, K., & Murtazaoglu, S. (2015). Application of fuzzy logic approach in predicting the lateral confinement coefficient for RC columns wrapped with CFRP. *Engineering Structures*, *88*, 74–91. doi:10.1016/j.engstruct.2015.01.039

Dunn, J. C. (1973). A Fuzzy Relative of the ISODATA Process and Its Use in Detecting Compact Well-Separated Clusters. *Journal of Cybernetics*, *3*(3), 32–57. doi:10.1080/01969727308546046

Fardis, M. N. (2005). *Designers' guide to EN 1998-1 and EN 1998-5 Eurocode 8: design of structures for earthquake resistance: general rules, seismic actions, design rules for buildings, foundations and retaining structures* (Vol. 8). Thomas Telford Services Limited.

Fasan, M., Amadio, C., Noè, S., Panza, G., Magrin, A., Romanelli, F., & Vaccari, F. (2015). *A new design strategy based on a deterministic definition of the seismic input to overcome the limits of design procedures based on probabilistic approaches.* arXiv preprint arXiv:1509.09119

FEMA356. P. (2000). Commentary for the Seismic Rehabilitation of Buildings. FEMA-356, Federal Emergency Management Agency.

Guclu, R., & Yazici, H. (2007). Fuzzy Logic Control of a Non-linear Structural System against Earthquake Induced Vibration. *Journal of Vibration and Control*, *13*(11), 1535–1551. doi:10.1177/1077546307077663

Hamburger, R. (1996). *Implementing performance-based seismic design in structural engineering practice.* Paper presented at the 11th World Conference on Earthquake Engineering, Acapulco, Mexico.

Holland, J. H. (1975). *Adaptation in natural and artificial systems: an introductory analysis with applications to biology, control, and artificial intelligence*. U Michigan Press.

Hong, N. K., Hong, S.-G., & Chang, S.-P. (2003). Computer-supported evaluation for seismic performance of existing buildings. *Advances in Engineering Software*, *34*(2), 87–101. doi:10.1016/S0965-9978(02)00106-0

Mamdani, E. H. (1976). Advances in the linguistic synthesis of fuzzy controllers. *International Journal of Man-Machine Studies*, *8*(6), 669–678. doi:10.1016/S0020-7373(76)80028-4

Mandelbrot, B. B. (1983). *The fractal geometry of nature/Revised and enlarged edition* (p. 1). New York: WH Freeman and Co.

McCulloch, W. S., & Pitts, W. (1943). A logical calculus of the ideas immanent in nervous activity. *The Bulletin of Mathematical Biophysics*, *5*(4), 115–133. doi:10.1007/BF02478259

Mistakidis, E. S., & Georgiou, D. N. (2003). Fuzzy sets in seismic inelastic analysis and design of reinforced concrete frames. *Advances in Engineering Software*, *34*(10), 589–599. doi:10.1016/S0965-9978(03)00115-7

Ozkul, S., Ayoub, A., & Altunkaynak, A. (2014). Fuzzy-logic based inelastic displacement ratios of degrading RC structures. *Engineering Structures*, *75*, 590–603. doi:10.1016/j.engstruct.2014.06.030

Pakdamar, F., & Guler, K. (2012). Evaluation of Flexible Performance of Reinforced Concrete Structures Using A Nonlinear Static Procedure Provided by Fuzzy Logic. *Advances in Structural Engineering*, *15*(12), 2173–2190. doi:10.1260/1369-4332.15.12.2173

Pekelnicky, R., & Poland, C. (2013). ASCE 41-13: Seismic Evaluation and Retrofit of Existing Buildings. *SEAOC 2012 Convention Proceedings*.

Sgambi, L. (2004). Fuzzy theory based approach for three-dimensional nonlinear analysis of reinforced concrete two-blade bridge piers. *Computers & Structures*, *82*(13-14), 1067–1076. doi:10.1016/j.compstruc.2004.03.016

Smith, R. (2016). Aristotle's Logic. *The Stanford Encyclopedia of Philosophy*. Retrieved from http://plato.stanford.edu/archives/spr2016/entries/aristotle-logic/

Tesfamaraim, S., & Saatcioglu, M. (2008). Seismic Risk Assessment of RC Buildings Using Fuzzy Synthetic Evaluation. *Journal of Earthquake Engineering*, *12*(7), 1157–1184. doi:10.1080/13632460802003785

TSC, M. o. P. W. a. S. G. o. R. o. T. (2007). *Specification for Buildings to be Built in Seismic Zones in Turkey*. TSC2007.

Zadeh, L. A. (1965). Fuzzy sets. *Information and Control*, *8*(3), 338–353. doi:10.1016/S0019-9958(65)90241-X

Zadeh, L. A. (1968). Probability measures of Fuzzy events. *Journal of Mathematical Analysis and Applications*, *23*(2), 421–427. doi:10.1016/0022-247X(68)90078-4

Zadeh, L. A. (1973). Outline of a New Approach to Analysis of Complex Systems and Decision Processes. *IEEE Transactions on Systems Man and Cybernetics, 3*(1), 28-44. doi:10.1109/Tsmc.1973.5408575

Chapter 10
A Framework and Case Study for the Resilience of Infrastructures

Ali Golara
NIGC, Iran

ABSTRACT

This chapter defines resilience in different contexts comprehensively, and organizes the mathematical theory of network resilience by providing a generalization in order to create a quantitative framework for resilience characterization of an infrastructure network. At this point, a new performance index measuring delivery importance was employed for an applied purpose and an industrial example using realistic data was solved to evaluate the resilience of the entire network. It can be utilized for any type of hazard which might lead to the disruption of the system. The principles and theory in this study can also be applied to other infrastructures that are interconnected and operate as a network, such as transporting systems, electrical power, water supply and distribution systems.

INTRODUCTION

Natural disasters always result to great damages and losses in natural and man-made environments. These damages are crystallized in different forms of Fatality, Economic, Social, Security, Cultural, and Military damages, and can as well lead to decay and disintegration of the society. The extent and severity of these damages are a function of environmental circumstances, the human understanding of risk and the risk-taking behaviour of the society. Thus, estimating the impact of the risks on the environment and human beings is a main prerequisite. Every potential and physical destructive event, phenomenon or human activity that may lead to loss of life or injury, destruction of buildings, disorder, impaired social, economical and infrastructural systems, may well contain the secret, hidden aspects and conditions, which may represent a sign of future threats.

The first difficulty in identifying the risk is acquiring and collecting reliable information, as well as classifying them and their impact on the various categories. Next is to identify possible ways in reduc-

DOI: 10.4018/978-1-5225-2089-4.ch010

ing the damages caused by these risks. The linking of technical findings in the risk with the increasing levels of awareness and risk warnings can be referred to other challenges. Furthermore, risk-taking is an increasingly global concern and its impact and activities in one area can affect the risk-taking in other areas. Providing a system with layers of information for risk factors, the vulnerable components exposed to the disasters, amount of damages derived from the risk factors can be a step in reducing and controlling the risk-taking.

Lack of detailed knowledge of risk factors or lack of adequate information may possibly affect the risk information of the built environment and change vulnerability. For example, in the case of earthquake risk, the first step is the gathering of information of identified active faults. However, there are considerable uncertainties in this step, since unidentified and hidden faults may possibly lead to future major incidents. Hence, the gathering and classification of such information in order to predict damages due to the event, is vital. Also, this information must be collected from the most detailed analysis such as the response of all components of a complicated system against the incidents to most general information such as the culture of society and population density. Therefore, there are lot of uncertainties in dealing with risk assessment methodologies.

Contrary to risk analysis, which begins with the identification of hazards and characterization of probabilities (Golara, 2014), resilience analysis enhances the system response to surprise events. Due to the tight coupling of complex systems, the adverse impacts of a failure in one engineered system may propagate, and possibly amplify, through a number of other connected systems (Vespignani, 2010). Recently, several unprecedented accidents, including the Deepwater Horizon oil spill in 2010 and the Fukushima Daiichi nuclear disaster in 2011, have prompted the need for resilience of infrastructure systems and communities (Alderson et al., 2014). The April 2013 attack on the Pacific Gas and Electric Metcalf electricity substation in California (Parfomak, 2014) as well as the explosion of a high-pressure pipeline that carries natural gas from Iran to Turkey in 2015, serve as a reminder that deliberate threats to infrastructures still exist. Scholars or institutes from various fields have defined the term "Resilience", in different ways (e.g. Holling, 1973; Haimes, 2009; Aven, 2011; Bruneau et al., 2003; Kahan et al., 2009; and Vugrin et al., 2010).

Resilience is a broad concept that presents different definitions for different sciences and applications. This concept was first introduced by Holling (1973) as a descriptive term in ecology. Since the publication in 1973, it has been widely used in various scientific fields such as Psychological, Economical, Urban, Disaster management alongside national and international infrastructures and relevant issues. Several studies have been carried out on the characteristics of resilient communities, which can be summarized as follows: Resilient systems should have a variety of similar functional components in order to avoid a total system failure when one component fails. On the other hand, the number of components of different functions should be able to protect the system against different threats. This resilient system has internal autonomy and can also resist against attack from outside the system (it has strength). Moreover, resilient system components are linked in such a way that they support each other (correlated) and have the necessary flexibility to adopt against changes.

In addition, resilience has also been raised in discussions of sociology. The vulnerability is first defined in these discussions. Vulnerability is a function of exposure and sensitivity of a system. For instance, communities that exist within the boundaries of islands are much more vulnerable to the destruction of coastal storms and floods than inner cities. The vulnerability of the built environment also depends on its vicinity with the source of danger or threat. Non-consolidated infrastructures and buildings, inadequate public infrastructures, as well as industrial and commercial development, increase the vulnerability of the built environment in the societies. The density of the built environment is another important indicator of vulnerability in the society. Another significant parameter of resilience is the economic health of the society that is closely related to its industrial and commercial development. For instance, societies that have a uniaxial economy, based only on tourism industry, are much more vulnerable than societies that are economically diverse. Finally, social and demographic characteristics of residents are the basic parameters of the vulnerability of a community. Significant social factors include:

- Age,
- Gender,
- Race,
- Socio-economic status,
- Populations with special needs (physical and mental problems, homelessness and passersby),
- Non-natives, and
- Seasonal tourism.

Crisis management is directly related to planning and urban management topics. The use of urbanism principles and applying concepts like form, texture and structure of the city, urban land use, communication networks and urban infrastructures can greatly decrease the effects of disasters. Statistical analysis of population growth in the world based on data from UN-Habitat (2016) shows that in 1950, an equivalent of 68% of the world's population, have lived in developing and underdeveloped countries and by 2030, the corresponding rate will be 85%. The population in developed countries in 1950 was equivalent to 32%, moreover by 2030; this number will be 15% of the world's total population. In 1800, the urban population was equal to 2%, 30% in 1950, and 47% in 2000, little more than 50% in 2008 and it is anticipated that by 2030, 60% of the world's population will live in cities. To describe the conditions of growth and development of urbanization in the world and with regards to the adaptation to vulnerability issues of urban settlements, two approaches were discussed: First, planning to decrease the concentration of urban population, especially in cities located in high-risk areas and secondly, planning to create a safe urban environment based on the creation of resistant urban environments, buildings and infrastructures, as well as giving more information to the citizens. Probably in the current situation, for making a resilient city, the second approach to urban planning should be chosen. This is due to the fact that urban growth and development is the function of the processes, and factors that requires more cities and longtime have been formed in their natural contexts.

The term of resiliency in the discussion of urban management is like bridging the gaps between the disaster risk reduction and adaptation to environmental changes. Literally, the term urban resilience refers to a concept that can easily communicate with all stages and aspects of disaster and crisis management as well as using it to achieve solutions in addressing conceptual complexities and answering researchers' questions. Comprehensive strategy for disaster risk management and increased resilience in the metropolitan area with the agreement of 168 countries on the basis of Hyogo document (Hyogo Framework for Action 2005:2015) is based on the following:

1. To identify, assess and control the risk;
2. Reduce risk through preventive measures;
3. Insurance and financing of disaster risk;

4. Readiness;
5. Call and response after the wake of the crisis, recovery and reconstruction aimed at decreasing the risk of future disasters.

In critical electrical networks, the issue of reliable and stable supply of electricity has particular importance. Due to the increasing growth of electricity distribution networks of medium and low voltage, their resilience in incidents such as severe wind, storms, cold and frost, flood and earthquake must be considered in every phase of the design. Factors affecting air and ground distribution network resilience can be summarized as follows: Using standard and high-quality products, good design while focusing on proper operation, maintenance and repair with the attitude toward easy implementation.

The impact of events such as earthquakes on transportation network elements can be counted as fragmentation of base and pavement layers, unstable embankments, retaining walls, bridges and tunnels also in addition to secondary damages caused by obstruction passages through damaged buildings and urban facilities like the failure of electricity, gas and water or sewage pipes, etc. Therefore, due to the geographic extent of the network and interaction with other vital arteries and urban elements, identifying their risk and vulnerability in contrast to buildings and concentrated installations, is not practical. It should also be noted that the transportation system, even in non-critical conditions before the events, faces challenges like traffic congestion, providing convenient access and creating appropriate sponsoring organization. Consequently, analyzing the resilience of transportation network in major cities requires various methods and studies.

Based on the previously mentioned definition of resilience in different contexts, over the past decade, several researchers have developed a variety of methodologies and also focused on the study of interdependent infrastructures. According to Zio and Sansavini (2011) and Rinaldi (2004), these modeling and simulation techniques can be grouped into six broad categories:

1. Aggregate supply and demand tools, which evaluate the total demand for infrastructure services in a region and the ability to supply those services (e.g. Apostolakis & Lemon, 2005; Adachi & Ellingwood, 2008; Piwowar et al., 2009; Helseth & Holen, 2009);
2. Dynamic Simulations, which examine infrastructure operations, the effects of disruptions, and the associated downstream consequences (e.g. Duenas-Osorio, 2007; Svendsen & Wolthusen, 2007; Ouyang et al., 2009);
3. Agent-based models, which allow the analysis of the operational characteristics and physical states of infrastructures (e.g. Casalicchio et al., 2007; Schläpfer et al., 2008);
4. Physics-based models, which analyze physical aspects of infrastructures with standard engineering techniques, for example, power flow and stability analyses for electric power grids or hydraulic analyses on pipeline systems (e.g. Chen & McCalley, 2005);
5. Population mobility models, which examine the movement of entities through geographical regions (e.g. Germann et al., 2006); and
6. Leontief input-output models, which basically provide a linear, aggregated, time-independent analysis of the generation, flow, and consumption of various commodities among infrastructure sectors (e.g. Jiang & Haimes, 2004; Reed et al., 2009; Cagno et al., 2010).

Most studies have described the concept of resilience and determine its dimensions as well as its different characteristics. Nevertheless, quite little research has been carried out in creating tools in measuring and evaluating the resiliency. Thus, we need a method that can easily utilize resiliency in all types of vital structure networks regardless of the type of risk. In this study, a new performance index measuring the delivery importance of high-pressure gas network is proposed to evaluate the resilience index of the entire network. Furthermore, it can be employed for various natural disasters like earthquakes and fires, or man-made hazard like deliberate attacks which might lead to the disruption of the system. The principles and theory applied to the natural gas supply and distribution network in this study can also be applied to other infrastructures that are interconnected and operate as a network, including electrical power and water supply and distribution systems. Thus, a comprehensive method based on the concept of network is presented in this chapter.

The differences between 'Resilience' and 'Reliability and Delivery Importance' are worth clarifying in this study. The primary difference is that, resilience applies to the functions or services performed by the network, while reliability is usually employed to describe the failure and repair behavior of individual network elements. A primary indicator of network resilience is the reliability of the services that run on the network; however, service reliability is not the only reasonable delivery function for network resilience characterization (Tortorella, 2005). All but one of several useful delivery functions describe the intended purpose of a network. As defined by Tortorella (2010), delivery importance represents a generalized idea of reliability importance (Ramirez-Marquez et al., 2006). Thus, delivery importance of a subset of the network is then defined in terms of the change in the value of the delivery function, while considering conditions of subset change (for instance, increase or decrease in capacity or removal from service). So, delivery importance represents a generalized idea of reliability importance (Birnbaum et al., 1961). Network resilience is defined as the degree to which the network is sensitive to changes in its infrastructure or elements. So, a network is resilient if there are few subsets that have high delivery importance.

Methodology

This chapter utilizes the most commonly used network reliability theory. Hence, in order to capture the notion that such network performs a specific desired function, a delivery function associated with a network is defined to specify the expected service of the network. For a commodity network like a gas/water/power, etc. transport network, the delivery function may be the quantity of commodity delivered to specified destination nodes over a stated period of time. Golara and Esmaeily (2016) discussed network resilience in quantitative terms and defined 'scalar network resilience' R_S,for a deterministic capacitated network at a nominal capacity matrix C_0 as the largest of the absolute delivery importance values in the negative direction (capacities decrease) across all subsets of the elements (nodes and links) in the network. This study only considers the links:

$$R_S = 1 - (Max\left|DI\_(Lij; C_0)\right|) \qquad where\, i, j, =1 \text{ to } N \text{ and } i \neq j \tag{1}$$

where i and j indicate the number of nodes, N is the total number of nodes in the network. Lij is the link between node i and node j, while $DI\_(Lij; C_0)$ is the delivery importance for that link and it is defined, in R_S methodology, as the total shortage in the whole network, when this link is not available on the network as a result of any type of hazard, which in this case is:

$$DI\_(Lij;C_0) = \frac{\text{Shortage of Natural Gas in the Whole Network (in Volume)}}{\text{Total Natural Gas Injected into the Network (in Volume)}} \tag{2}$$

The first essential idea of network resilience is that, a resilient network is one that has few subsets with large delivery importance, or conversely, most subsets have small delivery importance. This expresses the idea that removal, or decrease in capacity of most network elements has little impact on the delivery function of the network if the network is resilient. Scalar network resilience was employed to calculate resilience of the entire network in this methodology.

A broad measure of resilience that captures time parameter can be expressed, in general terms, by the concepts shown in Figure 1. Once a failure occurs, resilience is the inherent ability of a system to survive and recover from this disturbance. Consequently, resilience is represented as a combination of survivability and recoverability. This notional representation is widely utilized in the literature (Tierney & Bruneau, 2007; Castet & Saleh, 2012; Ayyub, 2013) to illustrate the fundamental ideas behind resilience. The approach in this study is based on the notion that a measure, $Q(t)$, which varies with time, is defined for the quality of the infrastructure of a community. Specifically, performance can range from 0 to 100%, where 100% means no degradation in service and 0% means no service is available. If an event occurs at time T_{0E}, it could cause sufficient damage to the infrastructure such that the quality is immediately reduced. Restoration of the infrastructure is expected to occur over time, as indicated in the figure, until time T_{RE} when it is completely repaired.

In a dynamic network, scalar network resilience (Equation 1) is a random variable, and a function of time. In a dynamic stochastic capacitated network, the capacities are represented by stochastic processes with time as their parameter space. These networks model realistic networks having network element capacities that change at random due to failures and repairs to parts of the network element. For instance, the capacity of a network element in a deterministic capacitated network could be modeled with a continuous-time, discrete-state semi-Markov process (Puterman, 2009).

The link capacity $(L_{AB};C_{AB})$ in a continuum network that delivers gas/water/power etc. from node A to node B can be modeled as continuous time Gaussian processes, and it indicates the change in capacity of the network elements from time to time as result of failures and repairs, degradation, corrosion, earthquake, explosion, terrorist attack, etc. Here, the delivery function is a specific quality of network

Figure 1. Schematic representation of resilience

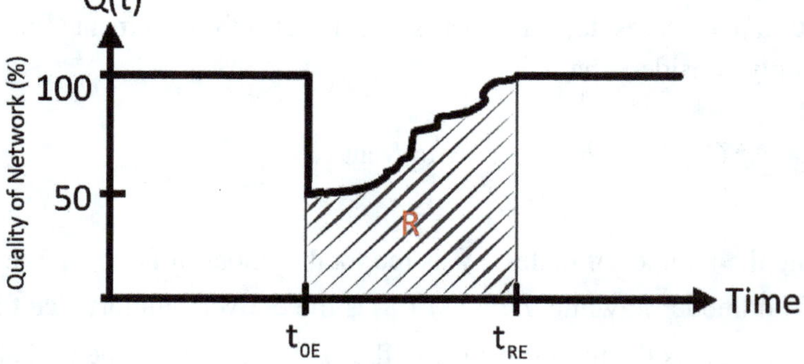

$Q(t)$. For instance, it can be volume/flow/power delivered from node A to node B over a period of time [t_{0E}, t_{RE}] where t is the time; t_{0E} is the initial time of the extreme event E; t_{EN} is the recovery time from event E; therefore, $t_{RE} = t_{0E} + t_{EN}$. Node A and B can be considered as both the ends of a single pipeline or arbitrary main nodes in a network.

The capacity of a gas/water/power network element is the maximum volume/flow/power permitted in that network element. Where $F(t)$ indicates the volume/flow/power in the network at time t assuming a maximum-flow scenario (Ford & Fulkerson, 1962), and $F_{AB}(t)$ indicates flow from node A to node B under this scenario. Then the delivery function is consistent with this model:

$$T(X;t_{EN}) = \int_{t_{0E}}^{t_{RE}} Q(t)dt = \int_{t_{0E}}^{t_{RE}} F_{AB}(t)dt \tag{3}$$

And the delivery importance of a subset of network elements is given by the random variable:

$$DI\_(L_{ij};X) = \sum \int_{t_{0E}}^{t_{RE}} \frac{\partial F_{AB}}{\partial x_{ij}}(t)dt \tag{4}$$

Thus, the 'dynamic network resilience' with respect to the given delivery function is:

$$R_D = \sum \int_{t_{0E}}^{t_{RE}} (100 - \frac{\partial F_{AB}}{\partial x_{ij}}(t)dt) \tag{5}$$

There is no significant difference between Equation (1) and Equation (5). Nevertheless, the methodology summarized in Equation (5) is more general than the one proposed by Equation (1), since within the framework of Equation (1), only the presence or absence of a link was considered, whereas within the framework of Equation (5), other uncertainties including time of disruption of event, repair time to bring the network to service, and network balancing time are considered.

An Applied and Industrial Example and Discussion

This chapter presents a comprehensive model of resilience quantification, which can be applied to complex systems of networks. It also provides a good example for examining the application of this methodology to a real problem. For this purpose, the Iranian high-pressure natural gas network is considered as a complex network. Initially, this complex network was divided into ten manageable areas with respect to flows and area nodes (Iranian gas supply system and ten gas areas of Iran are depicted in Figure 2). Apparently, the pipelines deliver natural gas from the processing facilities to consumers. These gas transfers are considered to occur within a "Node". Gas can as well flow between nodes. The flow between nodes occurs in the "Links". As shown in Figure 3, each area is considered as a node area. All natural gas pipelines that transport the gas between two adjacent states are combined and modeled as one link joining the two node areas (Table 1, first and second columns). Processing facilities are mod-

Figure 2. Iranian gas supply system, showing the location of main cities, refineries, compressor stations high-pressure gas pipe lines and ten gas areas of Iran

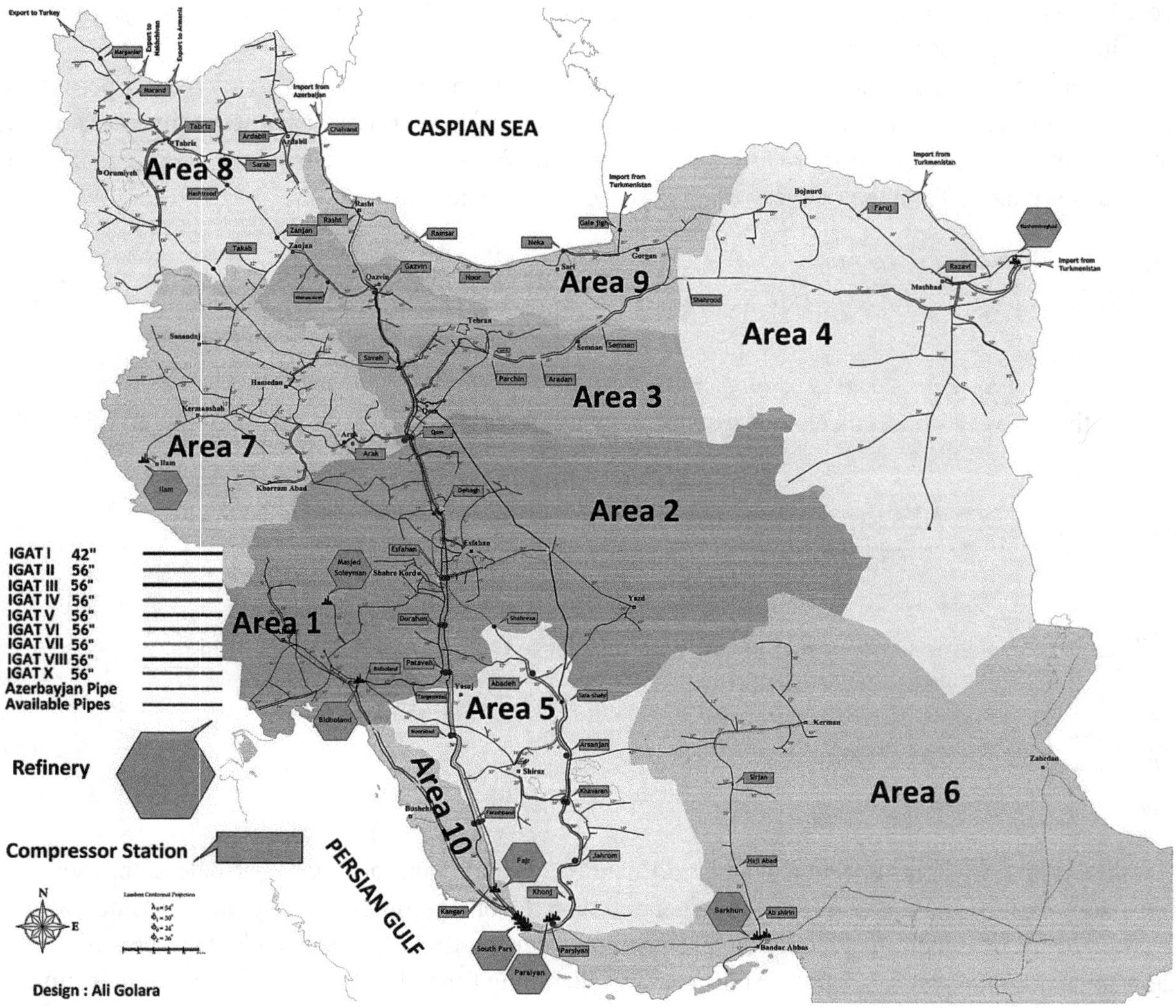

eled as source nodes in node area. Import terminals and Export terminals are modeled as source nodes and consumer nodes, respectively.

In the following discussion of damage scenarios as a result of any type of natural disasters, comparisons will be made with the worst case scenario. Knowing that the price of natural gas will rise after 2015, and the rise is projected to continue through 2040 (Annual Energy Outlook, 2013) and considering the fact that the maximum gas consumption in Iran occurs in January or early February (Golara & Esmaeily, 2016), the characteristics of the Iranian high-pressure gas supply system in January 2015 is modeled. It is obvious that when a link breaks in a network, the gas has to be transferred by other links. Nevertheless, due to the capacity limit of reaming pipelines and links, all needed gas in nodes/areas can't be transferred by available links in the network and gas shortages are therefore inevitable.

For January 2015, forecasting natural gas consumption, production and demand is calculated by equations derived from Golara et al. (2015) and Golora and Esmaeily (2016). Golora et al. (2015) ap-

Figure 3. Schematic model of Iranian high-pressure natural gas network

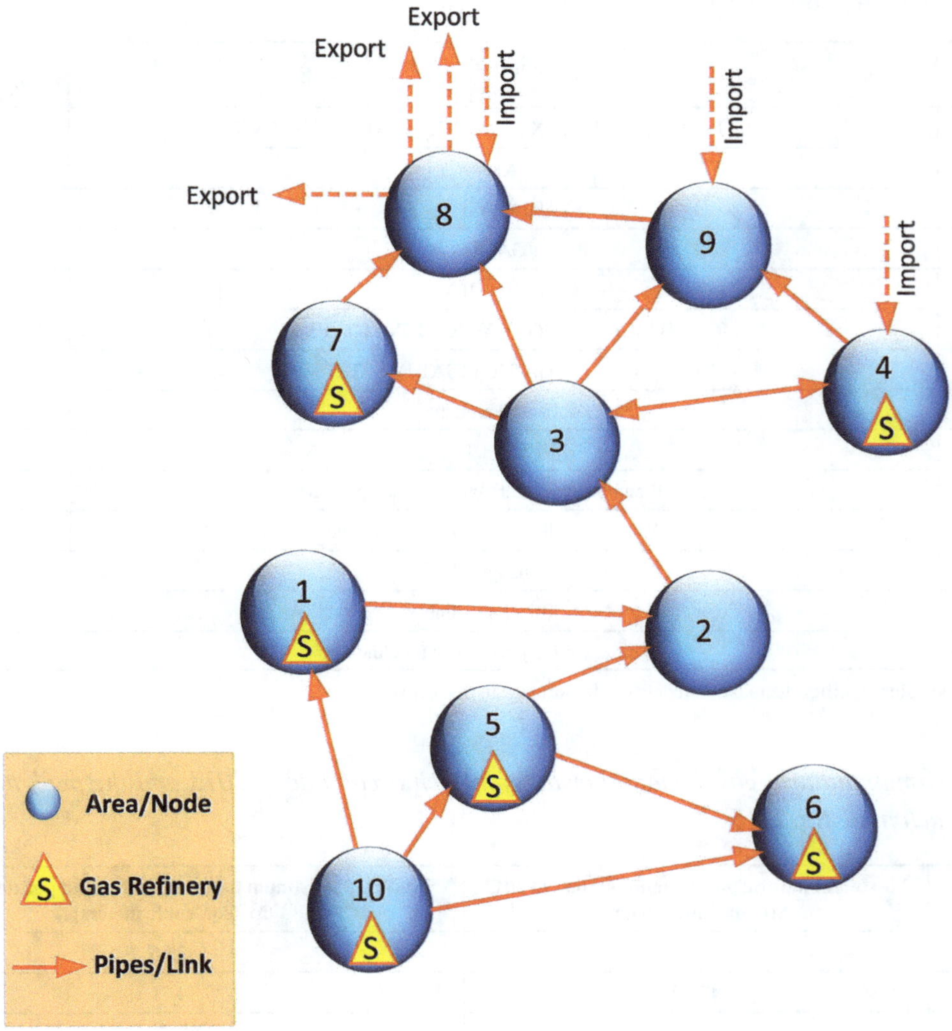

plied Generic Algorithm (GA) to cover uncertainties and develop the set of formulas for forecasting the future of Iran's natural gas production and consumption. The predicted maximum natural gas demand per day for each area/node in 2015 was determined by multiplying the factor of natural gas demand growth (29% per year) by the maximum natural gas demand per day in 2013. Table 2 shows the expected peak demand for each node in 2015. Gas demand data for each area node in 2013 were obtained from the NIGC, which is responsible for providing statistics and information relevant to energy production and demand.

The natural gas flow through each area node can be easily modeled either manually or with the aid of a simulation tool such as SIMONE (1993). To prevent the software from artificially restricting flow, the capacity of each area node is modeled as the sum of the maximum flows into the node. As mentioned earlier, the delivery importance for a link will be defined as the natural gas shortage in the whole network, when this link is not available on network because of any type of hazard. Disabling or removing a link does not allow the flow of gas downstream from a node. The delivery importance of each link is calculated employing Equation (2) and is also listed in the third column of Table 1.

Table 1. Main high-pressure gas pipelines that are combined and modeled as one link joining the two area nodes and calculated delivery importance of each link

Link	Pipeline	Delivery Importance
L10,5	IGAT II ; IGAT III ; IGAT X ; IGAT VII ; IGAT IV ; IGAT VIII	0.52
L10,1	IGAT VI	0.11
L10,6	IGAT VII	0.012
L1,2	IGAT I	0.08
L5,6	IGAT VII	0.011
L5,2	IGAT II ; IGAT III ; IGAT X ; IGAT IV ; IGAT VIII	0.47
L2,3	IGAT I ; IGAT II ; IGAT III ; IGAT IV ; IGAT VIII	**0.67**
L3,4	A 48-inch pipeline	0.081
L3,9	IGAT III	0.14
L3,8	IGAT II and Second Azerbayjan 48-inch pipelines	0.27
L3,7	A 30-inch pipeline and a 36-inch pipeline	0.1
L4,9	A 30-inch pipeline	0.06
L9,8	A 30-inch pipeline	0.07
L7,8	Third Azerbayjan 48-inch pipeline	0.069

Note: The bold value signifies the maximum delivery importance in this network.

Table 2. Maximum natural gas demand per day for each area/node in 2013 and the peak demand expected for each node for 2015

Area/Node	Maximum natural gas demand per day in 2013 (Million cubic meters)	Predicted maximum natural gas demand per day in 2015 (Million cubic meters)
1	27.19	42.96
2	47.29	74.71
3	94.15	148.75
4	34.08	53.85
5	16.12	25.46
6	11.01	17.39
7	39.55	62.48
8	42.60	67.31
9	32.66	51.60
10	15.83	25.02

As earlier mentioned in Equation 1, network resilience is calculated based on the largest of the absolute delivery importance values across all subsets of the elements. By considering only links in this study, the scalar resilience of the entire network is equal to:

$$R = 1 - (Max|DI\_(Lij;\ C_0)|) = 1 - 0.67 = 0.33 \text{ Or } 33\% \quad (6)$$

CONCLUSION

This chapter presents a comprehensive model of resilience quantification, which can be applied to complex systems of networks. It also provides a good example for examining the application of this methodology to a real-world problem. Like other critical infrastructures, the protection of the whole natural gas supply and distribution network is not feasible. So, a proper understanding of how this critical infrastructure behaves as a network and responds to potential changes, will help responsible institutes and government officials to identify important items and critical issues to be addressed in order to promote sound decisions. Regarding the case study discussed in this study, specific actions can be taken by the government, decision makers and designers to increase the resilience of the Iranian high-pressure gas network such as the ones listed below:

1. One of the most common ways to promote greater network resilience is "over provisioning", that is, making the capacities of the network elements larger than what would be required to just meet stated service reliability objectives. Therefore, several alternatives may be evaluated before a choice is made. Some of these alternatives are as follows.
    a. 1.1-Construction of gas storage facilities in area nodes 4, 7, 8 and 9
    b. 1.2-Construction of a new 56-inch transmission lines between area nodes 2 and 7
    c. 1.3-Construction of a new 56-inch transmission line between area nodes 2 and 4
    d. 1.4-Construction of a new 56-inch transmission line between area nodes 10 and 7
2. In the absence of a national strategic approach to protect the important links in the network, the natural gas supply and distribution network will likely evolve in a manner that diminishes its resilience and increases both the vulnerability and the consequences of natural failures and potential attacks. The network resilience will be improved if preservation instructions and plans are in place to prevent failures in network elements from cascading or propagating to other network elements. Therefore, it is necessary to provide an instruction for protecting most important links.
    a. 2.1- Providing a conservation and preservation instruction for link 2 to 3 (L2, 3) and for link 5 to 10 (L5, 10)
    b. 2.2- Sudden and unpredictable ruptures are due to events like natural disasters and deliberate attacks. These are usually large in scope and extended in time. In high-pressure natural gas networks, long experience helped codify a set of network management principles that were employed to mitigate the effects of losing large portions of a network due to, e. g., earthquakes, floods, and other natural disasters. Therefore, providing a network management instruction for crisis situations is vital.

REFERENCES

Adachi, T., & Ellingwood, B. R. (2008). Serviceability of earthquake-damaged water systems: Effects of electrical power availability and power backup systems on system vulnerability. *Reliability Engineering & System Safety*, *93*(1), 78–88. doi:10.1016/j.ress.2006.10.014

Alderson, D. L., Gerald, G. B., & Matthew, W. (2014). Assessing and improving operational resilience of critical infrastructures and other systems. *Stat*, *745*, 70.

Annual Energy Outlook, Energy Information Administration. (2013). Retrieved from http://www.eia.gov/forecasts/aeo/archive.cfm

Apostolakis, E. G., & Lemon, M. D. (2005). A screening methodology for the identification and ranking of infrastructure vulnerabilities due to terrorism. *Risk Analysis*, *25*(2), 361–376. doi:10.1111/j.1539-6924.2005.00595.x PMID:15876210

Aven, T. (2011). On some recent definitions and analysis frameworks for risk, vulnerability, and resilience. *Risk Analysis*, *31*(4), 515–522. doi:10.1111/j.1539-6924.2010.01528.x PMID:21077926

Ayyub, B. (2013). Systems Resilience for Multihazard Environments: Definition, Metrics, and Valuation for Decision-Making. *Risk Analysis*. doi:10.1111/risa.12093 PMID:23875704

Birnbaum, Z. W., Esary, J. D., & Saunders, S. C. (1961). Multi-Component Systems and Their Reliability. *Technometrics*, *3*(1), 55–77. doi:10.1080/00401706.1961.10489927

Bruneau, M., Chang, S., Eguchi, R., Lee, G., ORourke, T., Reinhorn, A. M., & Winterfelt, D. V. et al. (2003). A framework to quantitatively assess and enhance the seismic resilience of communities. *Earthquake Spectra*, *19*(4), 733–752. doi:10.1193/1.1623497

Cagno, E., De Ambroggi, M., Grande, O., & Trucco, P. (2010). Risk analysis of underground infrastructures in urban area: time-dependent interoperability analysis. *Reliability, Risk and Safety: Theory and Applications - Proc. ESREL 2009 Europe Annual Conf.*, 1899-1906.

Casalicchio, E., Galli, E., & Tucci, S. (2007). Federated agent-based modeling and simulation approach to study interdependencies in IT critical infrastructures. *11th IEEE Symposium on Distributed Simulation and Real-Time Applications*. doi:10.1109/DS-RT.2007.11

Castet, J. F., & Saleh, J. H. (2012). On the concept of survivability, with application to spacecraft and space-based networks. *Reliability Engineering & System Safety*, *99*, 123–138. doi:10.1016/j.ress.2011.11.011

Chen, Q., & McCalley, J. D. (2005). Identifying high risk N-k contingencies for online security assessment. *IEEE Transactions on Power Systems*, *20*(2), 823–834. doi:10.1109/TPWRS.2005.846065

Department of Homeland Security. (2013). *National Infrastructure protection plan*. Washington, DC: NIPP.

Duenas-Osorio, L., Craig, J. I., & Goodno, B. J. (2007). Seismic response of critical interdependent networks. *Earthquake Engineering & Structural Dynamics*, *36*(2), 285–306. doi:10.1002/eqe.626

Ford, L. R., & Fulkerson, E. (1962). *Flows in Networks*. Princeton, NJ: Princeton University Press.

Germann, T. C., Kadau, K., Longini, I. M., & Macken, C. A. (2006). Mitigation strategies for pandemic influenza in the United States. *Proceedings of the National Academy of Sciences of the United States of America*, *103*(15), 5935–5940. doi:10.1073/pnas.0601266103 PMID:16585506

Golara, A. (2014). Probabilistic seismic hazard analysis of interconnected infrastructure: A case of Iranian high-pressure gas supply system. *Natural Hazards*, *73*(2), 567–577. doi:10.1007/s11069-014-1087-6

Golara, A., Bonyad, H., & Omidvar, H. (2015). Forecasting Iran's Natural Gas Production, Consumption. *Pipeline & Gas Journal*, *242*(8), 24–30.

Golara, A., & Esmaeily, A. (2016). Quantification and Enhancement of the Resilience of Infrastructure Networks. *Journal of Pipeline Systems Engineering and Practice*. doi: 10.1061/(ASCE)PS.1949-1204.0000250

Haimes, Y. Y. (2009). On the definition of resilience in systems. *Risk Analysis*, *29*(4), 498–501. doi:10.1111/j.1539-6924.2009.01216.x PMID:19335545

Helseth, A., & Holen, A. T. (2009). Structural vulnerability of energy distribution systems: Incorporating infrastructural dependencies. *Electrical Power and Energy Systems*, *31*(9), 531–537. doi:10.1016/j.ijepes.2009.03.023

Holling, C. S. (1973). Resilience and stability of ecological systems. *Annual Review of Ecology and Systematics*, *4*(1), 1–23. doi:10.1146/annurev.es.04.110173.000245

Hyogo Farmework for Action. (2005-2015). Retrieved from https://www.unisdr.org/we/coordinate/hfa

Jiang, P., & Haimes, Y. Y. (2004). Risk management for Leontief-based interdependency systems. *Risk Analysis*, *24*(5), 1215–1229. doi:10.1111/j.0272-4332.2004.00520.x PMID:15563289

Kahan, J. H., Andrew, C. A., & Justin, K. G., (2009). An operational framework for resilience. *Journal of Homeland Security and Emergency Management, 6*(1).

Ouyang, M., Hong, L., Mao, Z.-J., Yu, M.-H., & Qi, F. (2009). A methodological approach to analyze vulnerability of interdependent infrastructures. *Simulation Modelling Practice and Theory*, *17*(5), 817–828. doi:10.1016/j.simpat.2009.02.001

Parfomak, P. W, (2014). *Physical Security of the US Power Grid: High-Voltage Transformer Substations*. Tech. rept. Congressional Research Service.

Piwowar, J., Chatelet, E., & Laclemence, P. (2009). An efficient process to reduce infrastructure vulnerabilities facing malevolence. *Reliability Engineering & System Safety*, *94*(11), 1869–1877. doi:10.1016/j.ress.2009.06.009

Puterman, M. L. (2009). *Markov decision processes: discrete stochastic dynamic programming* (Vol. 414). Wiley.

Ramirez-Marquez, J. E., Rocco, C. M., Gebre, D. A., Coit, D. W., & Tortorella, M. (2006). New Insights on Multi-State Component Criticality and Importance. *Reliability Engineering & System Safety*, *91*(8), 894–904. doi:10.1016/j.ress.2005.08.009

Reed, D. A., Kapur, K. C., & Christie, R. D. (2009). Methodology for assessing the resilience of networked infrastructure. *IEEE Systems Journal*, *3*(2), 174–180. doi:10.1109/JSYST.2009.2017396

Rinaldi, S. M. (2004). Modeling and simulating critical infrastructures and their interdependencies. *Proc. Thirty-Seventh Annual Hawaii International Conf. on System Sciences*. Computer Society Press doi:10.1109/HICSS.2004.1265180

Schläpfer, M., Kessler, T., & Kröger, W. (2008). Reliability analysis of electric power systems using an object-oriented hybrid modeling approach. *Proc. 16th Power Systems Computation Conf.*

Svendsen, N. K., & Wolthusen, S. D. (2007). Connectivity models of interdependency in mixed-type critical infrastructure networks. *Information Security Technical Report*, *12*(1), 44–55. doi:10.1016/j.istr.2007.02.005

Tierney, K., & Bruneau, M. (2007). Conceptualized and measuring resilience. *TR News*, *250*, 14–17.

Tortorella, M. (2005). Service reliability theory and engineering, II: Models and examples. *Quality Technology & Quantitative Management*, *2*(1), 17–37. doi:10.1080/16843703.2005.11673087

Tortorella, M. (2010). Network Resiliency. Rutgers Center for Operations Research (RUTCOR Research Report).

Tortorella, M. (2010). *Network resiliency*. RUTCOR Research Reports.

UN-Habitat. (2016). *The World Cities Report 2016, Urbanization and Development: Emerging Futures*. Author.

Vespignani, A. (2010). The fragility of interdependency. *Nature*, *464*(7291), 984–985. doi:10.1038/464984a PMID:20393545

Vugrin, E. D., Drake, E. W., Ehlen, M. A., & Camphouse, R. C. (2010).A framework for assessing the resilience of infrastructure and economic systems. Sustainable and Resilient Critical Infrastructure Systems, 77-116. doi:10.1007/978-3-642-11405-2_3

Záworka, J. (1993). Project SIMONE—achievements and running development. *Proceedings of the Second International Workshop SIMONE on Innovative Approaches to Modeling and Optimal Control of Large Scale Pipeline Networks*, 1–24.

Zio, E., & Giovanni, S. (2011). Modeling interdependent network systems for identifying cascade-safe operating margins. *Reliability, IEEE Transactions on*, *60*(1), 94–101. doi:10.1109/TR.2010.2104211

Chapter 11
Numerical Methods for the Seismic Performance Assessment of Reinforced Concrete Buildings

Ulgen Mert Tugsal
Gebze Technical University, Turkey

Beyza Taskin
Istanbul Technical University, Turkey

ABSTRACT

Considering the fact that similar structural and construction deficiencies which are exposed during the recent destructive earthquake events are existing in many southern European, Middle Eastern and west Asian countries settling on highly seismic zones, designating the seismic adequacy of the existing building stock for providing structural safety is a significant necessitation in the mitigation of losses during the future seismic events. In most of these regions, a clear majority of the building stock has not been adequately designed or constructed in terms of the seismic regulations of the design codes, while some have even not benefitted from engineering services. Post-earthquake site observations demonstrate the insufficient capacities in these buildings depending on different structural and construction deficiencies. Within the scope of this research, it is aimed to investigate and compare the analytically calculated structural performances of a building ensemble consists of 3~5 story structures with known damage level by utilizing different procedures.

INTRODUCTION

The need for evaluating the seismic adequacy of existing structures has come into focus following the damage and collapse of numerous reinforced concrete (RC) structures during recent earthquakes in many southern European, Middle Eastern and west Asian countries settling on highly seismic zones. The vulnerable building stock in these regions have similar structural and construction deficiencies because of

DOI: 10.4018/978-1-5225-2089-4.ch011

the fact that they have not been adequately designed or constructed in terms of the seismic regulations of the design codes, while some have even not benefitted from engineering services. In most of these regions, a clear majority of the zone of occupation has been exposed to strong ground shakes during recent earthquakes. Most particularly, the destructive earthquakes occurred in Turkey over the past two and a half decades, which caused significant amounts of casualties and economic losses, revealed the insufficient structural capacity of the existing building stock and emphasized the need for performance assessment to estimate the potential damage and to make required provisions against future earthquakes.

Site-observed damages in buildings, mostly consisting of low- and mid-rise RC structures, point out inadequate reinforcement detailing, lack of confinement zones, heavy and large-span cantilevers and indirect supporting preventing the formation of regular structural frames. Within the scope of this research, a building stock which represents the general characteristics of existing structures is taken into consideration. The stock consists of 3~5 story RC frame structural buildings which have experienced moderate damages during the M_w 6.2 1998 Ceyhan Earthquake. The aim of the study is to perform the structural performance assessment of the representative building stock by utilizing:

1. Linear elastic procedure described in Turkish Earthquake Code, TEC (2007);
2. Nonlinear dynamic analysis procedure under the effect of the recorded strong ground motion during 1998 Ceyhan Earthquake;
3. Appendix-2 of the Application Regulation of the law no: 6306 by Republic of Turkey, Ministry of Environment and Urbanization (2012); and
4. Japanese seismic index method (2001) and then compare the findings.

If there is a necessity for a detailed investigation, different procedures are recommended for performance analysis of buildings in many seismic codes that present two approaches to accomplish this task: linear elastic and nonlinear methods. Even though there is a broad agreement that the nonlinear based procedures are better tools for the implementation of the performance based seismic engineering, the linear elastic methods are and will continue to be used due to their simplicity in modeling and analysis. Independent from the approach, seismic performance of the building is defined to be related with the probable damage in individual structural members subjected and evaluated under the effect of the earthquake load and is classified considering damage regions such as:

- Minimum,
- Advanced,
- Extensive, and
- Collapse.

Thus, as the first step of this research, linear elastic procedure is applied for the entire building ensemble and structural performances throughout the entire structural elements are investigated subsequent to sectional performances by considering the targeted performance level.

Secondly, each building is investigated in details by means of nonlinear based method. The aim of this method to be used for determining the structural performances of the buildings under seismic effect is enabling the measurement of the plastic deformation demands regarding the ductile behavior and internal force demands concerning the brittle behavior under a given earthquake. Afterwards, the demands are compared with the element capacities and structural performance evaluation shall be conducted both at sectional and building level. In this sense, nonlinear dynamic analysis procedure is utilized under the effect of the recorded strong ground motion during 1998 Ceyhan Earthquake and seismic performance assessment of the building ensemble is carried out.

As a final analytical step, the building ensemble is investigated regarding the guidelines of Appendix-2 of the Application Regulation of the law no: 6306 by Republic of Turkey, Ministry of Environment and Urbanization (RBDG-2012). By utilizing this method, it is not aimed to figure out the structural performance of the stock but it is targeted to find out the risky ones which are expected to suffer probable heavy damage or collapse under the effect of the earthquakes.

When an efficient, fast and step by step assessment method to figure out the seismic performance of an existing building similar to the previous procedure is required, rather simplified analysis techniques, such as 3-staged Japanese seismic index method (2001) or equivalents can be used. By performing the first step of this method, it is not expected to find out exactly which buildings are structurally resistant or not under the effect of an expected earthquake, but it is aimed to point out which buildings are rather reliable and can be kept in service under its current condition and which ones need to be evaluated more detailed within the near future. Therefore, the building ensemble is also inspected by applying the first stage Japanese seismic index method.

Consequently, the analytical results concerned with the aforementioned procedures are comparatively discussed also by taking into account the site-observed damage status of the representative building stock.

STRUCTURAL PERFORMANCE EVALUATION PROCEDURES

According to the regulations of the TEC, seismic performance of buildings is concerned with the expected damage potential of the building under the effect of the earthquake loads and identified on the basis of four different damage stages. Seismic performance level of a building can be designated either by applying linear elastic or nonlinear static and dynamic analysis procedures and making a decision on structural elements' damage regions. Primarily, the acceleration spectrum to be used can be determined according to Table 1 regarding the target performance level of the inspected building for each procedure. For the reason that this study focuses on residential type of buildings, investigations are performed for

Table 1. Target Performance Levels of Buildings under the Effect of Different Earthquakes

The Purpose of Usage and Type of the Building	Probability of Exceedance		
	50% in 50 Years	10% in 50 Years	2% in 50 Years
The buildings that should be used after earthquakes	-	IO	LS
The buildings that people stay in for a long time period	-	IO	LS
The buildings that people visit densely and stay in for a short time period	IO	LS	-
The buildings that contain hazardous materials	-	IO	CP
Other buildings (Residential, etc.)	-	LS	-

*IO: Immediate Occupancy; LS: Life Safety; CP: Collapse Prevention

'Life Safety' performance level and internal force and displacement demands are calculated under the effect of the design spectrum (Contingency Level Earthquake) defined in the TEC, which has a probability of exceedance of 10% within 50 years.

Linear Elastic Procedure

Linear elastic procedure consists of three basic steps:

- Structural analysis,
- Determination of capacity and
- Demand and evaluation of structural performance.

In case of investigating the existing buildings, dissimilarly from to-be-built buildings, efficient bending stiffnesses $(EI)_e$ of the cracked sections shall be used for RC components under the effect of bending. According to the TEC, while a coefficient is indicated for columns and shear walls which is related with the axial force and cross-sectional properties, a constant reduction factor is defined for beams that are under negligible axial forces (Equations 1 & 2). In this equation, $(EI)_0$ is the bending stiffness of the uncracked section; N_D is the axial compression force; A_c is the gross section area of the column or shear wall; f_{cm} is the concrete compressive strength.

For beams:

$$(EI)_e = 0.40(EI)_0 \tag{1}$$

For columns and shear walls:

$$\begin{aligned} N_D / (A_c f_{cm}) &\leq 0.10 \;\Rightarrow\; (EI)_e = 0.40(EI)_0 \\ N_D / (A_c f_{cm}) &\geq 0.40 \;\Rightarrow\; (EI)_e = 0.80(EI)_0 \end{aligned} \tag{2}$$

According to the procedure, in order that beams, columns and shear walls to be considered as ductile elements, V_e shear force calculated in accordance with the bending capacity in critical sections should not exceed the shear capacity V_r which is calculated according to TS-500: Requirements for Design and Construction of RC Structures (2000). The structural elements are classified as 'brittle' in case they do not satisfy this ductility requirement. After defining the fracture type, demand/capacity ratios (r) have to be established for beams, columns and shear walls in order to calculate the damage levels of structural elements numerically. Demand/capacity ratios of structural elements can be determined by dividing the internal forces calculated under seismic excitation to the excess capacity of the section considering the structural behavior factor as unity (R_a=1). The excess capacity of the section is the difference between total capacity of the section and the internal forces calculated under gravity loads. In order to make a decision for the damage zone of the elements, a comparison of the values of demand/capacity ratios with limiting values given in the TEC according to element type has to be done. After defining the damage rank for each element, structural performance throughout the building can be inspected considering the target performance level.

Nonlinear Dynamic Analysis Procedure

The aim of nonlinear procedures to be used in determining the structural performances of the buildings under seismic effect is enabling the measurement of the plastic deformation demands regarding the ductile behavior and internal force demands concerning the brittle behavior for a given earthquake. Afterwards, the demands are compared with the deformation and internal force capacities that are defined in the TEC and structural performance evaluation shall be conducted both at sectional and building level.

The nonlinear analysis methods included in the scope of this regulation are:

- Static Incremental Equivalent Seismic Load Method,
- Incremental Mode Combination Method and
- Dynamic Time History Analysis Method.

During the initial step of each method, structural model of the building should be analyzed under $G+nQ$ loading and deformation, plastic rotation and internal force demands should be calculated. The plastic curvature demand depending on the plastic rotation demand shall be obtained in any section using Equation 3, in which L_p is the length of the plastic deformation zone that is generally accepted to be equal to the half of the section length (h) in the bending direction:

$$\varphi_p = \frac{\theta_p}{L_p} \tag{3}$$

The total curvature demand ϕ_t of the section shall be obtained by adding the equivalent yield curvature ϕ_y to the ϕ_p plastic curvature demand. ϕ_y can be defined with the bilinear moment-curvature relationship obtained from the analysis conducted under the axial force demand of the section by means of using an appropriate reinforcing steel model that considers the strain hardening together with a concrete model chosen in accordance with the aim. Afterwards, the strain demands of concrete and reinforcing bars shall be obtained by utilizing moment-curvature analysis in accordance with the predefined total curvature demand. Finally, these demands will be compared with the strain capacities defined in the TEC so that determination of the performance of the structural systems at each section level can be realized. On the basis of elements' evaluation, structural performance can be assessed according to the target performance level throughout the building.

The Principles of the Risky Buildings Detection Guideline (RBDG-2012)

Appendix-2 of the Application Regulation of the law no: 6306 by the Republic of Turkey, Ministry of Environment and Urbanization includes the principals and the methodology of the detection of risky buildings under the effect of earthquake excitation. In other words, this method is not applicable to figure out exactly which buildings are structurally resistant or not under the effect of an expected destructive earthquake. According to RBDG, the building will be defined as 'Risky' if it has the potential of being heavily damaged or collapse under the effect of the design earthquake (CLE) defined for the region. The analysis for determining the risky buildings shall be carried out for the critical story of the building by considering the existing structural system characteristics and utilizing linear elastic analysis procedure.

Elastic design spectrum defined in the TEC shall be applied as the input earthquake effect and the risk status of the building will be determined under $G+nQ+E$ loading, utilizing the analysis methods regulated in the code. Effective bending stiffnesses $(EI)_e$ of the cracked sections shall be used for RC components under the effect of bending, where two different constant reduction factors are defined for beams/shear walls and columns as given in Equations 4 and 5. In this equation, E_{cm} is the modulus of elasticity of concrete with the unit MPa and can be obtained by Equation 6.

Beams and Shear Walls:

$$(EI)_e = 0.30(E_{cm}I)_0 \tag{4}$$

Columns:

$$(EI)_e = 0.50(E_{cm}I)_0 \tag{5}$$

$$E_{cm} = 5000\left(f_{cm}\right)^{0.5} \tag{6}$$

During the evaluation, columns will be classified into 3 groups according to their shear force under earthquake loads to shear resistance ratios (V_e/V_r) and the transverse reinforcement detailing at confinement zones. A, B and C groups represent the elements subjected to flexural failure; flexural + shear failure and shear failure, respectively. Shear walls will be classified in two groups namely, brittle and ductile, depending on V_e/V_r and their height H_w to planar length l_w ratios.

According to the procedure, demand/capacity ratio ($m=M_{G+nQ+E}/M_K$) will be considered during decision making about the element's performance. Here M_{G+nQ+E} is the internal bending moment of a section subjected to the combined effects of gravitational and seismic loads. M_K will be computed by considering the interaction of the axial force N_K, which is calculated from the reduced load case $G+nQ\pm E/6$. m values of columns and shear walls remaining at the investigated critical story and drift ratio demands will be compared with the provided limit values in the guideline depending on the classification of structural element group.

As the final stage of the evaluation, axial compression stresses will be determined for the critical story under $G+nQ$ loading. In the case of obtaining the mean value of axial compression stresses to be greater than $0.6f_{cm}$, the building will be accepted as 'Risky' as if any of the columns or shear walls risk limit is exceeded. See Table 2.

Table 2. Limit values of Story Shear FoRCe Ratio according to RBDG-2012

Mean Axial Stress for Walls and Columns (=Total Stress for Walls and Columns / Number of Walls and Columns)	Limit Values of Story Shear Force Ratio
$\geq 0.6 f_{cm}$	0
$0.1 f_{cm} \geq$	0.30

Source: (RBDG, 2012)

Japanese Seismic Index Method (2001)

This standard is based on both site inspection and structural analysis to represent the seismic performance of a building in terms of seismic index of structure I_S and seismic index of non-structural elements I_N. Method can be applied during the seismic evaluation and the verification of seismic retrofitting of existing low-rise and mid-rise RC buildings. Three levels of screening procedures have been prepared in the standard and any level of the screening procedure may be used in accordance with the purpose of evaluation and the structural characteristics of the building.

According to the standard, first level inspection method should be conducted on the following investigation items, which are basically necessary for the calculation of the seismic index of structure:

1. Material strengths and cross-sectional dimensions for calculation of the strengths of structural members.
2. Crackings in concrete and deformations of structure for the evaluation of time index.
3. Building configuration for the evaluation of irregularity index.

The seismic index of structure I_S shall be calculated by Equation 7 at each story and in each principal horizontal direction of the building. The irregularity index S_D in the first screening and the time index T may be used commonly for all stories and directions.

$$I_S = E_0 S_D T \tag{7}$$

Here:

E_0= Basic seismic index of the structure.
S_D= Irregularity index.
T= Time index.

The basic seismic index of the structure E_0, which is to evaluate the basic seismic performance of the building by assuming other sub-indices as unity, shall be calculated for each story and each direction based on the ultimate strength, failure mode and ductility of the building.

Seismic safety of a building shall be judged by Equation 8:

$$I_S \geq I_{SO} \tag{8}$$

where:

I_S: Seismic index of structure.
I_{SO}: Seismic demand index of structure.

I_{SO} should be calculated by Equation 9 regardless of the story in the building:

$$I_{SO} = E_S \times Z \times G \times U \tag{9}$$

where:

E_S: Basic seismic demand index of structure (=0.8 for first level screening).
Z: Seismic Zone index.
G: Ground index.
U: Usage index.

If Equation 8 is satisfied, the building may be assessed to be 'Safe: The building possesses the seismic capacity required against the expected earthquake excitations'. Otherwise, the building should be assessed to be 'Uncertain' in terms of seismic safety.

Herein this study, first level inspection method is conducted regarding the critical story of each building and results are compared and discussed with nonlinear time history analysis findings.

INVESTIGATED RC BUILDINGS AND EARTHQUAKE ENSEMBLE

Structural Properties of the Building Stock

A set of 15 existing residential RC buildings, which have experienced moderate damage during Ceyhan Earthquake occurred in Turkey in 1998 are considered herein this paper. In the entire set number of stories for the buildings varies from 3-story to 5-story, therefore the set consists of low- to mid-rise structures that have relatively limited torsional irregularity in plan. All buildings with frame structural systems are investigated in details by means of structural material quality, reinforcement amount and detailing of bars and local site conditions. Concrete class is found to be varying from C7~C20 (f_{ck}=7.0~20.0 MPa), while the reinforcing steel is S220 (f_{yk}=220 MPa) class except for one 5-story building. Below Table 3 tabulates some characteristics of the buildings, where $\eta_{bi,max}$ is the coefficient of torsional irregularity and W is the total weight of the building.

Strong Ground Motion to be Considered

Ceyhan district of Adana province experienced a destructive earthquake of M_w 6.2 on June 27, 1998 with a maximum ground acceleration value of 0.28g. Although the focal distance to Ceyhan was 32 km, fault rupture was directed towards Ceyhan, causing heavy damage in the district because of the directivity effect (Tezcan & Boduroglu, 1998). Sucuoglu et al, 2000 indicated that there were a total number of 146 casualties; 12 RC buildings were collapsed and 120 were classified as moderately damaged in Ceyhan. On the other hand, the moderately damaged building ensemble considered herein this research settles in Yüreğir district which is 43 km far away from epicenter and 9km far away from the recording station. Figure 1 illustrates the recorded strong ground motions of Ceyhan Earthquake for each horizontal component and Figure 2 represents the comparison of elastic response spectra of EW component with the design spectra given in the TEC for local soil class of *Z2* and effective ground acceleration coefficient of A_0=0.3g.

Table 3. Structural Characteristics of the Inspected Buildings

Building Code	No of Stories	h_{Total} (m)	f_{ck} (Mpa)	f_{yk} (Mpa)	W (kN)	$\eta_{bi,max}$
L11	3	8.25	16.0	220	3950.7	1.14
P20	3	8.85	8.6	220	7676.8	1.03
P22	3	9.65	7.0	220	5151.4	1.07
P23	3	8.70	8.3	220	3872.7	1.10
P30	3	8.55	11.9	220	4009.2	1.04
P21	4	12.10	10.0	220	6382.8	1.13
P25	4	11.05	12.0	220	5890.4	1.21
P51	4	12.25	13.1	220	5731.8	1.25
SE03	4	13.30	12.3	220	8198.4	1.10
SL13	4	12.80	14.5	220	14147.1	1.56
CM5	5	14.75	20.0	420	19050.6	1.04
P06	5	13.75	13.3	220	8104.4	1.02
P62	5	14.75	10.0	220	15680.1	1.12
P79	5	13.75	11.5	220	8104.4	1.02
SE05	5	14.25	8.2	220	10587.2	1.23

Figure 1. Ground motion records of Ceyhan Earthquake

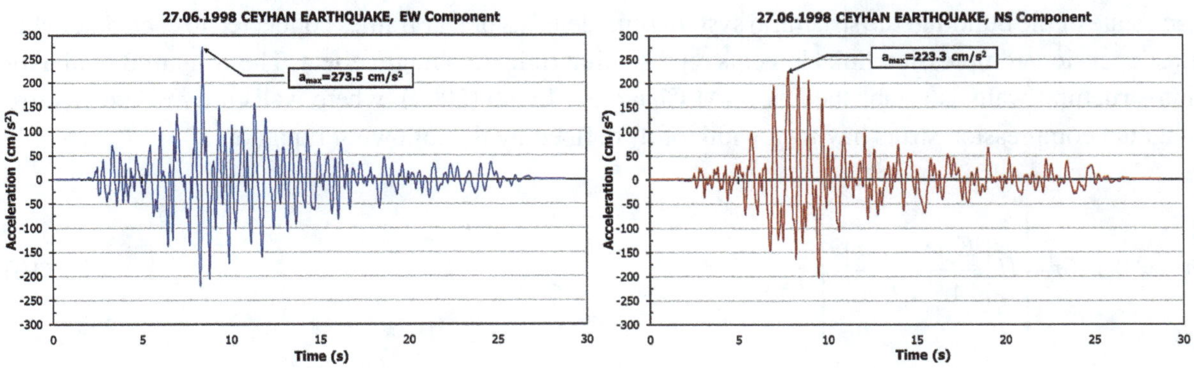

While performing nonlinear dynamic method, planar models of structures in the building ensemble are analyzed under the effect of EW component of Ceyhan Earthquake which has relatively maximum ground acceleration with a value of $0.28g$.

Analytical Modeling of the Structural Systems

Buildings are modeled utilizing DRAIN-2DX, however the modified version by Ascheim (2005) is preferred, which is capable of handling the stiffness and strength degradation through Takeda hysteretic model. Nonlinear behavior of columns and beams is assumed to be represented with a stiffness degrading

Figure 2. Comparison of elastic response spectra of each horizontal component and design spectra in the TEC.

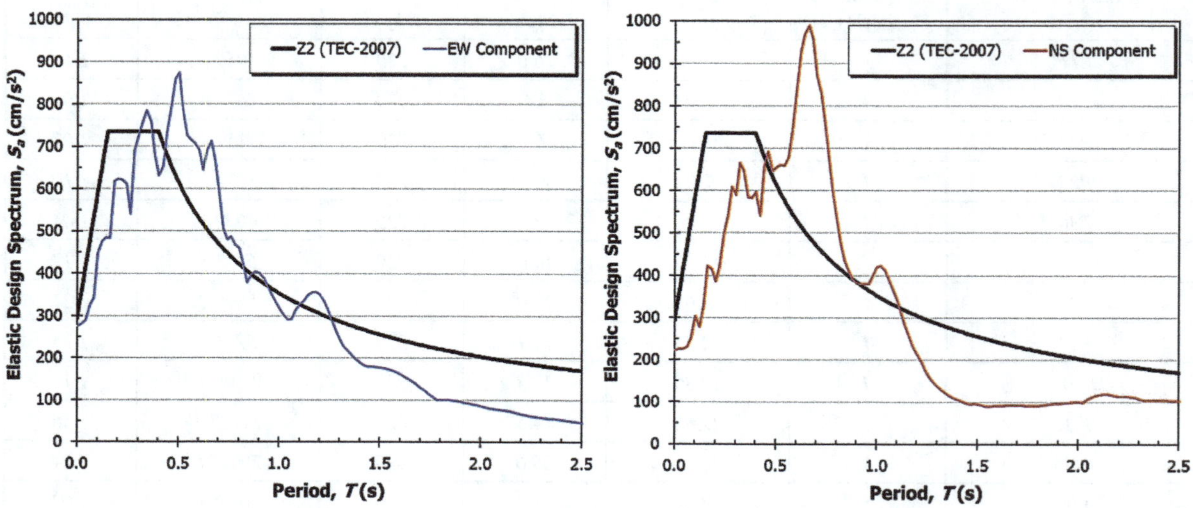

hysteresis model. A single earth-retaining basement wall in one of the buildings is modeled considering the pinching effects. For the entire building stock, a structural damping of 5% and a strain-hardening of 3% are taken into account.

Assigning element type-7 for stiffness-degrading elements and type-9 for the non-structural and reinforced concrete walls, the analytical model for DRAIN-2DX computer program is established. For each state of the building, the structural system is modeled as planar frames connected to each other with elastic tension/compression link elements representing rigid diaphragm effect. The structural model for non-structural walls is established as in Al-Chaar and Lamb (2002), where walls are modeled as two diagonal compression struts having a width of a_w defined by the following Equation 10:

$$a_w = 0.175 r_w \left\{ H \left(\frac{E_w t \sin 2\phi}{4 E_c I_c h_w} \right)^{\frac{1}{4}} \right\}^{-0.4} \quad (10)$$

Here t is the thickness; r_w is the diagonal length and ϕ is the diagonal slope angle of the wall. Considering the wall height as h_w; story height H; the moment of inertia of the neighboring columns as I_c and the modulus of elasticity for infill walls E_w=1000MPa and concrete E_c varies between 22600~28500MPa, strut widths a_w are computed for each non-structural wall. Figure 3 shows the modeling of infill walls.

Figure 4 shows the backbone curves for the hysteretic relations employed to the structural and non-structural elements during the nonlinear dynamic analysis. In the left figure, M_y and M_u represent the yield and ultimate moments and χ_y and χ_u are the corresponding curvatures. k_1 is the initial stiffness; k_2 and k_3 are the post-yield stiffnesses in the positive and negative directions where α is the strain hardening ratio and k_4 is the unloading stiffness. In the model for the infill walls, k_1 represents the axial stiffness of the uncracked section, u_1 and u_2 are the cracking and ultimate deformations, respectively.

Figure 3. Modeling of infill walls

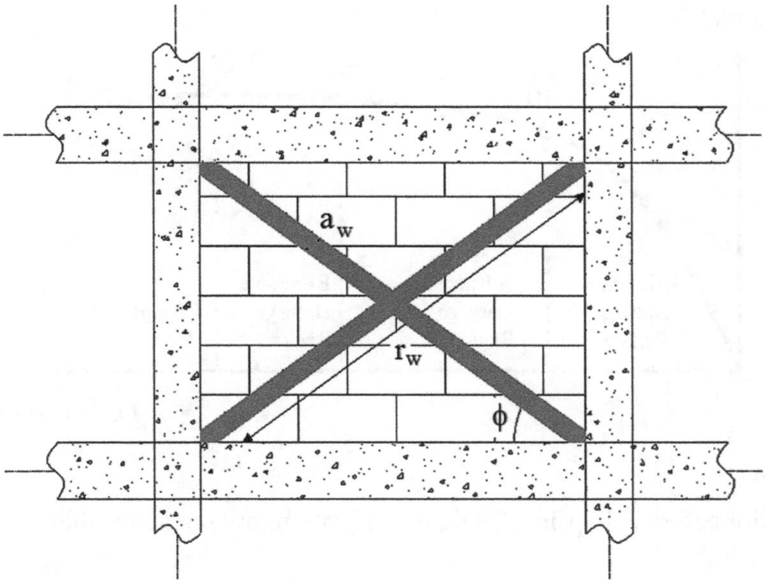

Figure 4. Hysteresis curves for structural (left) and non-structural elements (right)

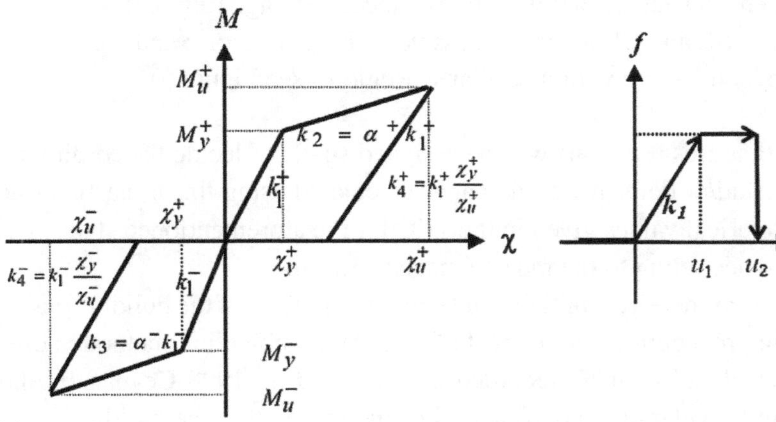

PERFORMANCE ASSESSMENT OF BUILDING ENSEMBLE

In the TEC, three limit states have been defined for ductile elements at section level which are:

- Minimum Damage State (MD),
- Safety Limit (SL), and
- Collapse Limit (CL) (see Figure 5).

MD state defines the beginning of the behavior beyond yielding; SL defines the limit of the behavior beyond yielding that the section is capable of safely ensuring the strength and CL defines the limit of

Figure 5. Section damage boundaries and regions defined in the TEC

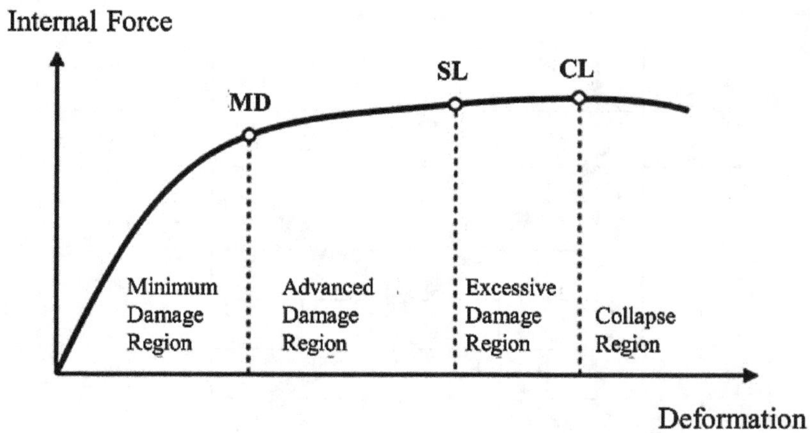

the structural behavior before collapsing. These classifications are not applicable for elements damaged in a brittle condition.

- Elements that the damages at critical sections do not reach MD are within the 'Minimum Damage Region';
- Those in-between MD and SL are within 'Advanced Damage Region';
- Those in-between SL and CL are in 'Excessive Damage Region'; and
- Those going beyond CL are within 'Collapse Region' (see Figure 5).

Damage regions that cross sections will be assigned shall be decided according to the comparison of the internal forces and/or deformation demands calculated using linear elastic or nonlinear analysis methods with the numerical values given in the TEC for the aforementioned states. Damage of the element shall be decided according to the most damaged section.

Within the scope of this research, initially, 3D structural models of the building ensemble are assessed utilizing linear elastic procedure given in the TEC. Modal combination method is used in the analyses and the maximum acceleration value recorded as 0.28*g* during 1998 Ceyhan Earthquake is defined as the effective ground acceleration coefficient. For the reason that the building ensemble consists of residential buildings, 'Life Safety' performance level is targeted during the analyses under the effect of the earthquake which has the probability of exceedance of 10% within 50 years. According to the results of linear elastic procedure, performance evaluation is achieved at section and building levels of all structures in the ensemble for each direction.

As the second step, internal force and deformation demands of each structural element in each building within the ensemble are obtained utilizing nonlinear time history procedure under the effect of EW component of Ceyhan Earthquake. Detailed investigations are performed for time histories of structural demands of column and beam elements that remain at the critical story. Total curvature demands are determined depending upon the absolute maximum value of the total rotation demands. Strains for concrete and reinforcing steel corresponding to total curvature for the section are obtained and finally these demands are compared with the limit values given in the code for the considered performance level in order to find out the performance status of the elements.

As a final analytical work, risk assessments of the building ensemble are performed utilizing the RBDG procedure (methodology). Similar to the linear analysis procedure given in the TEC, modal combination method is used in the analyses and 0.28*g* of maximum ground acceleration value is taken into account as the effective ground acceleration coefficient. Depending on the calculations in pursuant of RBDG, risky buildings are determined and besides, detailed investigations are performed in order to find out the elements that exceed the risk limit. It is assumed that, these elements that exceed the risk limit remain in collapse region.

Figure 6 illustrates the dominant vibration periods of the buildings in each direction utilizing 3 different analytical procedures. At the left side of Figure 6, the ratio of periods for 2D and 3D models of each building for non-damaged (virgin) state which are calculated by utilizing modal analysis is given. It is observed that, the periods obtained from 2D models derive 85% convergence when compared with 3D model findings. The difference may be revealed in consequence of considering the contribution of non-structural infill walls in 2D models. Furthermore, the right side of Figure 6 represents the ratio of dominant periods for 2D and 3D models of each building for damaged state (cracked section) which are determined during performance/risk evaluations. For this aim, modal analyses are renewed for the building ensemble after nonlinear dynamic analyses and vibration periods for the damaged states of 2D models of structures are recalculated. During the calculations performed for TEC and RBDG in which 3D models are considered, damaged state periods are determined considering the numerical values depending on effective bending stiffness. According to this, nonlinear dynamic analysis results indicate the significant stiffness degradation of buildings' structural systems.

Structural performance evaluations carried out for each directions of the building in the ensemble utilizing linear elastic and nonlinear procedures are given for beams elements in Figure 7. Damage regions illustrated in Figure 7 are the four different damage regions regulated in the TEC.

Damage distribution of beams determined according to nonlinear dynamic analysis procedure indicate lower values when compared with the findings of linear elastic procedure. Since risk assessment

Figure 6. Dominant vibration periods of the buildings according to the aforementioned procedures for virgin (left) and damaged (right) states

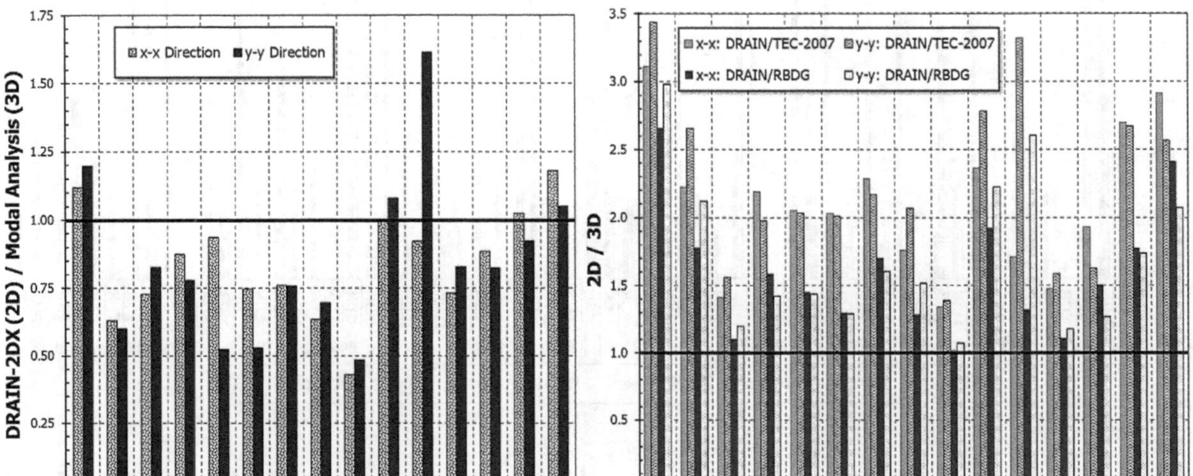

Figure 7. Damage distribution of beams according to the linear elastic and nonlinear dynamic analyses

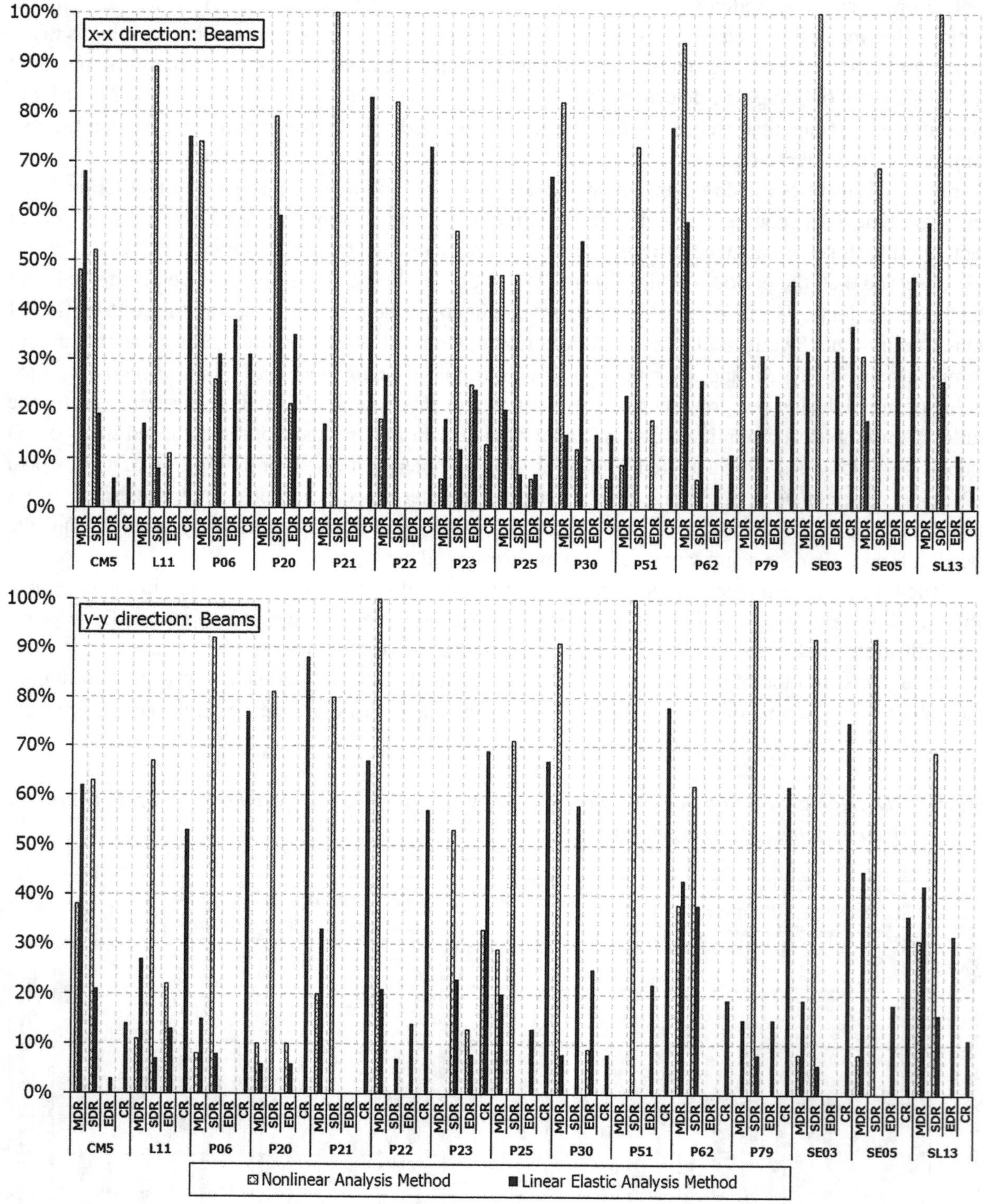

for beams is not required in RBDG, only the results of the two analytical methods within the context of TEC are compared in Figure 7.

A similar investigation is executed for columns according to the structural performance/risk assessment results for each analytical procedure as given in Figure 8. As mentioned before, according to regulations of RBDG, it is assumed that the elements that exceed the risk limit at the critical story will be considered in collapse region. For that reason, the comparison of three procedures in terms of columns is conducted only for elements that remain in this region.

When nonlinear dynamic analysis procedure is utilized, it is observed that a number of 10 building have structural elements in collapse region according to the performance results of columns in x-x direc-

Figure 8. Damage distribution of columns according to different analysis procedures

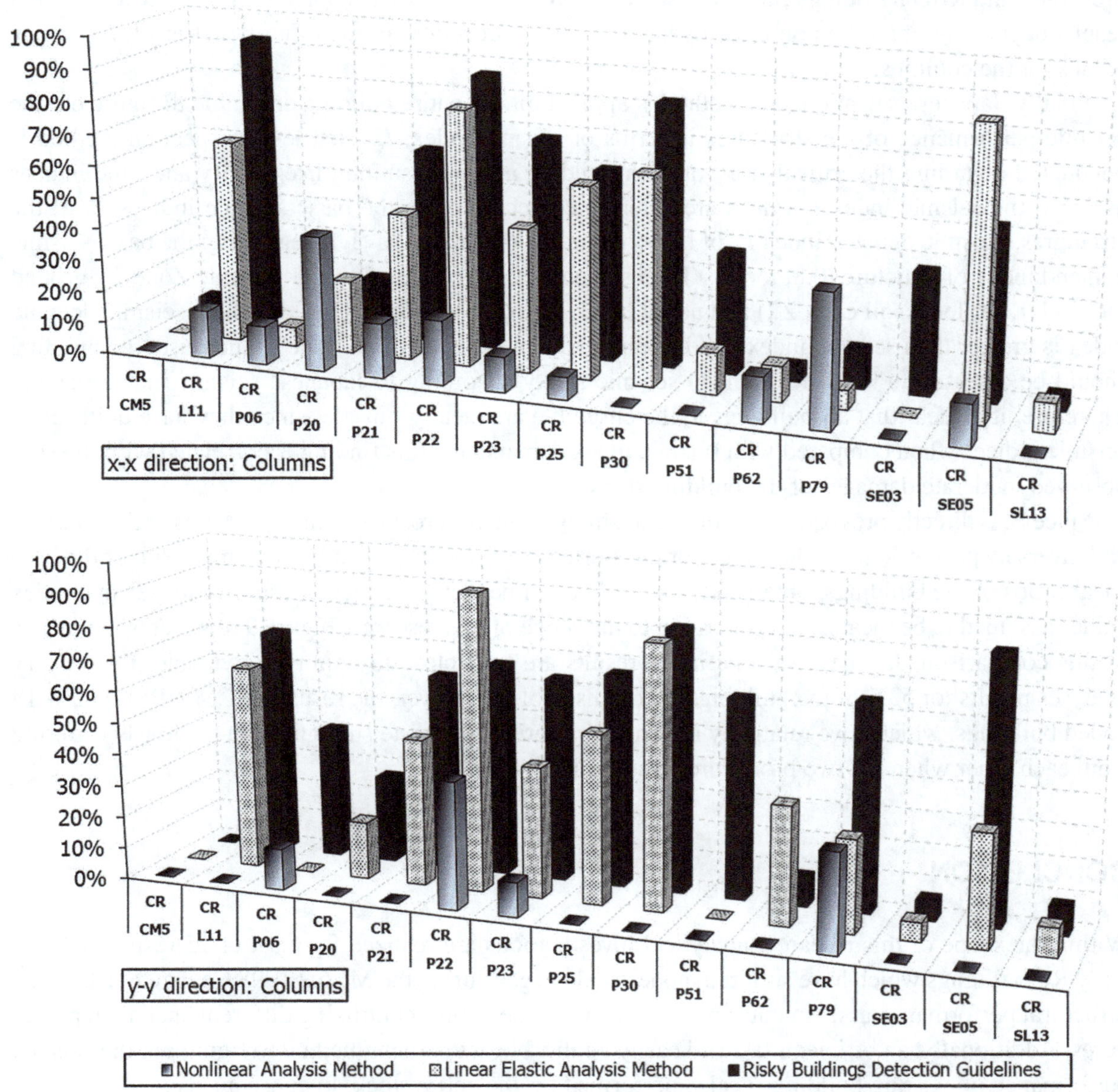

tion. For the calculations in the same direction, linear elastic procedure and RBDG come up with the results of 13 and 14 buildings, respectively. In case of performing such an investigation for y-y direction, the number of buildings that have collapsed columns are 4, 12 and 14 according to nonlinear dynamic procedure, linear elastic method and RBDG, respectively.

Even though the unfavorable evaluation results seem to be obtained from the findings of RBDG at a quick glance, the restriction regulated in the TEC for the ratio of the total shear force of the columns to story shear force, which is numerically 20%, has to be taken into consideration. In this sense, shear force ratios of the columns, which are tabulated in Table 4, are investigated in details for the purpose of comparing the three procedures accurately. When the linear elastic procedure results are examined, the shear force ratio of the columns in excessive damage region exceeds 20% limitation for 9 of the buildings in both *x-x* and *y-y* directions. Therefore, if these columns that are exposed to shear force over 20% are assumed to be collapsed regarding the restriction given in the TEC, in this instance linear elastic procedure and RBDG findings can be considered as consonant with each other. Performance assessment results depending upon the nonlinear dynamic analysis procedure present the lowest level of damage values for the columns.

Finally, Japanese seismic index method is applied for the entire building ensemble to figure out the seismic performance of the structures in terms of seismic index, I_S. First level inspection method is conducted regarding the critical story of each building and by assuming irregularity and time indices as unity, the seismic index I_S values are determined according to the basic seismic indices E_0 of the structures. Seismic demand index I_{S0} of the structures is calculated as '1.1' regarding: the basic seismic demand index of structure (0.8, for first level screening); zone index (1.0, for Seismic Zone-1); ground index (1.1, for local soil class Z2) and usage index (1.25, for residential buildings). If seismic demand index is greater than seismic index, the building may be assessed to be 'Safe'; otherwise, the building should be assessed to be 'Uncertain' in seismic safety according to Japanese seismic index method. Therefore, it is clear that all buildings in the ensemble are clearly 'unsafe' since they have quite small seismic indices when compared with seismic index demand and this finding is correlated with the site-observed moderate damages of the building stock.

Since E_0 is directly proportional to the shear strength and the cross-section dimensions of the columns and inversely proportional to the weight of the stories above the inspected level; comparison of the base shear ratios of the buildings, obtained from nonlinear time history analyses results, with seismic index values assumed to be meaningful according to the scope of this research. Figure 9 represents the results of this comparison. Japanese seismic index results are very close to nonlinear dynamic time history analyses results for 53% of the buildings in the ensemble. However, the results of CM5, P51 and SL13 coded buildings, which have relatively higher characteristic concrete strengths, are obviously differing from each other when the two procedures results are compared.

CONCLUSION

Within the scope of this research, analytical investigations are realized for a set of 15 residential 3~5 story RC buildings which have suffered moderate damages during the M_w 6.2 1998 Ceyhan Earthquake. Structural performance assessments are carried out for the entire set utilizing different analytical procedures. Additionally, an efficient, fast and easily applicable assessment method to figure out the seismic performance of existing buildings is also performed for the entire stock. In consequence:

Table 4. Shear foRCe ratios of the columns which are in EDR and CR at critical story according to the different evaluation procedures

Building	**Damage Region**	**Shear FoRCe Ratio of the Damaged Columns (%)**					
		Linear		**RBDG**		**Nonlinear**	
		X-X	*Y-Y*	*X-X*	*Y-Y*	*X-X*	*Y-Y*
L11	*EDR*	8%	15%			11%	12%
	CR	87%	85%	98%	88%	30%	-
P20	*EDR*	**21%**	**58%**			4%	9%
	CR	59%	32%	43%	51%	53%	-
P22	*EDR*	5%	3%			16%	42%
	CR	95%	97%	95%	80%	13%	26%
P23	*EDR*	**34%**	**34%**			3%	6%
	CR	54%	61%	86%	85%	20%	13%
P30	*EDR*	13%	9%			-	-
	CR	87%	91%	86%	90%	-	-
P21	*EDR*	**36%**	**28%**			28%	-
	CR	57%	67%	71%	69%	15%	-
P25	*EDR*	**25%**	**22%**			-	26%
	CR	72%	69%	73%	84%	4%	-
P51	*EDR*	**37%**	**38%**			-	-
	CR	22%	6%	62%	75%	-	-
SE03	*EDR*	**46%**	**26%**			-	-
	CR	-	32%	50%	31%	-	-
SL13	*EDR*	-	-			-	-
	CR	18%	11%	6%	7%	-	-
CM5	*EDR*	-	-			-	-
	CR	-	-	7%	-	-	-
P06	*EDR*	**34%**	**73%**			5%	-
	CR	42%	-	45%	48%	10%	18%
P62	*EDR*	**24%**	-			35%	17%
	CR	12%	38%	46%	7%	23%	-
P79	*EDR*	**45%**	**37%**			-	20%
	CR	42%	41%	45%	76%	36%	35%
SE05	*EDR*	-	**54%**			58%	-
	CR	98%	44%	63%	88%	26%	-

1. Linear elastic procedure described in the 2007 TEC;
2. Nonlinear dynamic analysis procedure under the effect of the recorded strong ground motion during 1998 Ceyhan Earthquake;

Figure 9. Comparison of the seismic indices with shear foRCe ratios

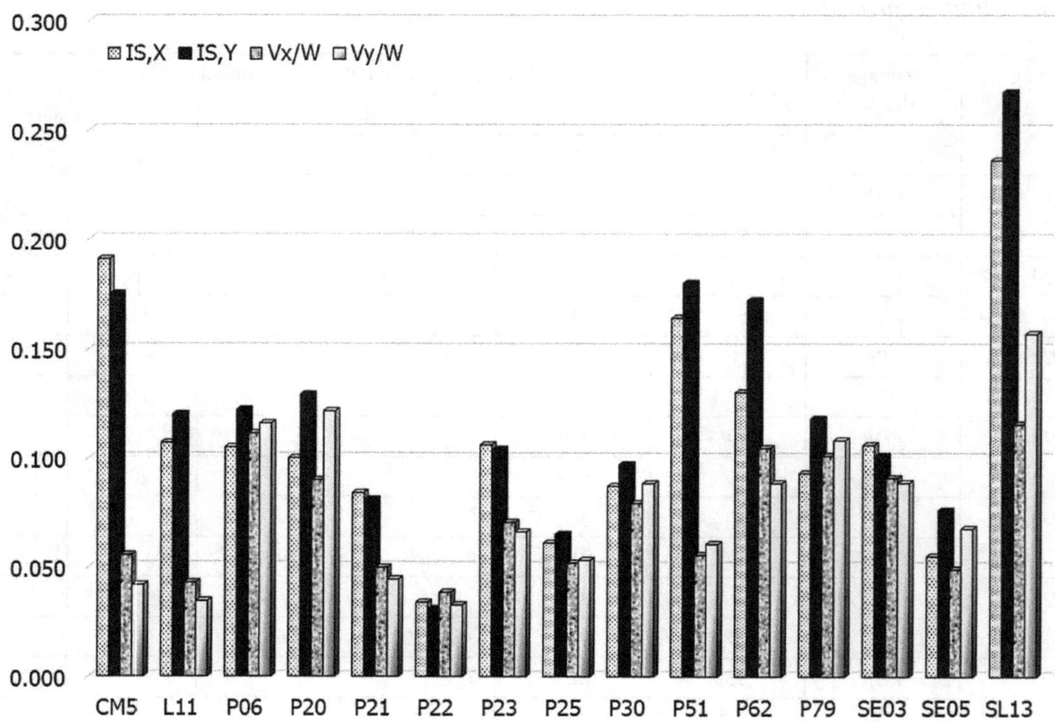

3. Appendix-2 of the Application Regulation of the law no: 6306 by Republic of Turkey, Ministry of Environment and Urbanization (RBDG); and
4. Japanese seismic index method findings are comparatively discussed also by taking into account the site-observed damage status of the representative building stock.

The outcomes of the study in the light of the evaluations are summarized below.

- The periods obtained from 2D and 3D models differ with a level of 15% and it is argued that the difference may be revealed in consequence of considering the contribution of non-structural infill walls in 2D models. Nevertheless, it is a known fact that the non-structural infill walls will be disengaged in 8~10 seconds during dynamic loading and it is presumed that in this circumstances significant discrepancy will not occur during structural performance assessment. The damaged state periods of the buildings which are recalculated after earthquake loading are found out to be longer when compared with the periods calculated by considering the numerical values depending on effective bending stiffness according to RBDG and TEC regulations.
- Damage distribution of beams determined according to nonlinear dynamic analysis procedure indicate lower values when compared with the findings of linear elastic procedure. Similar result is also valid for columns.
- The comparison of the three procedures in terms of columns is conducted only for elements that remain in collapse region. Performance assessment results depending upon the nonlinear dynamic

analysis procedure present the lowest level of damage values for the columns in terms of the ratio of columns in collapse region to the total number of columns in the story, while RBDG presents the highest ratios.

- When nonlinear dynamic analysis procedure is utilized, it is observed that a number of 10 building have structural elements in collapse region according to the performance results of columns in *x-x* direction. For the calculations in the same direction, linear elastic procedure and RBDG come up with the results of 13 and 14 buildings, respectively. In case of performing such an investigation for *y-y* direction, the number of buildings that have collapsed columns are 4, 12 and 14 according to nonlinear dynamic procedure, linear elastic method and RBDG, respectively.
- On the other hand, TEC regulates a restriction for the ratio of total shear force of the columns to story shear force, which is numerically 20%, in case of 'Life Safety' performance level target. Therefore, even though the unfavorable evaluation results seem to be obtained from the findings of RBDG at first sight, it has to be considered that the shear force ratio of the columns in excessive damage region exceeds 20% limitation for 9 of the buildings in both *x-x* and *y-y* directions according to RBDG. In this in-stance, linear elastic procedure and RBDG findings can be considered as consonant with each other.
- Nonlinear dynamic time history analyses results are observed to be very close to Japanese seismic index results for 53% of the buildings in the ensemble. However, the results of CM5, P51 and SL13 coded buildings, which have relatively higher characteristic concrete strengths, are obviously differing from each other.

ACKNOWLEDGMENT

Authors would like to thank Mr. İsmail Hakkı Besler and Mr. Faruk Saka of İdeYAPI Inc. for donating the structural analysis program ideCAD. Their support is sincerely acknowledged.

REFERENCES

Al-Chaar, G., & Lamb, G. (2002). Design of fiber-reinfoRCed polymer materials for seismic rehabilitation of infilled concrete structures. ERDC/CERL TR-02-33. US Army Corps of Engineers.

Ascheim, M. (2005). Retrieved from http://nisee.berkeley.edu/software/drain2dx

Prakash, V., Powell, G. H., & Campbell, S. (1993). *DRAIN-2DX base program description and user guide, SEMM Report, UCB/SEMM-93/17*. University of California-Berkeley.

RBDG. (2012). *Ministry of Environment and Urbanization: The guidelines regarding the detection of risky buildings, Republic of Turkey*. Ankara, Turkey: Council of Ministers. (in Turkish)

Standard for Seismic Evaluation of Existing ReinfoRCed Concrete Buildings. (2001). The Japan Building Disaster Prevention Association.

Sucuoğlu, H., Gür, T., & Gülkan, P. (2000). The Adana-Ceyhan Earthquake of 27 June 1998: Seis-mic retrofit of 120 RC buildings. *Proceedings of 12th World Conference on Earthquake Engineering*.

TS500. (2000). *Requirements for Design and Construction of RC Structures*. Ankara, Turkey: Turkish Standards Institute.

TEC. (2007). *Turkish Earthquake Resistant Design Code*. Ankara, Turkey: Ministry of Publicworks and Settle-ment.

Tezcan, S. S., & Boduroglu, H. (1998). *A reconnaissance report- The June 27, 1998 Adana-Ceyhan Earthquake*. Turkish Earthquake Foundation. Retrieved from www.depremvakfi.org

Compilation of References

Abarbanel, H. D. I., Brown, R., & Kadtke, J. B. (1990). Prediction in chaotic nonlinear systems: Methods for time series with broadband Fourier spectra. *Physical Review A.*, *41*, 1742–1807. PMID:9903289

Abrahamson, N. A. (1992). Non-stationary spectral matching. *Seismological Research Letters, 63*(1), 30.

Abrahamson, N. A., & W. J. Silva (1997). Empirical Response Spectral Attenuation Relations for Shallow Crustal Earthquakes. *Seismological Research Letters, 68*(1).

Adachi, T., & Ellingwood, B. R. (2008). Serviceability of earthquake-damaged water systems: Effects of electrical power availability and power backup systems on system vulnerability. *Reliability Engineering & System Safety*, *93*(1), 78–88. doi:10.1016/j.ress.2006.10.014

Adekristi, A., & Matthew, R. E. (2015). Time-Domain Spectral Matching of Earthquake Ground Motions using Broyden Updating. *Journal of Earthquake Engineering*, 1–20.

Aki, K. (1968). Seismic displacements near a fault. *Journal of Geophysical Research*, *73*(16), 5359–5376. doi:10.1029/JB073i016p05359

Al Atik, L., & Norman A. (2010). An improved method for nonstationary spectral matching. *Earthquake Spectra, 26*(3), 601-617.

Al-Chaar, G., & Lamb, G. (2002). Design of fiber-reinfoRCed polymer materials for seismic rehabilitation of infilled concrete structures. ERDC/CERL TR-02-33. US Army Corps of Engineers.

Alderson, D. L., Gerald, G. B., & Matthew, W. (2014). Assessing and improving operational resilience of critical infrastructures and other systems. *Stat*, *745*, 70.

Ambraseys, N., & Douglas, J. (2000). *Reappraisal of the effect of vertical ground motions on response*. ESEE Report No. 00-4. Imperial College, Civil and Environmental Engineering Dept., London.

American Society of Civil Engineers. (2010). *Minimum design loads for buildings and other structures*. Amer Society of Civil Engineers.

Annual Energy Outlook, Energy Information Administration. (2013). Retrieved from http://www.eia.gov/forecasts/aeo/archive.cfm

Antoniou, S., & Pinho, R. (2004a). Advantages and limitations of adaptive and non adaptive force-based pushover Procedures. *Journal of Earthquake Engineering*, *8*(4), 497–522. doi:10.1080/13632460409350498

Antoniou, S., & Pinho, R. (2004b). Development and verification of a displacement-based adaptive pushover Procedure. *Journal of Earthquake Engineering*, *8*(5), 643–661. doi:10.1080/13632460409350504

Antonopoulos, C. P., & Triantafillou, T. C. (2002). Analysis of FRP-strengthened RC beam-column joints. *Journal of Composites for Construction*, *6*(1), 41–51. doi:10.1061/(ASCE)1090-0268(2002)6:1(41)

Apostolakis, E. G., & Lemon, M. D. (2005). A screening methodology for the identification and ranking of infrastructure vulnerabilities due to terrorism. *Risk Analysis*, *25*(2), 361–376. doi:10.1111/j.1539-6924.2005.00595.x PMID:15876210

Arnold, V. I. (1978). *Mathematical methods of classical mechanics*. New York: Springer.

ASCE/SEI 41-06. (2006). *Seismic rehabilitation of existing buildings*. American Society of Civil Engineers.

Ascheim, M. (2005). Retrieved from http://nisee.berkeley.edu/software/drain2dx

Asteris, P. G., & Plevris, V. (Eds.). (2015). *Handbook of Research on Seismic Assessment and Rehabilitation of Historic Structures*. IGI Global; doi:10.4018/978-1-4666-8286-3

ATC (Applied Technology Council). (1996). *Seismic evaluation and retrofit of concrete buildings*. ATC (Report No. ATC-40).

ATC. (1996). *Seismic evaluation and retrofit of concrete buildings*. Report ATC-40. Applied Technology Council.

Atkinson, G. M., & Boore, D. M. (2006). Earthquake ground-motion prediction equations for Eastern North America. *Bulletin of the Seismological Society of America*, *96*(6), 2181–2205. doi:10.1785/0120050245

Aven, T. (2011). On some recent definitions and analysis frameworks for risk, vulnerability, and resilience. *Risk Analysis*, *31*(4), 515–522. doi:10.1111/j.1539-6924.2010.01528.x PMID:21077926

Ay, B. Ö., & Akkar, S. (2014). Evaluation of a recently proposed record selection and scaling procedure for low-rise to mid-rise reinforced concrete buildings and its use for probabilistic risk assessment studies. *Earthquake Engineering & Structural Dynamics*, *43*(6), 889–908. doi:10.1002/eqe.2378

Aydinoğlu, M. N. (2003). An incremental response spectrum analysis procedure based on inelastic spectral displacements for multi-mode seismic performance evaluation. *Bull Earthquake Eng*, *1*, 3-36.

Aydinoğlu, M. N. (2004). *An improved pushover procedure for engineering practice: Incremental response spectrum analysis procedure (IRSA)*. International workshop on performance-based seismic design: concept and implementation, Bled, Slovenia.

Aydinoğlu, M. N. (2007). A response spectrum-based nonlinear assessment tool for practice: incremental response spectrum analysis (IRSA). *ISET Journal of Earthquake Technology*, *44*(1), 169–192.

Ayyub, B. (2013). Systems Resilience for Multihazard Environments: Definition, Metrics, and Valuation for Decision-Making. *Risk Analysis*. doi:10.1111/risa.12093 PMID:23875704

Baker, J. W. (2010). Conditional mean spectrum: Tool for ground-motion selection. *Journal of Structural Engineering*, *137*(3), 322–331. doi:10.1061/(ASCE)ST.1943-541X.0000215

Baker, J. W., & Cornell, C. A. (2006). Spectral shape, epsilon and record selection. *Earthquake Engineering & Structural Dynamics*, *35*(9), 1077–1095. doi:10.1002/eqe.571

Baker, J. W., & Cornell, C. A. (2006b). Correlation of Response Spectral Values for Multicomponent Ground Motions. *Bulletin of the Seismological Society of America*, *96*(1), 215–227. doi:10.1785/0120050060

Baker, J. W., & Jayaram, N. (2008, February). Correlation of Spectral Acceleration Values from NGA Ground Motion Models. *Earthquake Spectra*, *24*(1), 299–317. doi:10.1193/1.2857544

Baros, D. K., & Anagnastopoulos, S. A. (2008) An assessment of static nonlinear pushover analyses in 2-D and 3-D applications, In 3D Pushover 2008 - Proceedings of nonlinear static methods for design/assessment of 3D structures. IST Press.

Beresnev, I. A., & Atkinson, G. M. (1997). Modeling finite-fault radiation from the ωn spectrum. *Bulletin of the Seismological Society of America*, *87*(1), 67–84.

Beresnev, I. A., & Atkinson, G. M. (1998). FINSIMa FORTRAN program for simulating stochastic acceleration time histories from finite faults. *Seismological Research Letters*, *69*(1), 27–32. doi:10.1785/gssrl.69.1.27

Bezdek, J. C., Ehrlich, R., & Full, W. (1984). FCM: The fuzzy c-means clustering algorithm. *Computers & Geosciences*, *10*(2), 191–203. doi:10.1016/0098-3004(84)90020-7

Birnbaum, Z. W., Esary, J. D., & Saunders, S. C. (1961). Multi-Component Systems and Their Reliability. *Technometrics*, *3*(1), 55–77. doi:10.1080/00401706.1961.10489927

Bischof, C. R. (2002). ACI Manual of Concrete Practice, 2010. *Practice, 552*, 2002–2002.

Biskinis, D. E., Roupakias, G. K., & Fardis, M. N. (2004). Degradation Of Shear Strength Of Reinforced Concrete Members With Inelastic Cyclic Displacements. *ACI Structural Journal*, (101): 773–783.

Bolotin, V. V. (2013). *Random vibrations of elastic systems*. Springer.

Bommer, J. J., & Acevedo, A. B. (2004). The use of real earthquake accelerograms as input to dynamic analysis. *Journal of Earthquake Engineering*, *1*(1), 43–91. doi:10.1080/13632460409350521

Bommer, J. J., Acevedo, A. B., & Douglas, J. (2003). The Selection and Scaling of Real Earthquake Accelerograms for Use in Seismic Design and Assessment.*Proceedings of ACI International Conference on Seismic Bridge Design and Retrofit*.

Bommer, J. J., Scott, S. G., & Sarma, S. K. (2000). Hazard-consistent earthquake scenarios. *Soil Dynamics and Earthquake Engineering*, *19*(4), 219–231. doi:10.1016/S0267-7261(00)00012-9

Boore, D. M. (2000). *SMSIM – FORTRAN Programs for Simulating Ground Motions from Earthquakes: Version 2.0 – A Revision of OFR 96-80-A*. USGS Open File Report OF 00-509.

Boore, D. M. (1983). Stochastic simulation of high-frequency ground motions based on seismological models of the radiated spectra. *Bulletin of the Seismological Society of America*, *73*, 1865–1894.

Boore, D. M. (2003). Simulation of ground motion using the stochastic method. *Pure and Applied Geophysics*, *160*(3), 635–676. doi:10.1007/PL00012553

Boore, D. M. (2009). Comparing stochastic point-source and finite-source ground-motion simulations: SMSIM and EXSIM. *Bulletin of the Seismological Society of America*, *99*(6), 3202–3216. doi:10.1785/0120090056

Bouchon, M., & Aki, K. (1977). Discrete wave-number representation of seismic wavefields. *Bulletin of the Seismological Society of America*, *67*, 259–277.

Bozorgnia, Y., & Campbell, K. W. (2004). The vertical-to-horizontal response spectral ratio and tentative procedures for developing simplified V/H and vertical design spectra. *Journal of Earthquake Engineering*, *8*(2), 175–207.

Bozorgnia, Y., Niazi, M., & Campbell, K. W. (1995). Characteristics of free-field vertical ground motion during the Northridge earthquake. *Earthquake Spectra*, *11*, 515–525.

Broderick, B. M., & Elnashai, A. S. (1995). Analysis of the failure of Interstate 10 freeway ramp during the Northridge earthquake of 17 January 1994. *Earthquake Engineering & Structural Dynamics*, *24*, 189–208.

Bruneau, M., Chang, S., Eguchi, R., Lee, G., ORourke, T., Reinhorn, A. M., & Winterfelt, D. V. et al. (2003). A framework to quantitatively assess and enhance the seismic resilience of communities. *Earthquake Spectra*, *19*(4), 733–752. doi:10.1193/1.1623497

Bruneau, M., & Saatciolu, M. (1993). Performance of Structures during the 1992 Erzincan Earthquake. *Canadian Journal of Civil Engineering*, *20*(2).

Brune, J. N. (1970). Tectonic stress and the spectra of seismic shear waves from earthquakes. *Journal of Geophysical Research*, *75*(26), 4997–5009. doi:10.1029/JB075i026p04997

Building Seismic Safety Council (BSSC). (2000). *Pre-standard and commentary for the seismic rehabilitation of buildings. FEMA-356*. Washington, DC: Federal Emergency Management Agency.

Burton, T. D. (1982). Non-linear oscillator limit cycle analysis using a time transformation approach. *International Journal of Non-linear Mechanics*, *17*(1), 7–19.

Cagno, E., De Ambroggi, M., Grande, O., & Trucco, P. (2010). Risk analysis of underground infrastructures in urban area: time-dependent interoperability analysis. *Reliability, Risk and Safety: Theory and Applications - Proc. ESREL 2009 Europe Annual Conf.*, 1899-1906.

Canbay, E., Ersoy, U., & Ozcebe, G. (2003). Contribution of reinforced concrete infills to seismic behavior of structural systems. *ACI Structural Journal*, *100*(5), 637–643.

Carr, A. J. (2006). *Ruaumoko Program for Inelastic Dynamic Analysis*. Christchurch, New Zealand: Department of Civil Engineering, University of Canterbury.

Carydis, P., Castiglioni, C., Lekkas, E., Kostaki, I., Lebesis, N., & Drei, A. (2012). The Emilia Romagna, May 2012 earthquake sequence. The influence of the vertical earthquake component and related geoscientific and engineering aspects. *Ingegneria Sismica*, *29*(2-3), 31–58.

Casalicchio, E., Galli, E., & Tucci, S. (2007). Federated agent-based modeling and simulation approach to study interdependencies in IT critical infrastructures. *11th IEEE Symposium on Distributed Simulation and Real-Time Applications*. doi:10.1109/DS-RT.2007.11

Casarotti, C., & Pinho, R. (2007). An adaptive capacity spectrum method for assessment of bridges subjected to earthquake action. *Bulletin of Earthquake Engineering*, *5*(3), 377–390. doi:10.1007/s10518-007-9031-8

Castet, J. F., & Saleh, J. H. (2012). On the concept of survivability, with application to spacecraft and space-based networks. *Reliability Engineering & System Safety*, *99*, 123–138. doi:10.1016/j.ress.2011.11.011

CEB-FIB (Comité Européen du Béton - Fédération Internationale du Béton). (2003). *Seismic assessment and retrofit of reinforced concrete buildings*. CEB-FIB Bulletin no. 24. International federation for structural concrete, task group 7.1.

CEN (Comité Européen de Normalisation). (2005a). *Eurocode 2 - Design of concrete structures Part 1-1: General rules and rules for buildings - EN 1992-1-1*. CEN.

CEN (Comité Européen de Normalisation). (2005b). *Eurocode 8 - Design of structures for earthquake resistance - Part 3: Assessment and retrofitting of buildings - EN 1998-3*. CEN.

CEN. (2004a). *European Standard EN 1992-1-1. Eurocode 2: Design of concrete structures - Part 1-1: General rules and rules for buildings*. Brussels: European Committee for Standardization.

CEN. (2004b). *European Standard EN 1998-1. Eurocode 8: Design of structures for earthquake resistance. Part 1: General rules, seismic action and rules for buildings*. Brussels: European Committee for Standardization.

CEN. (2005). *European Standard EN 1998-3. Eurocode 8: Design of structures for earthquake resistance. Part 3: Assessment and retrofitting of buildings*. Brussels: European Committee for Standardization.

Chang, F. K., & Krinitzsky, E. L. (1977). *Duration, spectral content, and predominant period of strong motion earthquake records from western United States*. US Army Engineer Waterways Experiment Station.

Charney, F. A. (2008). Unintended consequences of modelling damping in structures. *Journal of Structural Engineering*, *134*(4), 581–592. doi:10.1061/(ASCE)0733-9445(2008)134:4(581)

Chen, Q., & McCalley, J. D. (2005). Identifying high risk N-k contingencies for online security assessment. *IEEE Transactions on Power Systems*, *20*(2), 823–834. doi:10.1109/TPWRS.2005.846065

Chopra, A. K. (2007). *Dynamics of Structures* (3rd ed.). Prentice Hall.

Chopra, A. K., & Chinatanapakdee, C. (2004). Inelastic deformation ratios for design and evaluation of structures: Single-degree-of-freedom bilinear systems. *Journal of Structural Engineering*, *130*(9), 1309–1319. doi:10.1061/(ASCE)0733-9445(2004)130:9(1309)

Chopra, A. K., & Goel, R. K. (2001). *A modal pushover analysis procedure to estimate seismic demands for buildings: Theory and Preliminary Evaluation. PEER Report 2001/03*. Pacific Earthquake Engineering Center, University of California Berkeley.

Chopra, A. K., & Goel, R. K. (2002). A modal pushover analysis procedure for estimating seismic demands for buildings. *Earthquake Engineering & Structural Dynamics*, *31*(3), 561–582. doi:10.1002/eqe.144

Chossat, P., & Lauterbach, R. (2000). *Methods in equivalent bifurcations and dynamical systems*. Singapore: World Scientific Publishing.

Circolare Ministeriale n. 617. (2009). *Cons. Sup. LL. PP., 'Istruzioni per l'applicazione delle Nuove Norme Tecniche per le Costruzioni' di cui al decreto ministeriale del 14.01.2008*. G.U. del 26.02.2009 n. 47, supplemento ordinario n. 27. (in Italian).

Clementi, F., Quagliarini, E., Maracchini, G., & Lenci, S. (2015). Post-World War II Italian school buildings: Typical and specific seismic vulnerabilities. *Journal of Building Engineering*, *4*, 152–166. doi:10.1016/j.jobe.2015.09.008

Clementi, F., Scalbi, A., & Lenci, S. (2016). Seismic performance of precast reinforced concrete buildings with dowel pin connections. *Journal of Building Engineering*, *7*, 224–238. doi:10.1016/j.jobe.2016.06.013

Code, T. E. (2007). *Specifications for buildings to be built in seismic areas*. Ankara, Turkey: Ministry of Public Works and Settlement. (in Turkish)

CodeT. E. (2016). http://www.deprem.gov.tr/belgeler2016/tbdy.pdf

Committee, S. V. (1995). *Performance-based seismic engineering*. Sacramento, CA: Structural Engineers Association of California.

Computers, Structures Incorporated (CSI). (2007). *SAP 2000 NL*. Author.

Consiglio Nazionale delle Ricerche (CNR). (2013). *CNR-DT 2012/2013 - Istruzioni per la Valutazione Affidabilistica della Sicurezza Sismica di Edifici Esistenti*. Retrieved from http://www.cnr.it/documenti/norme/IstruzioniCNR_DT212_2013.pdf

Constantinou, M. C., Soong, T. T., & Dargush, G. F. (1998). Passive Energy Dissipation Systems for Structural Design and Retrofit, MCEER Monograph No. 1.

Cornell, C. A. (1968). Engineering seismic risk analysis. *Bulletin of the Seismological Society of America*, *58*(5), 1583–1606.

Department of Homeland Security. (2013). *National Infrastructure protection plan*. Washington, DC: NIPP.

Di Sarno, L., & Elnashai, A. S. (2009). Bracing systems for seismic retrofitting of steel frames. *Journal of Constructional Steel Research, 65*(2), 452–465. doi:10.1016/j.jcsr.2008.02.013

Doran, B., Yetilmezsoy, K., & Murtazaoglu, S. (2015). Application of fuzzy logic approach in predicting the lateral confinement coefficient for RC columns wrapped with CFRP. *Engineering Structures, 88*, 74–91. doi:10.1016/j.engstruct.2015.01.039

Duenas-Osorio, L., Craig, J. I., & Goodno, B. J. (2007). Seismic response of critical interdependent networks. *Earthquake Engineering & Structural Dynamics, 36*(2), 285–306. doi:10.1002/eqe.626

Dunn, J. C. (1973). A Fuzzy Relative of the ISODATA Process and Its Use in Detecting Compact Well-Separated Clusters. *Journal of Cybernetics, 3*(3), 32–57. doi:10.1080/01969727308546046

EasyOver. (2016). *Analysis Manual*. Author.

Elnashai, A. S. (2000). Advanced inelastic (pushover) analysis for seismic design and assessment. *The G.Penelis Symposium*.

Elnashai, A. S., & Di Sarno, L. (2008). Fundamentals of Earthquake Engineering. John Wiley & Sons. http://doi.org/doi:10.1002/9780470024867

Elnashai, A. S. (2001). Advanced inelastic static (pushover) analysis for earthquake applications. *Structural Engineering & Mechanics, 12*(1), 51–69. doi:10.12989/sem.2001.12.1.051

Elnashai, A. S., & Papazoglou, A. J. (1996). Analytical and field evidence of the damaging effect of vertical earthquake ground motion. *Earthquake Engineering & Structural Dynamics, 25*, 1109–1137.

Elwood, K. J., & Moehle, J. P. (2005). Axial Capacity Model for Shear-Damaged Columns. *ACI Structural Journal, 102*(4), 578–587.

Ergun, M., & Sevket, A. (2014). Comparing the effects of scaled and real earthquake records on structural response. *Earthquakes and Structures, 6*(4), 375–392. doi:10.12989/eas.2014.6.4.375

Eurocode 8: Design of structures for earthquake resistance. (2005).

Fahjan, Y. (2008). Selection and scaling of real earthquake accelerograms to fit the Turkish design spectra.[In Turkish with Extended Summary]. *Technical Journal of Turkish Chamber of Civil Engineers, 3*(19), 4423–4444.

Fahjan, Y. M. (2008). Selection and scaling of real earthquake accelerograms to fit the Turkish design spectra. *Teknik Dergi, 19*(3), 4423–4444.

Fajfar, P. (1999). Capacity spectrum method based on inelastic demand spectra. *Earthquake Engineering & Structural Dynamics, 28*(9), 979–993. doi:10.1002/(SICI)1096-9845(199909)28:9<979::AID-EQE850>3.0.CO;2-1

Fajfar, P. (2000). A nonlinear analysis method for performance-based seismic design. *Earthquake Spectra, 16*(3), 573–592. doi:10.1193/1.1586128

Fajfar, P., & Gašperšič, P. (1996). The N2 method for the seismic damage analysis of rc buildings. *Earthquake Engineering & Structural Dynamics, 25*(1), 31–46. doi:10.1002/(SICI)1096-9845(199601)25:1<31::AID-EQE534>3.0.CO;2-V

Fajfar, P., Marušic, D., & Peruš, I. (2005). Torsional effects in the pushover-based seismic analysis of buildings. *Journal of Earthquake Engineering, 9*(6), 831–854. doi:10.1080/13632460509350568

Fardis, M. (2009). *Seismic Design, Assessment and Retrofitting of Concrete Buildings* (3rd ed.). New York: Springer. doi:10.1007/978-1-4020-9842-0

Fardis, M. N. (2005). *Designers' guide to EN 1998-1 and EN 1998-5 Eurocode 8: design of structures for earthquake resistance: general rules, seismic actions, design rules for buildings, foundations and retaining structures* (Vol. 8). Thomas Telford Services Limited.

Fardis, M. N., Schetakis, A., & Strepelias, E. (2013). RC buildings retrofitted by converting frame bays into RC walls. *Bulletin of Earthquake Engineering, 11*(5), 1541–1561. doi:10.1007/s10518-013-9435-6

Fasan, M., Amadio, C., Noè, S., Panza, G., Magrin, A., Romanelli, F., & Vaccari, F. (2015). *A new design strategy based on a deterministic definition of the seismic input to overcome the limits of design procedures based on probabilistic approaches.* arXiv preprint arXiv:1509.09119

FEMA (Federal Emergency Management Agency). (2000). *FEMA 356: Pre-Standard and Commentary for the Seismic Rehabilitation of Buildings.* Washington, DC: FEMA.

FEMA. (2000). *356. Pre-standard and Commentary for the Seismic Rehabilitation of Buildings - Applied Technology Council.* Washington, DC: ATC.

FEMA356. P. (2000). Commentary for the Seismic Rehabilitation of Buildings. FEMA-356, Federal Emergency Management Agency.

Ferraioli, M. (2015). Case study of seismic performance assessment of irregular RC buildings: Hospital structure of Avezzano. *Earthquake Engineering and Engineering Vibration, 14*(1), 141–156. doi:10.1007/s11803-015-0012-7

fib. T.G. 7.1. (2003). *Seismic Assessment and Retrofit of Reinforced Concrete Buildings: State-of-the-art report.* Bulletin 24, Sprint-Digital-Druck.

Fiore, A., Porco, F., Uva, G., & Mezzina, M. (2013). On the dispersion of data collected by in situ diagnostic of the existing concrete. *Construction & Building Materials, 47*, 208–217. doi:10.1016/j.conbuildmat.2013.05.001

Ford, L. R., & Fulkerson, E. (1962). *Flows in Networks.* Princeton, NJ: Princeton University Press.

Foti, D., & Mongelli, M. (2011). *Isolatori sismici per edifice esistenti e di nuova costruzione.* Dario Flaccovio Editore. (in Italian)

Fujita, T. (1999), *Demonstration of Effectiveness of Seismic Isolation in the Hanshin-Awaji Earthquake and Progress of Applications of Base-Isolated Buildings.* Report on 1995 Kobe Earthquake by INCEDE, ERC and KOBEnet.

Fuller, K., Lim, C., Loo, S., Melkumyan, M., & Muniandy, K. (2000). *Design and Testing of High Damping Rubber Earthquake Bearings for Retrofit Project in Armenia.* In S. Balassanian, A. Cisternas, & M. Melkumyan (Eds.), *Earthquake Hazard and Seismic Risk Reduction* (pp. 379–385). Kluwer Academic Publishers. doi:10.1007/978-94-015-9544-5_40

Garevski, M. (2010). Development, Production and Implementation of Low Cost Rubber Bearings. Springer. doi:10.1007/978-90-481-9544-2_17

Germann, T. C., Kadau, K., Longini, I. M., & Macken, C. A. (2006). Mitigation strategies for pandemic influenza in the United States. *Proceedings of the National Academy of Sciences of the United States of America, 103*(15), 5935–5940. doi:10.1073/pnas.0601266103 PMID:16585506

Giannuzzi, D., Ballarini, R., Hucklebridge, A., Pollino, M., & Valente, M. (2014). Braced ductile shear panel: New seismic-resistant framing system. *Journal of Structural Engineering, 140*(2), 1–11. doi:10.1061/(ASCE)ST.1943-541X.0000814

Goel, R. K., & Chopra, A. K. (2005). Extension of Modal Pushover Analysis to Compute Member Forces. *Earthquake Spectra, 21*(1), 125–139. doi:10.1193/1.1851545

Golara, A., & Esmaeily, A. (2016). Quantification and Enhancement of the Resilience of Infrastructure Networks. *Journal of Pipeline Systems Engineering and Practice*. doi: 10.1061/(ASCE)PS.1949-1204.0000250

Golara, A. (2014). Probabilistic seismic hazard analysis of interconnected infrastructure: A case of Iranian high-pressure gas supply system. *Natural Hazards*, *73*(2), 567–577. doi:10.1007/s11069-014-1087-6

Golara, A., Bonyad, H., & Omidvar, H. (2015). Forecasting Iran's Natural Gas Production, Consumption. *Pipeline & Gas Journal*, *242*(8), 24–30.

Görgülü, O., & Taskin, B. (2015). Numerical simulation of RC infill walls under cyclic loading and calibration with widely used hysteretic models and experiments. *B EarthqEng*, *13*(9), 2591-2610. doi:10.1007/s10518-015-9739-9

Guckenheimer, J., & Holmes, P. (1983). *Nonlinear oscillations, dynamical systems and bifurcations of vector fields*. New York: Springer.

Guclu, R., & Yazici, H. (2007). Fuzzy Logic Control of a Non-linear Structural System against Earthquake Induced Vibration. *Journal of Vibration and Control*, *13*(11), 1535–1551. doi:10.1177/1077546307077663

Gupta, B., & Kunnath, S. K. (2000). Adaptive spectra-based pushover procedure for seismic evaluation of structures. *Earthquake Spectra*, *16*(2), 367–391. doi:10.1193/1.1586117

Hadley, D. M., & Helmberger, D. V. (1980). Simulation of strong ground motions. *Bulletin of the Seismological Society of America*, *70*(2), 617–630.

Haimes, Y. Y. (2009). On the definition of resilience in systems. *Risk Analysis*, *29*(4), 498–501. doi:10.1111/j.1539-6924.2009.01216.x PMID:19335545

Hamburger, R. (1996). *Implementing performance-based seismic design in structural engineering practice*. Paper presented at the 11th World Conference on Earthquake Engineering, Acapulco, Mexico.

Hancock, J., & Bommer, J. J. (2006). A state-of-knowledge review of the influence of strong-motion duration on structural damage. *Earthquake Spectra*, *22*(3), 827–845. doi:10.1193/1.2220576

Hartzell, S. H. (1978). Earthquake aftershocks as Greens functions. *Geophysical Research Letters*, *5*(1), 1–4. doi:10.1029/GL005i001p00001

Haselton, C. B., Liel, A. B., & Lange, S. T. (2008). Beam-Column Element Model Calibrated for Predicting Flexural Response Leading to Global Collapse of RC Frame Buildings. *Peer 2007, 3*.

Haselton, C. B., Baker, J. W., Liel, A. B., & Deierlein, G. G. (2011). Accounting for ground-motion spectral shape characteristics in structural collapse assessment through an adjustment for epsilon. *Journal of Structural Engineering*, *137*(3), 332–344. doi:10.1061/(ASCE)ST.1943-541X.0000103

Haselton, C. B., Liel, A. B., Deierlein, G. G., Dean, B. S., & Chou, J. H. (2011). Seismic Collapse Safety of Reinforced Concrete Buildings. I: Assessment of Ductile Moment Frames. *Journal of Structural Engineering*, *137*(4), 481–491. doi:10.1061/(ASCE)ST.1943-541X.0000318

Hatwal, H., Mallik, A. K., & Ghosh, A. (1983). Forced nonlinear oscillations of an autoparametric system. *ASME Journal of Applied Mechanics*, *50*(3), 657–662.

Haxton, R. S., & Barr, A. D. S. (1974). The autoparametric vibration absorber. *ASME Journal of Applied Mechanics*, *94*, 119–125.

Helseth, A., & Holen, A. T. (2009). Structural vulnerability of energy distribution systems: Incorporating infrastructural dependencies. *Electrical Power and Energy Systems*, *31*(9), 531–537. doi:10.1016/j.ijepes.2009.03.023

Holland, J. H. (1975). *Adaptation in natural and artificial systems: an introductory analysis with applications to biology, control, and artificial intelligence*. U Michigan Press.

Holling, C. S. (1973). Resilience and stability of ecological systems. *Annual Review of Ecology and Systematics*, *4*(1), 1–23. doi:10.1146/annurev.es.04.110173.000245

Hong, N. K., Hong, S.-G., & Chang, S.-P. (2003). Computer-supported evaluation for seismic performance of existing buildings. *Advances in Engineering Software*, *34*(2), 87–101. doi:10.1016/S0965-9978(02)00106-0

Hutchıngs, L. (1991). Prediction" of strong ground motion for the 1989 Loma Prieta earthquake using empirical Green's functions. *Bulletin of the Seismological Society of America*, *81*, 88–121.

Hutchıngs, L., & Wu, F. (1990). Empirical Greens functions from small earthquakes: A waveform study of locally recorded aftershocks of the San Fernando earthquake. *Journal of Geophysical Research*, *95*(B2), 1187–1214. doi:10.1029/JB095iB02p01187

Hyogo Farmework for Action. (2005-2015). Retrieved from https://www.unisdr.org/we/coordinate/hfa

Ibarra, L. F., Medina, R. A., & Krawinkler, H. (2005). Hysteretic models that incorporate strength and stiffness deterioration. *Earthquake Engineering & Structural Dynamics*, *34*(12), 1489–1511. doi:10.1002/eqe.495

IdeCAD 3D Integrated analyses, design and detailing software for reinforced concrete structure. (n.d.). IdeYAPI Ltd. Retrieved from www.idecad.com

Iervolino, I., & Cornell, C. A. (2005). Record selection for nonlinear seismic analysis of structures. *Earthquake Spectra*, *21*(3), 685–713. doi:10.1193/1.1990199

Iooss, G., & Adelmeyer, M. (1992). *Topics in bifurcation theory and applications*. London: World Scientific Publishing.

Irikura, K. (1983). Semi-empirical estimation of strong ground motions during large earthquakes. *Bull. Disaster Prevention. Res. Inst Kyoto Univ*, *33*, 63–104.

Jan, T. S., Liu, M. W., & Kao, Y. C. (2004). An upper-bound pushover analysis procedure for estimating seismic demands of high-rise buildings. *Engineering Structures*, *26*(1), 117–128. doi:10.1016/j.engstruct.2003.09.003

Jiang, P., & Haimes, Y. Y. (2004). Risk management for Leontief-based interdependency systems. *Risk Analysis*, *24*(5), 1215–1229. doi:10.1111/j.0272-4332.2004.00520.x PMID:15563289

Kahan, J. H., Andrew, C. A., & Justin, K. G., (2009). An operational framework for resilience. *Journal of Homeland Security and Emergency Management, 6*(1).

Kalkan, E., Adalier, K., & Pamuk, A. (2004). Near source effects and engineering implications of recent earthquakes in Turkey.*Fifth International Conference on Case Histories in Geotechnical Engineering*, 13-17.

Kalkan, E., & Chopra, A. K. (2010). *Practical Guidelines to Select and Scale Earthquake Records for Nonlinear Response History Analysis of Structures*. U.S. Geological Survey Open-File Report.

Kalkan, E., & Chopra, A. K. (2011). Modal-Pushover-based Ground Motion Scaling Procedure. *Journal of Structural Engineering*, *137*(3), 298–310. doi:10.1061/(ASCE)ST.1943-541X.0000308

Kalkan, E., & Kunnath, S. K. (2006). Adaptive modal combination procedure for nonlinear static analysis of building structures. *Journal of Structural Engineering*, *132*(11), 1721–1731. doi:10.1061/(ASCE)0733-9445(2006)132:11(1721)

Kapitsa, P. L. (1951). Dynamic stability of a pendulum with a vibrating point of suspension. Journal of Experimental and Theoretical Physics, 21, 588–598. (In Russian)

Kasai, K., & Popov, E. P. (1986). *A study of seismically resistant eccentrically braced steel framesystems. Rep. No. UCB/EERC-86/01*. Berkeley, CA: Earthquake Engineering Research Center, University of California.

Kaul, M. K. (1978). Spectrum-consistent time-history generation. *Journal of the Engineering Mechanics Division, 104*(4), 781–788.

Kramer, S. L. (1996). *Geotechnical Earthquake Engineering*. Prentice Hall.

Krinitzsky, E. L., & Chang, F. K. (1977). *Specifying Peak Motions for Design Earthquakes, State-of the-Art for Assessing Earthquake Hazards in the United States, Report 7, Miscellaneous Paper S-73-1*. Vicksburg, MS: US Army Corps of Engineers.

Kuran, F., Demir, C., Koroglu, O., Kocaman, C., & İlki, A. (2007). Seismic Safety Analysis of an Existing 1502 Type Disaster Building Using New Version of Turkish Seismic Design Code.*Proceedings of the ECCOMAS Thematic Conference on COMPDYN*.

Lee, W. K., & Hsu, C. S. (1994). A global analysis of an harmonically excited spring-pendulum system with internal resonance. *Journal of Sound and Vibration, 171*(3), 335–359.

Liang, Z., Lee, G. C., Dargush, G. F., & Song, J. (2012). *Structural damping – Applications in seismic response modification*. Boca Raton, FL: CRC Press – Taylor & Francis.

Liel, A. B., Haselton, C. B., & Deierlein, G. G. (2011). Seismic Collapse Safety of Reinforced Concrete Buildings. II: Comparative Assessment of Nonductile and Ductile Moment Frames. *Journal of Structural Engineering, 137*(4), 492–502. doi:10.1061/(ASCE)ST.1943-541X.0000275

Lilhanand, K., Tseng, W. S. (1987). Generation of synthetic time histories compatible with multiple-damping design response spectra. *Structural Mechanics in Reactor Technology*.

Lilhanand, K., & Tseng, W. S. (1987). Development and application of realistic earthquake time histories compatible with multiple-damping design spectra. *Development, 3*, 7–8.

Lin, Y. K., & Cai, G. Q. (1995). *Probabilistic structural dynamics - Advanced theory and applications*. New York: McGraw-Hill.

Lopez, O. A., Hernandez, J. J., & Puig, J. (2004). Seismic Risk in Schools: The Venezuelan Project. *Proceedings of the Keeping schools safe in Earthquakes, Ad Hoc Experts' Group Meeting on Earthquake Safety in Schools*.

Luco, N., & Bazzurro, P. (2007). Does amplitude scaling of ground motion records result in biased nonlinear structural drift responses? *Earthquake Engineering & Structural Dynamics, 36*(13), 1813–1835. doi:10.1002/eqe.695

Magliulo, G., Maddaloni, G., & Cosenza, E. (2007). Comparison between non-linear dynamic analysis performed according to EC8 and elastic and non-linear static analyses. *Engineering Structures, 29*(11), 2893–2900. doi:10.1016/j.engstruct.2007.01.027

Mamdani, E. H. (1976). Advances in the linguistic synthesis of fuzzy controllers. *International Journal of Man-Machine Studies, 8*(6), 669–678. doi:10.1016/S0020-7373(76)80028-4

Mandelbrot, B. B. (1983). *The fractal geometry of nature/Revised and enlarged edition* (p. 1). New York: WH Freeman and Co.

Mander, J. B., Priestley, M. J. N., & Park, R. (1988). Theoretical Stress-Strain Model for Confined Concrete. *Journal of Structural Engineering, 114*(8), 1804–1826. doi:10.1061/(ASCE)0733-9445(1988)114:8(1804)

Martelli, A., Forni, M., Arato, G.-B., & Spadoni, B. (2001). Overview and Summary of the 7th International Seminar on Seismic Isolation, Passive Energy Dissipation and Active Control of Vibrations of Structures. *Proceedings of the 7th Int. Seminar*, i-xxxvii.

Martelli, A., & Forni, M. (2011). Seismic retrofit of existing buildings by means of seismic isolation: some remarks on the Italian experience and new projects. *Proceedings of the 3rd International Conference on Computational Methods in Structural Dynamics and Earthquake Engineering, COMPDYN 2011.*

Martelli, A., Forni, M., & Rizzo, S. (2008). Seismic isolation: present application and perspectives. *Proceedings of the ASSISi International Workshop on Base Isolated High-Rise Buildings*, 1 - 26.

Martinez-Rueda, J. E. (1998). Scaling procedure for natural accelerograms based on a system of spectrum intensity scales. *Earthquake Spectra*, *14*(1), 135–152. doi:10.1193/1.1585992

Martinez-Rueda, J. E., & Elnashai, A. S. (1997). Confined concrete model under cyclic load. *Materials and Structures*, *30*(197), 139–147. doi:10.1007/BF02486385

Masi, A. (2003). Seismic vulnerability assessment of gravity load designed RC frames. *Bulletin of Earthquake Engineering*, *1*(3), 371–395. doi:10.1023/B:BEEE.0000021426.31223.60

Masi, A., & Chiauzzi, L. (2013). An experimental study on the within-member variability of in situ concrete strength in RC building structures. *Construction & Building Materials*, *47*, 951–961. doi:10.1016/j.conbuildmat.2013.05.102

Masi, A., & Vona, M. (2012). Vulnerability assessment of gravity-load designed RC buildings: Evaluation of seismic capacity through non-linear dynamic analyses. *Engineering Structures*, *45*, 257–269. doi:10.1016/j.engstruct.2012.06.043

Mazzolani, F. (2008). Innovative metal systems for seismic upgrading of RC structures. *Journal of Constructional Steel Research*, *64*(7-8), 882–895. doi:10.1016/j.jcsr.2007.12.017

McCulloch, W. S., & Pitts, W. (1943). A logical calculus of the ideas immanent in nervous activity. *The Bulletin of Mathematical Biophysics*, *5*(4), 115–133. doi:10.1007/BF02478259

Melkumyan, M. (2006). Armenia is one of the world leaders in development and application of base isolation technologies. *MENSHIN Journal of the Japan Society of Seismic Isolation*, *54*, 38 - 41.

Melkumyan, M. (2001). The state of the art in development of testing facilities and execution of tests on isolation and bridge bearings in Armenia. *Proceedings 5th World Congress on Joints, Bearings and Seismic Systems for Concrete Structures.*

Melkumyan, M. (2004). Recent Applications of Seismic Isolation in Civil Buildings in Armenia. *Proceedings of the 13th World Conference on Earthquake Engineering.*

Melkumyan, M. (2005). Current Situation in Application of Seismic Isolation Technologies in Armenia. *Proceedings of the International Conference dedicated to the 250th anniversary of the 1755 Lisbon Earthquake*, 493-500.

Melkumyan, M. (2007). Seismic isolation of civil structures in Armenia - development and application of innovative structural concepts. *Proceedings ECCOMAS Thematic Conference on Computational Methods in Structural Dynamics and Earthquake Engineering.*

Melkumyan, M. (2009). Armenian seismic isolation technologies for civil structures - example on application of innovative structural concepts, R&D and design rules for developing countries. *Proceedings 11th World Conference on Seismic Isolation, Energy Dissipation and Active Vibration Control of Structures.*

Melkumyan, M. (2011a). Assessment of the Seismic Risk in the City of Yerevan and its Mitigation by Application of Innovative Seismic Isolation Technologies. *J. Pure and Applied Geophysics, 168*(3-4), 695–730. doi:10.1007/s00024-010-0110-4

Melkumyan, M. (2011b). *New Solutions in Seismic Isolation*. Yerevan: LUSABATS.

Melkumyan, M. (2013). Comparison of the analyses results of seismic isolated buildings by the design code and time histories *Journal of Civil Engineering and Science, 2*(3), 184–192.

Melkumyan, M. (2014a). Structural concept and analysis of the 15-story base isolated apartment building "Avan". *International Journal of Engineering Research and Management, 1*(7), 157–161.

Melkumyan, M. (2014b). Structural concept and analysis of the 17-story base isolated apartment building "Sevak". *International Journal of Engineering and Applied Sciences, 1*(3), 13–17.

Melkumyan, M., & Gevorgyan, E. (2008). Structural concept and analysis of a 17-story multifunctional residential complex with and without seismic isolation system. *Proceedings 2008 Seismic Engineering Conference Commemorating the 1908 Messina and Reggio Calabria Earthquake*, 1425 - 1432. doi:10.1063/1.2963766

Melkumyan, M., & Gevorgyan, E. (2010). Structural concept and analysis of 18-story residential complex "Northern Ray" with and without base isolation system. *Proceedings of the 14th European Conference on Earthquake Engineering.*

Melkumyan, M., Gevorgyan, E., & Hovhannisyan, H. (2005). Application of Base Isolation to Multifunctional Multistory Buildings in Yerevan, Armenia. *Proceedings of the 9th World Seminar on Seismic Isolation, Energy Dissipation and Active Vibration Control of Structures*, 2, 119-127.

Melkumyan, M., & Hakobyan, A. (2005). Testing of Seismic Isolation Rubber Bearings for Different Structures in Armenia. *Proceedings of the 9th World Seminar on Seismic Isolation, Energy Dissipation and Active Vibration Control of Structures*, 2, 439-446.

Melkumyan, M., & Hovhannisyan, H. (2006). New approaches in analysis and design of base isolated multistory multifunctional buildings. *Proceedings 1st European Conference on Earthquake Engineering and Seismology (a joint event of the 13th ECEE & 30th General Assembly of the ESC).*

Menegotto, M., & Pinto, P. E. (1973). Method of analysis for cyclically loaded R.C. plane frames including changes in geometry and non-elastic behavior of elements under combined normal force and bending. Preliminary Report. *IABSE, Zurich, 13*, 15–22.

Midas GEN. (2015). Analysis Manual.

Milani, G., & Valente, M. (2015a). Comparative pushover and limit analyses on seven masonry churches damaged by the 2012 Emilia-Romagna (Italy) seismic events: Possibilities of non-linear Finite Elements compared with pre-assigned failure mechanisms. *Engineering Failure Analysis, 47*, 129–161. doi:10.1016/j.engfailanal.2014.09.016

Milani, G., & Valente, M. (2015b). Failure analysis of seven masonry churches severely damaged during the 2012 Emilia-Romagna (Italy) earthquake: Non-linear dynamic analyses vs conventional static approaches. *Engineering Failure Analysis, 54*, 13–56. doi:10.1016/j.engfailanal.2015.03.016

Mistakidis, E. S., & Georgiou, D. N. (2003). Fuzzy sets in seismic inelastic analysis and design of reinforced concrete frames. *Advances in Engineering Software, 34*(10), 589–599. doi:10.1016/S0965-9978(03)00115-7

Mitchell, D., DeVall, R. H., Saatcioglu, M., Simpson, R., Tinawi, R., & Tremblay, R. (1995). Damage to concrete structures due to the 1994 Northridge earthquake. *Canadian Journal of Civil Engineering, 22*(1).

Miyamoto, K., & Gilani, A. (2007). Base Isolation for Seismic Retrofit of Structures: Application to a Historic Building in Romania. *Seismic Risk Reduction,Proceedings of the International Symposium*, 585-592.

Motazedian, D., Atkinson, G. M. (2005). Stochastic finite-fault modeling based on a dynamic corner frequency. *Bulletin of the Seismological Society of America, 95*(3), 995-1010.

Mpampatsikos, V., Nascimbene, R., & Petrini, L. (2008). A Critical Review of the R.C. Frame Existing Building Assessment Procedure According to Eurocode 8 and Italian Seismic Code. *Journal of Earthquake Engineering, 12*(sup1), 52–82. http://doi.org/10.1080/13632460801925020

Murphy, K. D., Bayly, P. V., Virgin, L. N., & Gottwald, J. A. (1994). Measuring the stability of periodic attractors using perturbation-induced transients: Applications to two non-linear oscillators. *Journal of Sound and Vibration, 172*(1), 85–102.

Nabergoj, R., & Tondl, A. (1994). A simulation of parametric ship rolling: Effects of hull bending and torsional elasticity. *Nonlinear Dynamics, 6*, 265–284.

Nabergoj, R., Tondl, A., & Virag, Z. (1994). Autoparametric resonance in an externally excited system. *Chaos, Solitons, and Fractals, 4*, 263–273.

Naeim, F., & Kelly, J. M. (1999). *Design of Seismic Isolated Structures: From Theory to Practice*. John Wiley & Sons. doi:10.1002/9780470172742

Naeim, F., & Lew, M. (1995). On the use of design spectrum compatible time histories. *Earthquake Spectra, 11*(1), 111–127. doi:10.1193/1.1585805

Náprstek, J., & Fischer, C. (2008). Non-linear auto-parametric stability loss of a slender structure due to random non-stationary seismic excitation. In *Proceedings of 14th Word Conference on Earthquake Engineering*. Chinese Assoc. Earthq. Eng.

Náprstek, J., & Fischer, C. (2012). Dynamic stability of a non-linear continuous system subjected to vertical seismic excitation. In B. H. V. Topping (Ed.), *Proceedings of the Eleventh International Conference on Computational Structures Technology* (paper 179). Civil-Comp Press.

Náprstek, J. (1996). Stochastic exponential and asymptotic stability of simple non-linear systems. *International Journal of Non-linear Mechanics, 31*(5), 693–705.

Náprstek, J. (1998). Non-linear self-excited random vibrations and stability of an SDOF system with parametric noises. *Meccanica, 33*, 267–277.

Náprstek, J. (2015b). Stochastic resonance - Challenges to engineering dynamics. In B. H. V. Topping & J. Kruis (Eds.), *Computational technology reviews* (pp. 53–101). Civil-Comp.

Náprstek, J., & Fischer, C. (2009). Auto-parametric semi-trivial and post-critical response of a spherical pendulum damper. *Computers & Structures, 87*, 1204–1215.

Náprstek, J., & Fischer, C. (2011). Auto-parametric stability loss and post-critical behavior of a three degrees of freedom system. In M. Papadrakakis et al. (Eds.), *Computational methods in stochastic dynamics* (pp. 267–289). Springer.

Náprstek, J., & Král, R. (2014). Finite element method analysis of Fokker-Plank equation in stationary and evolutionary versions. *Advances in Engineering Software, 72*, 28–38.

Náprstek, J., & Pirner, M. (2002). Non-linear behaviour and dynamic stability of a vibration spherical absorber. In A. Smyth et al. (Eds.), *Proceedings of 15th ASCE Engineering Mechanics Division Conference*. New York: Columbia University.

Nayfeh, A. H., & Mook, D. T. (2008). *Nonlinear Oscillations*. New York: John Wiley & Sons.

Nayfeh, A. H., Mook, D. T., & Marshall, L. R. (1973). Nonlinear coupling of pitch and roll modes in ship motion. *Journal of Hydronautics*, *7*, 145–152.

Newmark, N. M., Blume, J. A., & Kapur, K. K. (1973). Seismic design spectra for nuclear power plants. *Journal of the Power Division*, *99*, 287–303.

Newmark, N. M., & Hall, W. J. (1978). *Development of criteria for seismic review of selected nuclear power plants. NUREG/CR-0098.* Nuclear Regulatory Commission.

Newmark, N. M., & Hall, W. J. (1982). *Earthquake Spectra and Design, monograph. Earthquake Engineering Research Institute.* Berkeley, CA: EERI. Print

NTC. (2008). *Decreto Ministeriale 14/1/2008. Norme tecniche per le costruzioni.* Ministry of Infrastructures and Transportations, G.U. S.O. n. 29 on 2/4/2008.

Otani, S. (1981). Hysteretic models of reinforced concrete for earthquake response analysis. *Journal of the Faculty of Engineering*, *36*(2), 125–159.

Ouyang, M., Hong, L., Mao, Z.-J., Yu, M.-H., & Qi, F. (2009). A methodological approach to analyze vulnerability of interdependent infrastructures. *Simulation Modelling Practice and Theory*, *17*(5), 817–828. doi:10.1016/j.simpat.2009.02.001

Ozkul, S., Ayoub, A., & Altunkaynak, A. (2014). Fuzzy-logic based inelastic displacement ratios of degrading RC structures. *Engineering Structures*, *75*, 590–603. doi:10.1016/j.engstruct.2014.06.030

Pakdamar, F., & Guler, K. (2012). Evaluation of Flexible Performance of Reinforced Concrete Structures Using A Nonlinear Static Procedure Provided by Fuzzy Logic. *Advances in Structural Engineering*, *15*(12), 2173–2190. doi:10.1260/1369-4332.15.12.2173

Pall, A. S., & Marsh, C. (1982). Seismic Response of Friction Damped Braced Frames. *Journal of Structural Division, ASCE, St.*, *9*, 108.

Pall, A., & Pall, R. T. (2004). Performance-Based Design using Pall Friction Dampers – An economical design solution. *13th World Conference on Earthquake Engineering*, Vancouver, Canada.

Parfomak, P. W, (2014). *Physical Security of the US Power Grid: High-Voltage Transformer Substations.* Tech. rept. Congressional Research Service.

Pekelnicky, R., & Poland, C. (2013). ASCE 41-13: Seismic Evaluation and Retrofit of Existing Buildings. *SEAOC 2012 Convention Proceedings.*

Perform 3D Components & Elements. (2006). *For Perform-3D and Perform-Collapse.* Computer & Structures Inc.

Pierdicca, A., Clementi, F., Maracci, M., Isidori, D., & Lenci, S. (2015). Vibration-Based SHM of Ordinary Buildings: Detection and Quantification of Structural Damage. *ASME 2015 International Design Engineering Technical Conferences and Computers and Information in Engineering Conference*, 8. http://doi.org/ doi:10.1115/DETC2015-46763

Pinto, A., Verzeletti, G., Molina, J., Varum, H., Pinho, R., & Coelho, E. (2002). *Pseudo-dynamic tests on non-seismic resisting RC frames (bare and selective retrofit frames). EUR Report 20244 EN, ELSA.* Ispra, Italy: Joint Research Centre.

Piwowar, J., Chatelet, E., & Laclemence, P. (2009). An efficient process to reduce infrastructure vulnerabilities facing malevolence. *Reliability Engineering & System Safety*, *94*(11), 1869–1877. doi:10.1016/j.ress.2009.06.009

Plevris, V., Mitropoulou, C. C., & Lagaros, N. D. (Eds.). (2012). *Structural Seismic Design Optimization and Earthquake Engineering.* IGI Global. doi:10.4018/978-1-4666-1640-0

Postelnicu, T., Popa, V., Cotofana, D., Chesca, B., Ionescu, R., & Vacarenu, R. (2005) Study on seismic performance of existing buildings in Romania. Technical Report, Technical University of Civil Engineering, Bucharest, and Building Research Institute.

Powell, P. D. (2011). *Calculating determinants of block matrices*. arXiv:1112.4379 [math.RA]

Prakash, V., Powell, G. H., & Campbell, S. (1993). *DRAIN-2DX base program description and user guide, SEMM Report, UCB/SEMM-93/17*. University of California-Berkeley.

Preumont, A. (1984). The generation of spectrum compatible accelerograms for the design of nuclear power plants. *Earthquake Engineering & Structural Dynamics*, *12*(4), 481–497. doi:10.1002/eqe.4290120405

Priestley, M. J. N., Calvi, G. M., & Kowalsky, M. J. (2007). *Displacement-Based Seismic Design of Structures*. Pavia, Italy: IUSS Press.

Pugachev, V. S., & Sinitsyn, I. N. (1987). *Stochastic differential systems - Analysis and filtering*. Chichester, UK: J. Wiley.

Puterman, M. L. (2009). *Markov decision processes: discrete stochastic dynamic programming* (Vol. 414). Wiley.

Ramirez-Marquez, J. E., Rocco, C. M., Gebre, D. A., Coit, D. W., & Tortorella, M. (2006). New Insights on Multi-State Component Criticality and Importance. *Reliability Engineering & System Safety*, *91*(8), 894–904. doi:10.1016/j.ress.2005.08.009

RBDG. (2012). *Ministry of Environment and Urbanization: The guidelines regarding the detection of risky buildings, Republic of Turkey*. Ankara, Turkey: Council of Ministers. (in Turkish)

Reed, D. A., Kapur, K. C., & Christie, R. D. (2009). Methodology for assessing the resilience of networked infrastructure. *IEEE Systems Journal*, *3*(2), 174–180. doi:10.1109/JSYST.2009.2017396

Reiter, L. (1990). *Earthquake Hazard Analysis: Issues and Insights*. Columbia University Press.

Ren, Y., & Beards, C. F. (1994). A new receptance-based perturbative multi-harmonic balance method for the calculation of the steady state response of non-linear systems. *Journal of Sound and Vibration*, *172*(5), 593–604.

Retamales, R., & Boroschek, R. (2014). State-of -the-Art of Seismic Isolation and Energy Dissipation Applications in Chile. *MENSHIN Journal of the Japan Society of Seismic Isolation*, *84*, 55-69.

Reyes, J. C., & Chopra, A. K. (2011). Three-Dimensional Modal Pushover Analysis of Buildings Subjected to Two Components of Ground Motion, including its Evaluation for Tall Buildings. *Earthquake Engineering & Structural Dynamics*, *40*(7), 789–806. doi:10.1002/eqe.1060

Reyes, J. C., & Kalkan, E. (2012). How many records should be used in an ASCE/SEI-7 ground motion scaling procedure? *Earthquake Spectra*, *28*(3), 1223–1242. doi:10.1193/1.4000066

Ricles, J. M., & Popov, E. P. (1987). *Dynamic analysis of seismically resistant eccentrically braced frames. Rep. No. UCB/EERC-87/07*. Berkeley, CA: Earthquake Engineering Research Center, University of California.

Rinaldi, S. M. (2004). Modeling and simulating critical infrastructures and their interdependencies. *Proc. Thirty-Seventh Annual Hawaii International Conf. on System Sciences*. Computer Society Press doi:10.1109/HICSS.2004.1265180

Rozman, M., & Fajfar, P. (2009). Seismic response of a RC frame building designed according to old and modern practices. *Bulletin of Earthquake Engineering*, *7*(3), 779–799. doi:10.1007/s10518-009-9119-4

Saatcioglu, M., Mitchell, D., Tinawi, R., Gardner, N. J., Gillies, A. G., Ghobarah, A., & Lau, D. et al. (2001). The August 17, 1999, Kocaeli (Turkey) Earthquake Damage to Structures. *Canadian Journal of Civil Engineering*, *28*(4).

Saito, T. (2006). In M. Higashino & S. Okamoto (Eds.), *Observed Response of Seismically Isolated Buildings. In Response Control and Seismic Isolation of Buildings* (pp. 63–88). Taylor & Francis.

Salazar, A. R., & Haldar, A. (2000). Structural responses considering the vertical component of earthquakes. *Computers & Structures*, 131–145.

Sandri, M. (1996). Numerical calculation of Lyapunov exponents. *Mathematica Journal*, *6*, 78–84.

Schläpfer, M., Kessler, T., & Kröger, W. (2008). Reliability analysis of electric power systems using an object-oriented hybrid modeling approach. *Proc. 16th Power Systems Computation Conf.*

SeismoSoft (2007). *SeismoStruct – A Computer Program for Static and Dynamic Nonlinear Analysis of Framed Structures*.

Sezen, H., & Moehle, J. P. (2004). Shear strength model for lightly reinforced concrete columns. *Journal of Structural Engineering*, *130*(11), 1692–1703. doi:10.1061/(ASCE)0733-9445(2004)130:11(1692)

Sgambi, L. (2004). Fuzzy theory based approach for three-dimensional nonlinear analysis of reinforced concrete two-blade bridge piers. *Computers & Structures*, *82*(13-14), 1067–1076. doi:10.1016/j.compstruc.2004.03.016

Shakeri, K., Shayanfar, M. A., & Kabeyasawa, T. (2010). A story shear-based adaptive pushover for estimating seismic demands of buildings. *Engineering Structures*, *32*(1), 174–183. doi:10.1016/j.engstruct.2009.09.004

Shih, T.-Y., & Lin, Y. K. (1982). Vertical seismic load effect on hysteretic columns. *ASCE Journal of Engineering Mechanics Div.*, *108*, 242–254.

Shome, N., Cornell, C. A., Bazzurro, P., & Carballo, J. E. (1998). Earthquakes, records, and nonlinear responses. *Earthquake Spectra*, *14*(3), 469–500. doi:10.1193/1.1586011

Silva, W. J. (1997). Characteristics of vertical strong ground motions for applications to engineering design. In I. M. Friedland et al. (Eds.) *Proceedings of the FHWA/NCEER Workshop on the National Representation of Seismic Ground Motion for New and Existing Highway Facilities* (pp. 205-252). Technical Report NCEER-97-0010.

Smith, R. (2016). Aristotle's Logic. *The Stanford Encyclopedia of Philosophy*. Retrieved from http://plato.stanford.edu/archives/spr2016/entries/aristotle-logic/

Smith, H. J. T., & Blackburn, J. A. (1992). Experimental Study of an Inverted Pendulum. *American Journal of Physics*, *60*, 909–911.

Smyrou, E., Priestley, M. J. N., & Carr, A. J. (2011). Modelling of Elastic Damping in Nonlinear Time-history Analyses of Cantilever RC Walls. *Bulletin of Earthquake Engineering*, *9*(5), 1559–1578. doi:10.1007/s10518-011-9286-y

Soong, T. T., & Dargush, G. F. (1999). *Passive Energy Dissipation and Active Control, Structural Engineering Handbook* (4th ed.). CRC Press.

Spoelstra, M., & Monti, G. (1999). FRP-confined concrete model. *Journal of Composites for Construction*, *3*(3), 143–150. doi:10.1061/(ASCE)1090-0268(1999)3:3(143)

Standard for Seismic Evaluation of Existing ReinfoRCed Concrete Buildings. (2001). The Japan Building Disaster Prevention Association.

Stephenson, A. (1908). On a new type of dynamical stability. *Memoirs of the Manchester Literary and Philosophical Society*, *52*, 1–10.

Stewart, J. P., Chiou, S. J., Bray, J. D., Graves, R. W., Somerville, P. G., & Abrahamson, N. A. (2001). *Ground Motion Evaluation Procedures for Performance-Based Design, PEER Report 2001/09*. Berkeley, CA: Pacific Earthquake Engineering Research Center, University of California.

STRAUS7. (2016). *Theoretical manual-theoretical background to the Strand7 finite element analysis system*. Author.

Strepelias, T., Fardis, M. N., Bousias, S., Palios, X., & Biskinis, D. (2012). *RC Frames infilled into RC walls for seismic retrofitting: Design, experimental behavior and modeling. Report Series in Structural and Earthquake Engineering*. Patras: University of Patras.

Sucuoğlu, H., Gür, T., & Gülkan, P. (2000). The Adana-Ceyhan Earthquake of 27 June 1998: Seis-mic retrofit of 120 RC buildings. *Proceedings of 12th World Conference on Earthquake Engineering*.

Surmeli, M., & Yuksel, E. (2015). A Variation of Modal Pushover Analyses (VMPA) Based on a Non-Incremental Procedure. *Bulletin of Earthquake Engineering*, *13*(11), 3353–3379. doi:10.1007/s10518-015-9785-3

Svendsen, N. K., & Wolthusen, S. D. (2007). Connectivity models of interdependency in mixed-type critical infrastructure networks. *Information Security Technical Report*, *12*(1), 44–55. doi:10.1016/j.istr.2007.02.005

Symans, M. D., Charney, F. A., Whittaker, A. S., Constantinou, M. C., Kircher, C. A., Johnson, M. W., & McNamara, R. J. (2008). Energy Dissipation Systems for Seismic Applications: Current Practice and Recent Developments. *Journal of Structural Engineering*, *134*(1), 3–21. doi:10.1061/(ASCE)0733-9445(2008)134:1(3)

Tasdemir, M. A., & Ozkul, M. (1999). Marmara depremi beton araştırması.[In Turkish]. *Hazır Beton Birligi*, *6*(36), 32–35.

Taskın, B., & Tugsal, U. M. (2014) Seismic performance evaluation of school buildings subjected to 2011 Van earthquakes.*Proceedings of the 10th NCEE*.

TEC. (2007). *Specification for Buildings to be Built in Seismic Zones 2007*. Ministry of Public Works and Settlement Government of Republic of Turkey.

TEC. (2007). *Specification for buildings to be built in seismic zones. Turkish Standards Institution. Ministry of public works and settlement, Ankara TS-498 (1997) Design loads for buildings. Turkish Standards Institute*. Ankara: Ministry of Public Works and Settlement.

TEC. (2007). *Turkish Earthquake Resistant Design Code*. Ankara, Turkey: Ministry of Publicworks and Settle-ment.

Tehranizadeh, M., & Moshref, A. (2011). Performance-based optimization of steel moment resisting frames. *Scientia Iranica*, *18*(2), 198–204. doi:10.1016/j.scient.2011.03.029

Tesfamaraim, S., & Saatcioglu, M. (2008). Seismic Risk Assessment of RC Buildings Using Fuzzy Synthetic Evaluation. *Journal of Earthquake Engineering*, *12*(7), 1157–1184. doi:10.1080/13632460802003785

Tezcan, S. S., & Boduroglu, H. (1998). *A reconnaissance report- The June 27, 1998 Adana-Ceyhan Earthquake*. Turkish Earthquake Foundation. Retrieved from www.depremvakfi.org

The World Bank Implementation Completion Report No. 17255. (1997). *Armenia Earthquake Reconstruction Project*. World Bank.

Thompson, J. M. T., & Stewart, H. B. (2002). *Nonlinear dynamics and chaos* (2nd ed.). Chichester, UK: Wiley.

Tierney, K., & Bruneau, M. (2007). Conceptualized and measuring resilience. *TR News*, *250*, 14–17.

Tondl, A. (1991). *Quenching of self-excited vibrations*. Prague: Academia.

Tondl, A. (1997). To the analysis of autoparametric systems. *Zeitschrift für Angewandte Mathematik und Physik*, *77*(6), 407–418.

Tondl, A., & Nabergoj, R. (1994). Non-periodic and chaotic vibrations in a flow induced systems. *Chaos, Solitons, and Fractals*, *4*, 2193–2202.

Tondl, A., Ruijgrok, T., Verhulst, F., & Nabergoj, R. (2000). *Autoparametric resonance in mechanical systems*. Cambridge, UK: Cambridge University Press.

Tortorella, M. (2010). *Network resiliency*. RUTCOR Research Reports.

Tortorella, M. (2010). Network Resiliency. Rutgers Center for Operations Research (RUTCOR Research Report).

Tortorella, M. (2005). Service reliability theory and engineering, II: Models and examples. *Quality Technology & Quantitative Management*, *2*(1), 17–37. doi:10.1080/16843703.2005.11673087

Trifunac, M. D., & Brady, A. G. (1975). A Study on the Duration of Strong Earthquake Ground Motion. *Bulletin of the Seismological Society of America*, *65*, 581–626.

TS500. (2000). *Requirements for Design and Construction of RC Structures*. Ankara, Turkey: Turkish Standards Institute.

TSC, M. o. P. W. a. S. G. o. R. o. T. (2007). *Specification for Buildings to be Built in Seismic Zones in Turkey*. TSC2007.

UN-Habitat. (2016). *The World Cities Report 2016, Urbanization and Development: Emerging Futures*. Author.

Vail, C., Hubbell, J., O'Connor, B., King, J., & Pall, A. (2004). *Seismic upgrade of the Boeing Commercial Airplane Factory at Everett, WA, USA*. The 13th World Conference on Earthquake Engineering, Vancouver, British Columbia, Canada.

Valente, M. (2011). Eccentric steel bracing for seismic retrofitting of non-ductile reinforced concrete frames.*The Thirteenth International Conference on Civil, Structural and Environmental Engineering Computing CC2011*, Crete, Greece. doi:10.4203/ccp.96.196

Valente, M. (2012). Seismic rehabilitation of a three-storey R/C flat-slab prototype structure using different techniques. *Applied Mechanics and Materials*, *193-194*, 1346–1351. doi:10.4028/www.scientific.net/AMM.193-194.1346

Valente, M. (2013a). Seismic upgrading strategies for non-ductile plan-wise irregular R/C structures. *Procedia Engineering*, *54*, 539–553. doi:10.1016/j.proeng.2013.03.049

Valente, M. (2013b). Seismic strengthening of non-ductile R/C structures using infill wall or ductile steel bracing. *Advanced Materials Research*, *602-604*, 1583–1587. doi:10.4028/www.scientific.net/AMR.602-604.1583

Valente, M., & Milani, G. (2016a). Seismic assessment of historical masonry towers by means of simplified approaches and standard FEM. *Construction & Building Materials*, *108*, 74–104. doi:10.1016/j.conbuildmat.2016.01.025

Valente, M., & Milani, G. (2016b). Non-linear dynamic and static analyses on eight historical masonry towers in the North-East of Italy. *Engineering Structures*, *114*, 241–270. doi:10.1016/j.engstruct.2016.02.004

van der Burgh, A. H. P. (1968). On the asymptotic solutions of the differential equations of the elastic pendulum. *Journal de Mécanique*, *7*(4), 507–520.

Vanmarcke, E. H., Cornell, C. A., Gasparini, D. A., & Hou, S. (1976). *SIMQKE: A program for artificial motion generation*. Civil Engineering Department, Massachusetts Institute of Technology.

Venture, N. C. J. (2011). NIST GCR 11-917-15–selecting and scaling earthquake ground motions for performing response-history analyses. US National Institute for Standards and Technology.

Verderame, G. M., De Luca, F., Ricci, P., & Manfredi, G. (2011). Preliminary analysis of a soft-storey mechanism after the 2009 LAquila earthquake. *Earthquake Engineering & Structural Dynamics*, *40*(8), 925–944. doi:10.1002/eqe.1069

Vespignani, A. (2010). The fragility of interdependency. *Nature*, *464*(7291), 984–985. doi:10.1038/464984a PMID:20393545

Vugrin, E. D., Drake, E. W., Ehlen, M. A., & Camphouse, R. C. (2010).A framework for assessing the resilience of infrastructure and economic systems. Sustainable and Resilient Critical Infrastructure Systems, 77-116. doi:10.1007/978-3-642-11405-2_3

Whittaker, D., & Robinson, W. (2009). Progress of Application and R&D for Seismic Isolation and Passive Energy Dissipation for Civil and Industrial Structures in New Zealand. *Proceedings of the 11th World Conference on Seismic Isolation, Energy Dissipation and Active Vibration Control of Structures.*

Wu, F. (1978). Prediction of strong ground motion using small earthquakes, *Proceedings of the 2nd International Conference on Microzonation, 2,*701-704.

Xu, Z., & Cheung, Y. K. (1994). Averaging method using generalized harmonic functions for strongly non-linear oscillators. *Journal of Sound and Vibration*, *174*(4), 563–576.

Yankelevsky, D. Z., & Reinhardt, H. W. (1989). Uniaxial behavior of concrete in cyclic tension. *Journal of Structural Engineering*, *115*(1), 166–182. doi:10.1061/(ASCE)0733-9445(1989)115:1(166)

Youssefa, M. A., Ghaffarzadehb, H., & Nehdia, M. (2007). Seismic performance of RC frames with concentric internal steel bracing. *Engineering Structures*, *29*(7), 1561–1568. doi:10.1016/j.engstruct.2006.08.027

Zadeh, L. A. (1973). Outline of a New Approach to Analysis of Complex Systems and Decision Processes. *IEEE Transactions on Systems Man and Cybernetics, 3*(1), 28-44. doi:10.1109/Tsmc.1973.5408575

Zadeh, L. A. (1965). Fuzzy sets. *Information and Control*, *8*(3), 338–353. doi:10.1016/S0019-9958(65)90241-X

Zadeh, L. A. (1968). Probability measures of Fuzzy events. *Journal of Mathematical Analysis and Applications*, *23*(2), 421–427. doi:10.1016/0022-247X(68)90078-4

Záworka, J. (1993). Project SIMONE—achievements and running development. *Proceedings of the Second International Workshop SIMONE on Innovative Approaches to Modeling and Optimal Control of Large Scale Pipeline Networks*, 1–24.

Zembaty, Z. (2011). How to model rockburst seismic loads for civil engineering purposes? *Bulletin of Earthquake Engineering*, *9*(5), 1403–1416.

Zembaty, Z. (2013). Numerical analyses of seismic ground rotations from the wave passage effects. *Geotechnical. Geological and Earthquake Engineering*, *24*, 15–28.

Zhou, F., Tan, P., Heisha, W., & Huang, X. (2009). Recent Development and Application on Seismic Isolation in China. *Proceedings of the JSSI 15th Anniversary International Symposium on Seismic Response Controlled Buildings for Sustainable Society.*

Zhu, L., Elwood, K. J., & Haukaas, T. (2007). Classification and Seismic Safety Evaluation of Existing Reinforced Concrete Columns. *Journal of Structural Engineering*, *133*(9), 1316–1330. doi:10.1061/(ASCE)0733-9445(2007)133:9(1316)

Zio, E., & Giovanni, S. (2011). Modeling interdependent network systems for identifying cascade-safe operating margins. *Reliability, IEEE Transactions on*, *60*(1), 94–101. doi:10.1109/TR.2010.2104211

About the Contributors

Vagelis Plevris is an Associate Professor at the Department of Civil Engineering and Energy Technology of the Oslo and Akershus University College of Applied Sciences (HiOA) in Oslo, Norway. He holds a 5-year Bachelor's Degree in Civil Engineering from the National Technical University of Athens (NTUA) with specialization in Structural Engineering. He also holds an MSc from NTUA on "Structural Design and Analysis of Structures", a Master in Business Administration (MBA) from Athens University of Economics and Business (AUEB) and a PhD from the School of Civil Engineering of NTUA. His research work focuses mainly on the Finite Element Method, Static and Dynamic Analysis of Structures, Earthquake Engineering, Optimum Design of Structures, Reliability and Probabilistic Analysis of Structures and Neural Networks and their Applications in Engineering. His published work includes more than 60 publications in peer-reviewed journals, conference proceedings, edited books and others.

Georgia Kremmyda is a Dr Civil Engineer, currently appointed as the Civil Engineering Degree Leader in the School of Engineering at The University of Warwick. Georgia received her PhD degree in Earthquake Engineering, an MSc degree in Tunnelling and Underground Structures and an MEng degree in Structural Engineering from the Faculty of Civil Engineering at the National Technical University of Athens (NTUA), Greece. In her current role she is responsible for the delivery and management of teaching at undergraduate and MSc level in Civil Engineering within the School of Engineering, providing leadership in curriculum development, administration, liaison with industry and other activities. Georgia also holds more than 12 years experience in the Industry on leading positions. Her teaching and research interests are oriented to: Earthquake Engineering; Reliability analysis and probability of failure of structures; Precast Concrete Structures; Structural Dynamics.

Yasin Fahjan is Professor of Earthquake Engineering at Gebze Technical University. Has strong background on computational methods for earthquake risk assessments, soil-structure interaction, fluid-structure Interactions. Developed several educational and commercial codes and algorithms related to structural analyses, rapid response and early warning systems, strong ground motion simulations and seismic hazard analysis. He is the main developer for AFAD-RED system operated by Turkish Disaster & Emergency Management Authority for real time estimation of earthquake damages on nationwide level. He also worked as earthquake engineering consultant for many pioneering engineering projects in Turkey and overseas. Prof. Fahjan worked as a team leader for many research projects in national and international level. He has considerable publication in the field (over 100 papers in refereed journals and conference proceedings).

* * *

Erkan Akpinar is Assistant Professor at Kocaeli University, Turkey, since 2011. He has received his MSc and PhD degrees in 2004 and 2010 respectively from Kocaeli University. He studied in University of Ottawa, Canada as a visiting researcher between 2009 and 2010. His areas of interest are analysis and strengthening of reinforced concrete, masonry and precast structures. He is author and co-author of several publications. He is also member of several international Task Groups.

Malik Atik received his Ph.D. in Structural Engineering from the University of Lille1 Science and Technology in 2013. Currently, he works at Brand Energy & Infrastructure Services. His research is mainly in dynamics and seismic analysis of structures. He published several papers in refereed journals and conference proceedings.

Francesco Clementi obtained his PhD in Architecture, Buildings and Structures at Polytechnic University of Marche (UNIVPM) during the academic year 2008–2009. He is an Assistant Professor in Structural Mechanics at the Faculty of Engineering of UNIVPM from 2012. He has developed research and teaching activities at the Universities of Ancona, Camerino, Lublin and Sao Paulo. His research interests focus on composite materials, dynamics of structures, seismic vulnerability, structural health monitoring, fracture mechanics.

Giovanni Di Sciascio, born in 1979, graduated in Mechanical Engineering from the University of Bologna in 2005, licensed engineer in the Italian Register of professional engineers since 2005. He works in structural design, numeric modelling of complex parts with dedicated procedures for resistance and stability verifications, programming of applications for structural calculations and CAD 3D modelling.

Sergio Di Sciascio, born in 1949, graduated in Mechanical Engineering from the University of Bologna in 1975, licensed engineer in the Italian Register of professional engineers since 1977. He is a structural engineer, supervisor of works and provides experienced assistance to the assembly of metal structures.

Seckin Ersin is master degree student at Kocaeli University, Turkey, where he has received his Bachelor Degree in Civil Engineering. He is currently working at OTS, Engineering Company, in Istanbul, Turkey.

Cyril Fischer received his MS (1993) and PhD. (2002) in Numerical Mathematics from the Charles University in Prague. His professional interests cover a wide area of applied numerical methods in linear and non-linear algebra (MS thesis: "Large scale eigenvalue problem"), stochastic dynamics (PhD. thesis: "Numerical methods of stochastic mechanics"), non-linear dynamics, earthquake engineering, wind engineering, etc. In 1994 he was hired as a research assistant by the Institute of Theoretical and Applied Mechanics (Prague, Czech Republic) and is currently head of the Centre for Computer Technology and Informatics in ITAM. He published more than 100 publications in international journals and proceedings of international conferences. He works as a reviewer of several scientific journals or conference series. He has been awarded by the Babuška Prize (2002).

Ali Golara was born in 1982. He is a highly skilled professional with proven track record over 10 years of broad experience within civil engineering and gas industry in the field of hazard, risk and resilience.

He is responsible for strategic planning studies and the director of rehabilitation and strengthening committee (Gas Standard Committee) in National Iranian Gas Company (NIGC) and an official member of International Gas Union (IGU). He also manages contracts with government and private companies in both areas: research and scientific-technological support. He is a lecturer at Azad University of Tehran, where he teaches structural analysis, financial decision making, details and elements and earthquake engineering courses. He has a diverse range of skills including map design, vulnerability evaluation and strengthening of structures, project planning and control procedures. He has published more than 20 papers and serves as reviewer in several scientific journals (e.g. ASCE, Springer and Elsevier) and he is also able to write computer programs.

Orkun Gorgulu is a Senior Structural Engineer and Design Group Leader in Tekfen Engineering Company in Istanbul where he has been a member since 2008. Orkun Gorgulu completed his Ph.D. at Istanbul Technical University in 2015. His research interests lie in the area of RC Structures; Earthquake Engineering; Engineering Seismology; Nonlinear Dynamics; Performance Based Design. He is the author of several articles published in scientific journals and member of different international working groups.

Fatma Ilknur Kara received her MS in Earthquake and Structural Science Department from Gebze Institute of Technology in 2004. Currently she is a Phd student in Earthquake Engineering Department of Istanbul Technical University and Research Assistant in Earthquake and Structural Science Department of Gebze Technical University. Her research interests has been concentrated on structural dynamics, performance based design, disaster and risk management. Se is the author of several articles published in scientific journals and member of different international working groups.

Stefano Lenci is a Full Professor of Structural Engineering at the Polytechnic University of Marche. He has authored more than 300 publications, among which 125 papers are published in international scientific journals. Currently he is the Head of the PhD program in Civil, Environmental, Building Engineering, and Architecture. His scientific interests are nonlinear dynamics, dynamics of structures, seismic vulnerability, structural health monitoring and mechanical behaviour of composites.

Mikayel G. Melkumyan was born on June 10, 1951. He started his scientific and practical activity in 1973, immediately after graduation from the Civil Engineering Department of Yerevan Polytechnic Institute, carrying out both design and experimental-theoretical research works. From April 1990 through March 1991 he conducted research at the Institute of Industrial Science, University of Tokyo, where on the basis of his experimental research works he created a new hysteresis model to describe the shear behavior of reinforced concrete structures (walls, diaphragms). During a short period of time in 1995-1996, devoting him to the challenge of increasing earthquake resistance of existing buildings, he developed and implemented two unique methods of protecting existing buildings from earthquakes through base and roof isolation without interrupting exploitation of the buildings. These works are unprecedented in the world practice of earthquake resistant construction. His works in the fields of both non-linear behavior of reinforced-concrete structures and seismic isolation are well known to the international professional community by the weighty contribution to the science and practice of earthquake resistant construction. He has authored and co-authored 205 scientific works, including 17 books (among them four monographs), 10 normative documents, and 12 inventions.

Aydın Mert is a Researcher at the Bogazici University. He obtained his MS in Earth System Science from the Istanbul Technical University in 2005 and his Ph.D. in Geophysical Engineering from the Istanbul University in 2011. His research interests has been concentrated on three main fields: Earthquake Ground Motion Simulation, Earthquake Hazard Assessments, and Earthquake Early Warning and Rapid Response Systems. He is the author of several articles published in reputed scientific journals and different international conferences. He is also member of different international working groups.

Ulgen Mert Tugsal is a Research Assistant in the Department of Civil Engineering at Istanbul Technical University where she has been a faculty member since 2007. Ulgen Mert Tugsal completed her Ph.D. at Istanbul Technical University in 2016. Her research interests lie in the area of RC Structures; Earthquake Engineering; Engineering Seismology; Nonlinear Dynamics; Performance Based Design.

Gabriele Milani is associate professor of "Scienza delle Costruzioni" (Strength of Materials) at the Technical University of Milan, Italy. So far, he has authored more than 120 papers on international journals and edited a book. He is the second author in Scopus under the keyword "masonry". He has been awarded a Telford Premium by ICE in 2012, a most cited author 2005-2008 by Computers & Structures, a Bathe award in 2014. He is EIC of a journal dedicated to masonry (International Journal of Masonry Research and Innovation) and co-editor of a generalist civil engineering journal (Open Civil Engineering Journal). He is in the editorial board of several renewed international journals, including Computers & Structures and Structural Engineering and Mechanics.

Jiří Náprstek received his MS in Civil Engineering from the Czech Technical University in Prague in 1966. He finished his PhD. studies in the Institute of Theoretical and Applied Mechanics (Prague, Czech Republic) in 1972 at the field of wave propagation in the visco-elastic continuum under moving inertial load with harmonic and random character. He became the Doctor of Sciences (DSc) in 1997 after a successful defense of thesis dealing with Stochastic Dynamics and Stability of linear and nonlinear systems. His scientific career is devoted to Nonlinear Dynamics, Dynamic Stability, Stochastic Dynamics, Computational Mechanics, Dynamics of Non-holonomic Systems, Wind- Earthquake- and Traffic-Engineering. During his life he published more than 300 papers in international journals and proceedings of international conferences. He works as reviewer of 24 scientific journals (Elsevier, Wiley, Springer, ASCE, ASME). He has been awarded by the State Prize of the Republic (1982), twice by the Prize of the Academy of Sciences (1983, 1986) and by Frantisek Krizik Medal (2007).

Ferhat Pakdamar obtained his PhD from Istanbul Technical University, Turkey in 2010. He studied the "Evaluation of Flexible Performance of Reinforced Concrete Structures Using a Nonlinear Static Procedure Provided by Fuzzy Logic" for his PhD. He has two high-profile journal papers one of it is based on flexible performance and the other is "Stress-strain modelling of high strength concrete by fuzzy logic approach" and several conference publication papers. He studied on new generation technics like fuzzy logic, neural networks, genetic algorithms and fractal geometry. In 2004, when he is studying English in UC Berkeley / USA, he negotiated some subjects with Prof. Lutfi Asker ZADEH who is the founder of Fuzzy Logic. He focused on the seismic behaviour of structures, reinforced concrete structures especially bridges and historical buildings. He worked as a seismic expert to describe the risk assessment and to retrofit the over 100 bridges and viaduct projects around Istanbul/Turkey. As a part of this work, he generated the seismic risk maps for Istanbul. He also worked in some very important historical monument projects. He has a few conference publications for these projects. He did many earthquake risk assessment projects for the buildings in Turkey as an expert. With this qualification he is not only theoretical expert but also a field expert. Since July 2014, He is a lecturer in the Department of Architecture at the Gebze Technical University (old name Gebze Institute of Technology). He is delivering 8 modules for undergraduate, 4 modules for graduate students and currently supervising several MSc and PhD dissertations.

Marwan Sadek obtained his Diploma in Civil Engineering (first in class with distinction) in 1998 from the Faculty of Engineering at the Lebanese University. He was awarded a distinction scholarship from the AUF international agency to join the Master of Engineering at the University of Lille. Upon completing his Master degree in 1999 (first in Class), he was awarded a distinction scholarship to study the PhD in Civil Engineering at the same institution. He worked as bridge design engineer between 2002 and 2005 at the consulting engineering Firm ACOGEC. In September 2005, he joined the Polytechnic University of Lille where he worked as Associate Professor in the Civil Engineering department. He was the director of the Master degree program in Civil Engineering for 10 years. He earned the HDR degree (Habilitation à diriger des recherches) in 2011. He was elected to membership of the Administrative and Scientific Councils of the Polytechnic University of Lille between 2010 and 2014. He received the French Award of Scientific Excellence (Prime d'Excellence Scientifique PES) for eight years. Currently, and after obtaining a secondment agreement from the Polyetchnic University of Lille, he is Full Professor of Civil Engineering at the Institute of Technology of the Lebanese University. He is author and co-author of more than 60 papers in refereed journals and conference proceedings. His research activity is mainly devoted to: soil dynamics and Earthquake Engineering, soil structure interaction and development of advanced numerical modeling. He was involved in several consultancies related to structural and seismic design of bridges and heavy structures.

Isam Shahrour is a distinguished Professor at the University of Lille1 in France. Currently, he is director of the Civil and Geo-Environment Engineering Laboratory and coordinator of the SunRise Smart City project "A large scale demonstrator of the Smart City". His research concerns soil-structure interaction, geo-environmental engineering and more recently Smart Cities. He published around 110 journal papers and supervised 70 PhD theses. He was Vice President for "Research and Doctoral Program" at the University of Lille1 during the period 2007 – 2012.

Beyza Taskin is an Associate Professor in the Department of Civil Engineering at Istanbul Technical University where she has been a faculty member since 1993. Beyza Taskin completed her Ph.D. at Istanbul Technical University in 2001. Her research interests lie in the area of RC Structures; Earthquake Engineering; Engineering Seismology; Nonlinear Dynamics; Performance Based Design; Repair & Strengthening of Structures; Formwork Systems.

Marco Valente is Assistant Professor at Politecnico di Milano where he obtained his Ph.D. degree in Earthquake Engineering. His research activity mainly focuses on seismic analyses, design and retrofitting of structures, innovative techniques for seismic protection of structures; seismic performance assessment of steel; reinforced concrete, steel-concrete composite, precast and masonry structures. He is the author of more than 160 papers published in journals and conference proceedings in the field of Earthquake Engineering.

Index

O

P

Q

R

S

T

V

W

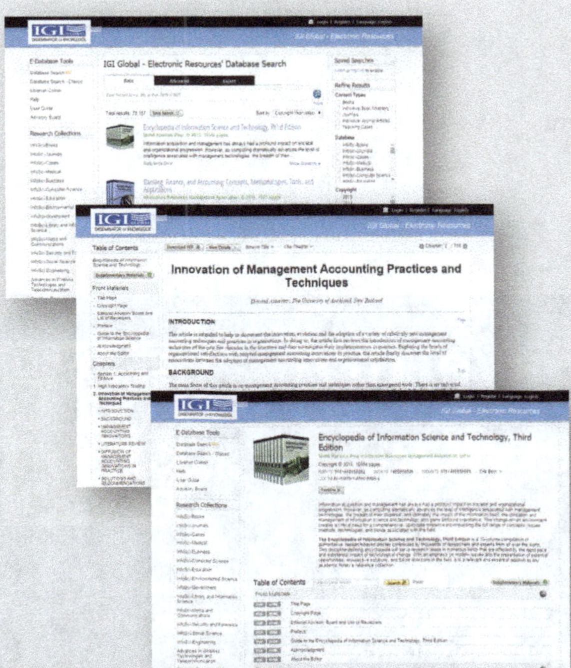

www.ingramcontent.com/pod-product-compliance
Lightning Source LLC
LaVergne TN
LVHW080536300926
849162LV00007B/1619